S0-DUV-990

The World of Natural Wine

The World of Natural Wine

WHAT IT IS, WHO MAKES IT, and WHY IT MATTERS

Aaron Ayscough

ARTISAN | NEW YORK

Library of Congress Cataloging-in-Publication Data is on file.

ISBN 978-1-57965-939-4

Cover and book design by Heitman-Ford + Co.

Artisan books are available at special discounts when purchased in bulk for premiums and sales promotions as well as for fund-raising or educational use. Special editions or book excerpts also can be created to specification. For details, contact the Special Sales Director at the address below, or send an e-mail to specialmarkets@workman.com.

For speaking engagements, contact speakersbureau@workman.com.

Published by Artisan
A division of Workman Publishing Co., Inc.
225 Varick Street
New York, NY 10014-4381
artisanbooks.com

Printed in China on responsibly sourced paper

First printing, August 2022

10 9 8 7 6 5 4 3 2 1

To my friends in the Beaujolais, where this book began:
Nicole, Marion, and Lambert

Contents

Preface

This is not a buying guide. Rather, this book is intended as a field guide—something you'd use if you were foraging in the woods, or crossing a desert. It's an attempt to describe, in practical terms, what the natural wine community is, how it came to be, and how to navigate it. For natural wine is more than a beverage: it's a counterculture.

The vignerons, chefs, restaurateurs, sommeliers, and wine drinkers that make up this international natural wine counterculture are united by a way of thinking about wine. It is a conscious approach that acknowledges that the aesthetics of wine are not merely subjective. How a wine tastes and smells depends on how it was farmed and how it was made. And how it is farmed and how it is made have effects on the health (both ecological and physical) of winemaking communities and wine drinkers.

It's easy to forget this, if, like most of us, you don't live in a winemaking community.

I grew up in rural Pennsylvania, where the closest one ever got to wine culture was state-run wine and spirits stores, places with all the passion and personality of dental clinics. By age twenty-three I had, rather by accident, become one of the biggest Italian wine buyers in Los Angeles, through my work as wine director of Nancy Silverton's Pizzeria Mozza, and later as a sommelier at the adjacent Osteria Mozza.

Two short years later, in 2009, I decided to move to Paris—not to work in wine, but to escape it. Like many young sommeliers, I was adept at memorizing wine books, grinning at guests, and inventing funny wine descriptions on the fly. But it all felt fake to me. I had no idea what I was talking about. No one around me seemed to know what they were talking about when they spoke about wine, either, and the fact that this was evidently no obstacle whatsoever to a successful career in wine made me want out of the field entirely.

Fortunately, when I arrived in Paris in 2009, I discovered a thriving wine subculture that countervailed everything I'd been taught. To my first Parisian wine mentors—people like restaurateur Guy Jeu, wine retailer Michel Moulherat, and the American expat restaurateur Kevin Blackwell—I must have seemed a slow student. I'd been indoctrinated to venerate Champagne, Bordeaux, and Burgundy, and for all I knew, biodynamic farming was as radical as it got. These Parisians never asked whether wines were biodynamic; what mattered to them was whether a wine was natural. The wines they venerated were, as often as not, Vin de France (without an appellation). They rarely opened Champagne or Burgundy, favoring *pétillant naturel* and Beaujolais. Bordeaux was unmentionable. Sulfites and filtration were not.

The proximity of Paris to France's wine regions afforded access to a different, more granular conversation about wine, one that felt so much more genuine than flavor descriptions and witty metaphors. I wanted in on it.

After a few years of frequenting the natural wine bistros of Paris and visiting vignerons in my spare time, I felt impelled to move to the Beaujolais in 2015. I had begun writing about natural wine by then, on a blog I called *Not Drinking Poison*. The conversation among Parisian natural wine aficionados no longer interested me; I was after the conversation among the natural vignerons themselves.

I worked a season of harvest and vinification with Fleurie vigneron Yvon Métras and his son Jules. The following year, I did the same with Morgon vigneron Guy Breton, known locally as "Le P'tit Max." When there was no more work in the cellar I did odd jobs for the rest of the region's natural vignerons—whatever I could to learn about winemaking alongside them. This book began in earnest then. In the years since, I've alternated work as a sommelier in Paris (at restaurants including Daniel Rose's Chez la Vieille and Bruno Verjus's Table) with work at wine estates (including Domaine Derain in Burgundy and Clos Fantine in the Languedoc) and writing projects (including two translations of the work of the wine scientist and natural wine progenitor Jules Chauvet). In 2020, I began studies at the Beaune Viticultural School in Burgundy to obtain a winemaking license.

Nothing has taught me as much as that first year in the Beaujolais. Living among vignerons and their families throws into sharp relief what is at stake in the nonuse of synthetic chemicals in farming. As a natural wine taster, I knew it to be a prerequisite for great winemaking. As an inhabitant of the herbicide-showered Beaujolais, I came to know it as an ethical imperative, as I witnessed an alarming rate of early-onset cancer in the families of friends and neighbors. This dynamic, in which the moral and the aesthetic intersect, undergirds the unusual fervor you encounter throughout the natural wine scene.

I went to the Beaujolais because of the central role its vignerons played in the formation of France's early natural wine community. I figured if I was going to learn about natural wine, I ought to begin at the beginning. This is also why this book focuses primarily on regions within France. Today, of course, natural wine is a global phenomenon, like hip-hop. Natural wine is made in Australia, Chile, Japan, and beyond, just as there are successful rappers in every language, all over the world. But the natural wine community was first and foremost a French phenomenon, just as hip-hop was an American one.

France was where a *natural wine culture*, distinct from a general wine culture, first developed among vignerons, bistro owners, retailers, journalists, and consumers. It coalesced in the 1990s and 2000s. Its roots began even earlier. (In 1978, as Grand Wizzard Theodore was inventing turntable scratching in the Bronx, Marcel Lapierre was conducting his first unsulfited vinification in the Beaujolais town of Villié-Morgon.) Understanding the natural wine community in France is key to appreciating the natural wine communities that have arisen around the world in its wake.

Today it can feel as though the universe of natural wine is expanding in all directions at a breathtaking rate. The phrase itself is an ideological battleground. For an understanding of the true spirit of natural wine, however, look to its origins: A handful of renegade Beaujolais vignerons and their friends in other regions. A côterie of thoughtful and idiosyncratic bistro owners in Paris. A flourishing underground of iconoclastic farmers and winemakers, working in the face of overwhelming economic forces to preserve an ideal of naturalness in wine production.

This book is a group portrait of this community—and an invitation to join it. A way to get to know the personalities behind natural wine, and how and why they do what they do. Grab a glass, and get inspired.

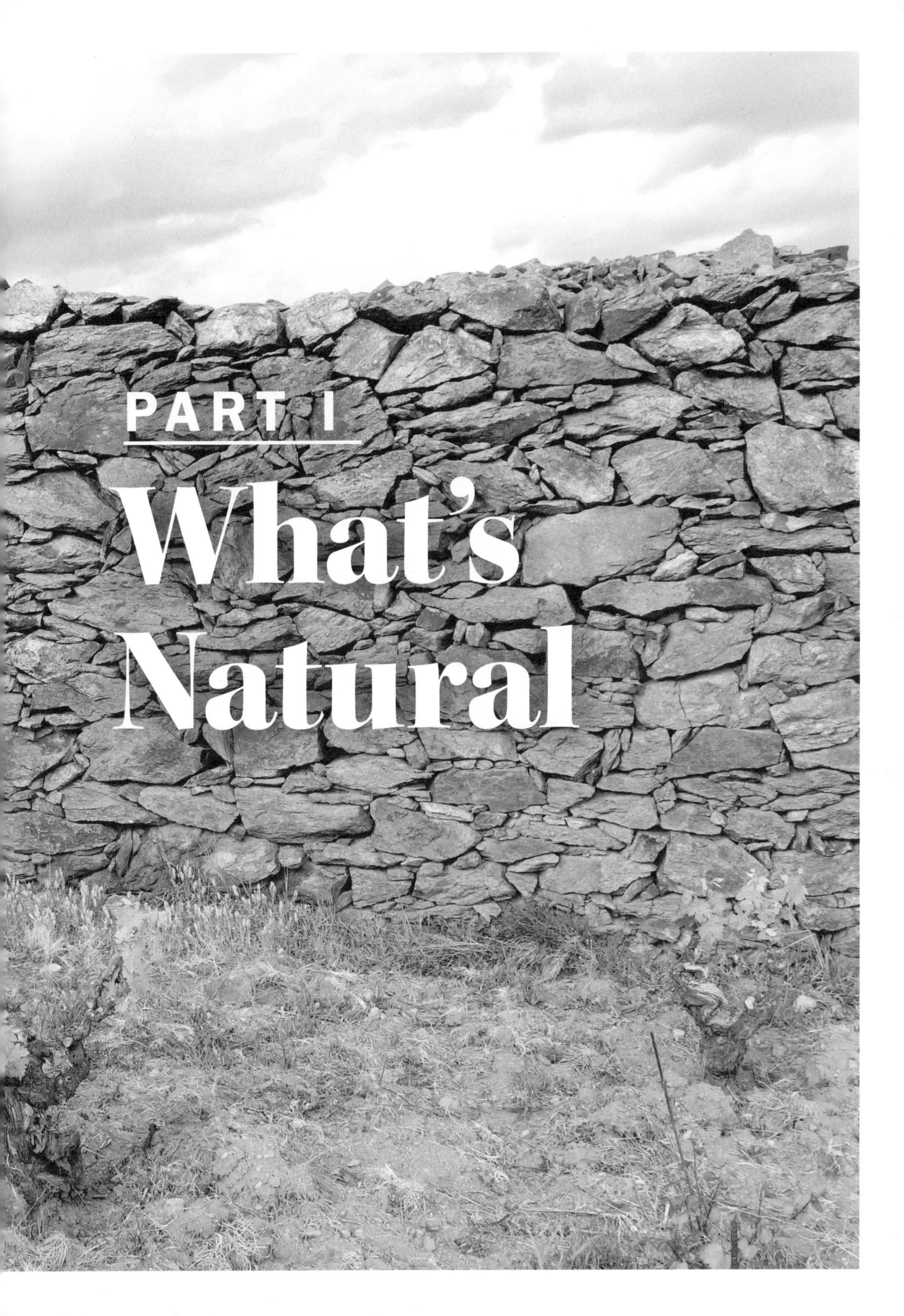

PART I

What's Natural

Previous pages: A shelter built into the dry stone wall surrounding a vineyard at Clos du Rouge Gorge in the Roussillon

Above left: A sign at the entry to the Courtois family estate in Soings-en-Sologne (Loir-et-Cher)

Above right: Claude Courtois in the vineyards beside his house

Left: The cellar at Les Cailloux du Paradis

Natural wine is at once a beverage and a culture. Efforts to define it in narrower terms, as simply a beverage, have so far been frustrated by the nonparticipation of the actors within natural wine culture, chiefly the natural winemakers themselves. Why?

It's because wine growing and winemaking are practices that require a high degree of sensitivity to local terrain and climate, as well as to local social and economic conditions. Few vignerons wish to adhere to a one-size-fits-all, internationalized rulebook for natural winemaking. Most are content with a definition along the lines of Supreme Court justice Potter Stewart's famous description of obscenity: "I'll know it when I see it."

That's fine for them. But how will *you* know natural wine when you see it? (And taste it, of course.)

It requires a familiarity with wine growing, winemaking, and the history of natural wine. This section shows how and why natural wine culture arose in France in the 1980s, 1990s, and early 2000s, and goes on to enumerate common practices within the contemporary idiom of natural wine growing and natural winemaking. It offers an empirical definition of natural wine—but it's less a rulebook than a description of a shared value system. Once you know the culture, you'll know the beverage.

A barrel of chenin aging in the cellar of vigneron and early natural wine proponent Jean-Pierre Robinot (Chahaignes, France)

1. A Way of Thinking About Wine

What Is Natural Wine?

A pumpover (see page 96) in progress at the cellar of southern Beaujolais vigneron Nicolas Dubost

Natural wine is wine with nothing to hide. Organically farmed grapes, fermented, aged, and bottled without the use of additives, filtration, or transformative corrective procedures. It's wine that maintains its traditional—and crucial—link to nature. This sets it apart from most wine on the market today, which has been stripped of its origins by intensive agriculture and invasive winemaking.

The link between wine and nature, of course, is the reason wine is unique in the first place. Wine has an astonishing capacity to reflect where it came from, the grapes it was made from, the land they were grown on, how that land was farmed, the weather that year. Natural wine is wine that tells the truth about these origins.

So you could say natural wine is the opposite of fraudulent wine. A fraudulent wine is a wine that purports to be something it's not. Most mass-market wines today are fraudulent in this sense, purporting to represent specific cultural and geographic origins while delivering an anonymous wine that could have been manufactured anywhere.

In the old days, wine fraud usually meant passing off a cheap wine from an unknown place as a more expensive wine from a famous place, or adding lots of sugar during fermentation to boost a wine's alcohol content, or adding colorants or distillates. These actions *denatured* a wine—they removed wine from its origins, preventing it from expressing the character of a vineyard and a vintage like it naturally would.

We're talking so much about natural wine today because wine lovers are becoming aware that wine fraud has become institutionalized. Natural wine begins with the bummer realization that nowadays, almost all wine on sale anywhere has been denatured, first by synthetic chemical farming, then by corrective winemaking.

Since the early twentieth century, agronomists and synthetic chemical treatments have come to dominate wine growing. Synthetic fertilizers artificially boost yields, destabilizing vineyard biomes and seeping into groundwater. Pesticides and herbicides applied to grapes devastate surrounding wildlife. They also find their way into wines and into our bodies. Over time, these treatments impoverish soils, a phenomenon that leads winegrowers to increase their use of synthetic fertilizers.

Meanwhile, since the 1980s, winemaking consultants called enologists and the transformative winemaking procedures they advocate have come to dominate winemaking. As long-distance export markets have grown in economic importance to old-world wine regions, the use of sulfites as a preservative and an antibacterial agent in winemaking has climbed steadily upward (see page 78). Excessive use of sulfites yields neutered, lifeless wines. Nor are sulfites the only common wine additive. Even organic winemakers are legally permitted to use seventy-two other additives, from tartaric acid to copper sulfate to artificial tannins to clarifying agents like pea protein. The list is longer for winemakers who don't seek organic certification.

Why do most winemakers add so much stuff to wine? Partly to compensate for the poor-quality grapes yielded by synthetic chemical farming. Partly to aid in the routinization of the winemaking process, making it less costly in terms of time and labor. Conventional wine production also creates a vicious cycle within the wine market. Nowadays, most drinkers expect all wines to conform to the debased, commodified wine styles that result from corrective winemaking.

Today's enologists employ sophisticated additives and machines that can make a wine from Burgundy taste like a wine from Sonoma, or vice versa. They can make a wine from a cold vintage taste like a hot vintage. They can add or subtract flavors and change alcohol levels. It's impressive, in a certain way. But in today's globalized wine marketplace, it has the not-so-miraculous effect of making most wines taste the same.

Your average wine label proudly displays an enticing place of origin: Provence, Rioja, Santa Barbara. But in the bottle, it's a faceless, anonymous wine that, thanks to transformative modern enology, could come from just about anywhere: Sicily, Burgundy, Mars. In other words, most wines purport to be something they're not: a wine with a true expression of a regional terroir. To prefer natural wine is to recognize that transformative modern enology is the new wine fraud. It plays the same role that blending foreign wines and sugar addition once played in correcting for, and enabling, destructive farming.

The good news is, wine drinkers are getting wise. They're demanding greater transparency and assurance about what's in the glass and how it got there. An exciting wine counterculture has arisen in support of this notion of natural wine. It began in France and has spread worldwide over the past two decades. To a large extent, it represents a rediscovery of the wisdom of traditional wine-growing and winemaking methods. To learn about

Vigneron Jules Métras during harvest at Fleurie's La Madone in the Beaujolais. On the left are his family's organically farmed vines. On the right are the herbicided vines of a neighbor.

these things, we turn away from agronomists and enologists and listen to winegrowers who make their own wines: vignerons.

Natural vignerons are at the center of an international natural wine community, united by a conviction that the goodness of a wine is inseparable from the naturalness of how it was made.

By the end of this book, you might begin to suspect that natural wine is *just wine*—wine as it's meant to be farmed and made and tasted and enjoyed, free from the distorting effects of the globalized wine market. You might begin to suspect that the majority of conventional wines are, in some sense, more like grape-based beverages. You'd be right.

NATURAL HISTORY

EIGHTEENTH-CENTURY WINE FRAUD

Throughout history, commerce laws and taxation have provided incentives to adulterate or dilute wine. It is in this sense that the word "natural" was first applied to wine: it signifies wine in its natural state, before any tampering by third parties. Fraud by wine dealers, transporters, hotels, and bars was rampant before the practice of estate bottling became normalized in the latter half of the twentieth century.

In 1789, the lawyer Jacques-Joseph Brac de la Perrière, in his pamphlet *Le Commerce des Vins, Réformé, Rectifié et Épuré* (*Wine Commerce, Reformed, Corrected, and Purified*), lamented: "The wine merchant of Paris never bought the wines of the Beaujolais to sell them in a natural state to the consumer; he only ever bought them to give energy to fat wines, flat and without virtue." Brac de la Perrière suggested that, were his plans to be enacted, "[Hotel owners and cabaret owners] would have natural wines that would distinguish them, and which would bring honor to their hotels and cabarets."

Parisian wine fraud had the effect of popularizing *guinguettes*, suburban cabarets found just outside the city limits (and thus the city's taxation of wine), described in 1790 by French National Assembly member Étienne Chevalier as places where poor families could take comfort in "natural wine at low prices, and comestibles infinitely less expensive than in Paris."

Today, on the contrary, natural wine is more abundant in Paris than in the rest of France. The city was home to the first retailers who began specializing in natural wine in the 1980s (see page 32).

An early-twentieth-century postcard featuring a guinguette in the Lyon suburb of Vaulx-en-Vélin

THE VIGNERON

When we talk about wine, words matter. Something as simple as the word "winemaker" can create misunderstandings.

In California and many other new-world regions, winemakers rarely farm vineyards they own, because vineyard land is extraordinarily expensive in these places. Instead, most winemakers purchase grapes from grape growers, usually called *winegrowers* to differentiate them from those who grow table grapes. These new-world winegrowers tend to farm on a scale much larger than their old-world counterparts, often subcontracting much of their farming. This has created a new-world wine culture in which wine growing is largely separated from winemaking.

This is a historical anathema. Great winemakers have always been farmers, not creative directors. You can't make great wines without a supreme sensitivity to growing great grapes.

Most old-world wine cultures have a word equivalent to the French *vigneron* (in Italian, *vignaiolo*; in Spanish, *viñador*), which means a winegrower who makes wine. It's as if the chef preparing your food is also the farmer who grew it. Historically, this is how wine has always been produced in older wine cultures, because farming decisions greatly inform the subsequent winemaking. There is a maxim in the wine world, no less true for being a cliché: "Great wines are made in the vines."

The English language has no word for vigneron. Happily, English is also a remarkably plastic language that allows for the adoption of useful foreign words. The first step to understanding natural wine is to make "vigneron" a part of your vocabulary.

Beaujolais vigneron Guy Breton in his over one-hundred-year-old vines in the Morgon hamlet of Saint Joseph

Beyond Organics and Biodynamics

You might have already heard of "organic wine" and "biodynamic wine" and wonder where natural wine fits in. Both organic wine and biodynamic wine are related to the idea of natural wine, because most natural wine is also organic and/or biodynamic.

But only a fraction of organic and biodynamic wines are also natural wines. Natural wines go significantly further in terms of purity in winemaking. (For more details, head straight to chapter 3.)

To be certified as organic (see page 53), a wine must be farmed without synthetic fertilizers, herbicides, or synthetic pesticides, which is to say, how small-scale farming was always conducted until well into the twentieth century. Legal definitions of "organic wine" vary slightly between the US and the EU.

Meanwhile, wine certified as biodynamic (see page 54) has been farmed according to a holistic system of agriculture based upon the ideas of the Austrian philosopher Rudolf Steiner, who emphasized the lunar calendar and biodiversity at both the farm level and the microbial level. Biodynamic-certified wine is necessarily also organic wine.

Natural wine is not a legal designation. It has no certification board (see page 116). It is a self-policing subculture of winemakers who advocate purity in winemaking as well as farming.

PERMITTED ADDITIVES AND WINEMAKING PROCESSES

Conventional Wine

Ammonium bisulfate | Ammonium sulfate | Arabic gum | Artificial tannins | Ascorbic acid | Bentonite | Beta glucanase enzymes | Calcium carbonate | Carboxymethylcellulose (CMC) | Casein | Cellulose gum | Chitosan | Citric acid | Concentrated must | Copper citrate | Copper sulfate | Diammonium phosphate | Dry yeasts | Electrodialysis | Electromembrane acidification treatment | Enological charcoal | Enrichment by reverse osmosis | Lactic acid | Lactic bacteria | Lysozyme | Malic acid | Metatartaric acid | Ovalbumen | Pea or wheat protein fining agents | Pectinase | Polyvinylpolypyrolidone (PVPP) | Potassium bicarbonate | Potassium caseinate | Potassium metabisulfate | Potassium metabisulfate | Silica gel | Sugar | Sulfur dioxide | Tangential filtration | Tartaric acid | Urease | Wood chips | Yeast nutrients | And more . . .

Organic Wine

Arabic gum | Artificial tannins | Ascorbic acid | Bentonite | Calcium carbonate | Casein | Citric acid | Concentrated must | Copper citrate | Copper sulfate | Diammonium phosphate | Dry yeasts | Enological charcoal | Enrichment by reverse osmosis | Lactic acid | Lactic bacteria | Malic acid | Metatartaric acid | Ovalbumen | Pea or wheat protein fining agents | Pectinase | Potassium bicarbonate | Potassium caseinate | Potassium metabisulfate | Silica gel | Sugar | Sulfur dioxide | Tangential filtration | Tartaric acid | Wood chips | Yeast nutrients

Biodynamic Wine

Bentonite

Enological charcoal

Olalbumen

Sugar

Sulfur dioxide

Tangential filtration

Yeast addition by special approval

Natural Wine

The more pragmatic factions accept:
Diatomaceous earth filtration
Plate filtration
Sulfur dioxide

The more radical factions accept:
No additives, fining, or filtration whatsoever

Maximum total SO_2 at analysis before bottling: Reds | Whites	Conventional Wine	Organic Wine	Biodynamic Wine	Natural Wine
	150mg/L | 200mg/L	**100mg/L | 150mg/L**	**70mg/L | 90mg/L**	**Traces | 30mg/L***

*Natural wine has no legal definition, so these total SO_2 figures refer to generally accepted norms within the French natural wine scene, not fixed legal limits.

NATURAL /LEX·I·CON/

THE PAYSAN

In English, the word "peasant" has negative connotations. Technically, it refers to a smallholder or agricultural laborer, but more often it is used in a derogatory way to allege a lack of sophistication. Such usage has arisen because peasants are almost extinct in Anglophone cultures. As the English polymath John Berger wrote in the introduction to his 1979 novel *Pig Earth*, "The British peasantry was destroyed (except in certain areas of Ireland and Scotland) well over a century ago. In the USA there have been no peasants in modern history because the rate of economic development based on monetary exchange was too rapid and too total."

In France, Italy, and other areas where peasant culture persists, the words for peasant have mostly positive connotations. A *paysan*—or *paisano*, in Italian—retains links to noncapitalist systems of exchange, e.g., bartering services for goods. It means someone connected to the land, someone without pretension, who cares strongly for the community. Sometimes it can connote a certain wariness toward outsiders. But many leading figures of the natural wine community are proud paysans, for peasant winemaking communities have become repositories for a wealth of unique regional knowledge.

Such communities provide a historical context for wine growing and winemaking that predates the modern globalized wine market. They retain a memory of how wine was made before its production was adapted for high volumes, long-distance export, and big profits.

Many old-world natural winemakers can recall the wine-growing and winemaking methods of their parents and grandparents. They and their families experienced the social upheaval and health and environmental costs of synthetic chemical agriculture. And, like some of their longtime wine clients, they perceived their wines changing for the worse with the arrival of synthetic fertilizers, herbicides, pesticides, and contemporary enology.

Fleurie vigneron Sylvain Chanudet outside his family's compound in Prion in 2015

Natural Wine in Five Basic Principles

The best way to think of naturalness in wine is as an ideal: wine made from grapes alone, with nothing added and nothing removed.

Sometimes it's difficult to tell if a wine is natural, especially when you're confronted with a wine you don't know. We're usually taught to ask how a wine *tastes*. Understanding natural wine requires asking yourself how a wine was *made*.

Many wines turn out to be natural in some ways, but not so natural in other ways. Even among natural winemakers, there are levels of naturalness. It's not black-and-white. So here are some general questions for determining how natural a given wine is.

1. Organic Farming

Is the wine produced from organically farmed grapes?

To be considered natural, it should be.

The exceptions: Some natural winemakers decline organic certification, out of frustration with the paperwork required or the perceived hypocrisy of other organic winemakers (see page 55). Other natural winemakers purchase chemically farmed fruit, a practice they often justify with parallel efforts to incentivize better agricultural practices among their grape providers.

2. Native Yeasts

Is the wine fermented on its own native yeasts rather than exogenous (often called "commercial" or "select") yeasts?

It should be. Native yeasts are considered critical for expressing the character of a vineyard and a vintage—and for fully realizing a wine's aromatic and structural potential.

The exception: Almost all Champagnes (along with other Champagne-method wines) receive an exogenous yeast addition before the bubbles are produced (see page 332).

3. Low or Zero Sulfite Addition

Are sulfites added to the wine? How much? And when?

Sulfites are used as a preservative and antioxidant in conventional winemaking. The best natural winemakers manage to make great wines with no sulfite addition whatsoever, while others add only tiny doses in the range of 10–30mg/L.

The exceptions... are legion. The great divide within the natural wine world is between winemakers who add small amounts of sulfites, and those who never add any at all. In the French natural wine scene, wines typically clock in between 0 to 30 milligrams per liter of total sulfites at final analysis. (For more on sulfite addition, see page 80.)

4. No Other Additives or Transformative Processes

Has the winemaker used other winemaking additives? Has the wine undergone a transformative process like thermovinification or reverse osmosis?

No other additives or transformative processes should be used in wine considered natural. (See pages 84–87 for explanations of winemaking additives and transformative processes.)

The exceptions: Some natural winemakers will resort to limited micro-oxygenation. Some natural winemakers resort to adding yeast nutrients and/or the antibacterial enzyme lysozyme, but both practices are considered taboo.

5. No Fining or Filtration

Has the wine been fined or filtered at any stage of the winemaking process?

If so, that's a pity. Fining and filtration strip wines of their character. They don't pollute a wine, like wine additives do, but a wine that has been fined and filtered is no longer a living wine. (See pages 108–109 for more on wine fining and filtration.)

The exceptions: Many otherwise natural winemakers filter their white wines and rosés before bottling. The best don't.

A French Phenomenon

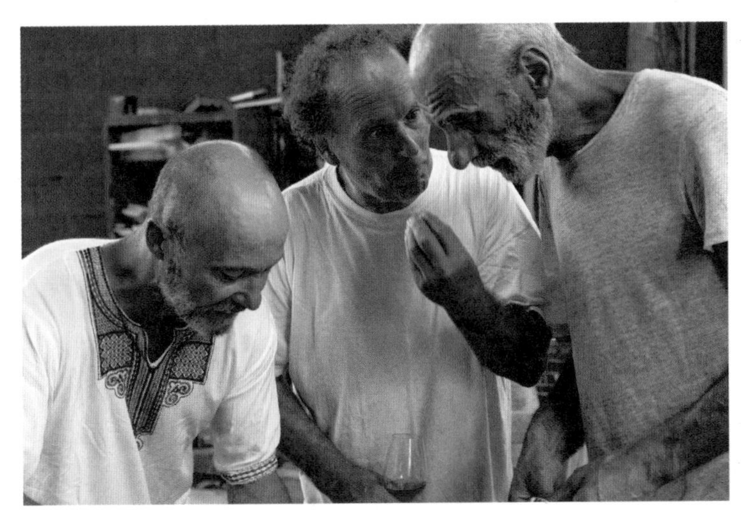

Ardèche vigneron Gérald Oustric (*center*) among friends opening oysters

Natural wine as a movement—among vignerons and consumers alike—began in France in the 1980s.

But why France?

Partly it's due to Paris, the world's most visited city (35.4 million tourists in 2019) and a hegemonic force within France. The French capital accounts for 25 percent of the country's economy, and all the train lines run through it. In wine terms, this means that vignerons from throughout France have historically all passed through Paris, where they've been able to meet and exchange ideas. (Italy and Spain, both old-world nations with ancient peasant winemaking traditions, are more fragmented both economically and culturally.) When resistance to wine industrialization began to take root in Paris in the 1980s (see page 32), it quickly spread to the rest of the nation.

France owes some of its primacy in the natural wine world today to Napoleon, who privatized church lands and created the Napoleonic Code, a system of succession that divided landownings equally among heirs. This gave rise to communities of agricultural smallholders throughout France. Statistics are not kept regarding the number of natural wine estates in a given nation, due to the elastic, unenforceable nature of the term "natural." A look at a nation's rate of organic viticulture, however, does offer an indication of its potential for natural winemaking, since natural wine is, in practice, a small subcategory of organic wine. As of 2019, France contains a quarter of the world's surface of organic vineyards. The US, by comparison, possesses just 4 percent.

Today we find the largest communities of natural winemakers in regions that share certain social and economic circumstances. There is usually a history of smallholding or sharecropping (as in the Beaujolais). Vineyard prices are low (as in Anjou), and a relatively generous social safety net exists (as in France at large). These circumstances allow vignerons to take on the inherent financial risks of natural winemaking.

Such circumstances exist often in old-world countries and almost never in the New World. It's possible to produce excellent natural wine in California, Oregon, or South Africa, for example. But cultural, historical, and economic factors make it uncommon enough that these and other new-world regions exert, for now, marginal influence on the wider landscape of natural wine. France remains natural wine's ground zero, setting the norms for production values on the international scene.

The Beaujolais, Cradle of Natural Wine

The origins of natural wine involve a patchwork of disparate cultures and individual stories. But the late Beaujolais vigneron Marcel Lapierre has attained preeminence in natural wine lore because, more than any other natural wine pioneer, Lapierre was able—and inclined—to inspire and organize wine buyers and fellow vignerons to share his commitment to purity in winemaking.

Natural winemaking in the Beaujolais began as a collaboration in the early 1980s between Lapierre; *négociant* (wine dealer), winemaker, and wine scientist Jules Chauvet; and itinerant winemaking consultant Jacques Néauport, who was then working as Chauvet's assistant. Together they began to produce a portion of Lapierre's wines without sulfites and without filtration. Throughout the 1980s, their circle expanded to include many of Lapierre's friends in the Beaujolais, along with a handful of peers in other regions.

Think of these winemakers as a subculture circumscribed in place and time, like punk in New York in the late 1970s. Since then, the word "punk" has been applied liberally: to everything from post-punk to pop-punk to dance-punk; to high fashion, to filmmaking, to art. We gauge the authenticity of later derivations of punk by their proximity—social or spiritual—to the original punk subculture. A similar technique works for assessing successive generations of natural winemakers.

So who were these original punks? Who is the Richard Hell, who is the Patti Smith, who is the Joey Ramone of natural wine?

In the Beaujolais, there was Jean Foillard (see page 132), Guy Breton (see page 128), Jean-Paul Thévenet (see page 133), Yvon Métras (see page 134), and Jean-Claude Chanudet (see page 129). The Mâconnais vigneron Jean-Jacques Robert (see page 207) was a peer in that early era, too. In the Jura, there was Pierre Overnoy (see page 180), brought into the circle by Jacques Néauport. In the Drôme, there was Philippe Laurent and Michèle Aubery of Domaine Gramenon (see page 230). In the northern Rhône, there was René-Jean Dard and François Ribo (see page 232), who maintained a certain distance from the Lapierre circle. Finally, there was François Dutheil of Château Sainte Anne (see page 322). Together they comprised the tiny network of French vignerons working to produce what was then called *vin sans soufre* in the 1980s and early 1990s.

The original Marcel Lapierre gang gathered around a table during a trip to the US with Bruce Neyers of Kermit Lynch Wine Merchant in 1993. *From left:* Jean Foillard, Yvon Métras, Marcel Lapierre, Guy Breton, Jean-Paul Thévenet, and Jean-Jacques Robert.

A NATURAL BEAUJOLAIS TIMELINE

1951: Jules Chauvet experiments with producing his Beaujolais-Villages with zero added sulfites.

1970: Jacques Néauport, then an English student in Lyon, accompanies his classmate's father to harvest in the Beaujolais. He meets Marcel Lapierre, a young winemaker at a neighboring estate.

1973: Lapierre takes over his family estate following the death of his father.

1978: Lapierre conducts his first attempts at winemaking without added sulfites.

1980: Néauport wins a contest, sponsored by *La Revue du Vin de France*, for Best Amateur of wine in France. He enrolls in the Lycée Viticole de Beaune to study winemaking.

1981: Néauport begins an internship with Chauvet. He introduces Lapierre to Chauvet. Lapierre embraces organic farming principles. Chauvet, Néauport, and Lapierre begin collaborating.

1984: Lapierre begins producing a significant portion of his wine without added sulfites or filtration.

Guy Breton (*left*) and Marcel Lapierre at one of the latter's famous Bastille Day parties, circa 1990

1985: François Morel, of Paris wine bistro Les Envierges, begins purchasing Lapierre's wines. Lapierre's neighbors Jean Foillard and Jean-Paul Thévenet begin producing natural wine following similar principles.

1988: Lapierre's neighbors Guy Breton and Georges Descombes begin producing natural wine.

1989: Jules Chauvet dies.

1992: Lapierre's neighbor Yvon Métras begins producing natural wine.

1995: Lapierre partners with Jean-Claude Chanudet and the latter's father-in-law, Joseph Chamonard, to purchase Beaujolais estate Château Cambon. A rift arises between Lapierre and Néauport, in part because the latter apparently felt left out of this new venture.

2006: Lapierre and friends organize La Beaujoloise, a tasting of natural wines from the Beaujolais that continues to this day, under the name Bien Boire en Beaujolais (see page 137).

2010: Marcel Lapierre dies. His son Mathieu Lapierre assumes responsibility at the estate.

2014: Marcel Lapierre's eldest daughter, Camille, joins Mathieu at the estate.

MARCEL LAPIERRE: THE GREAT COMMUNICATOR

1950–2010

In 1973, when Marcel Lapierre took over his family estate, the fad for Beaujolais Nouveau was taking a grave toll on farming and winemaking in the region. To ensure Nouveau wines were ready by early November, Beaujolais vignerons would harvest too early. This would oblige them to add sugar during fermentation, boosting the alcoholic potential of the underripe, overcropped harvest. (Inoculation with commercial yeasts would become widespread by the early 1980s.) Heavy sulfite dosage and filtration were used to eliminate any remaining yeast activity in the young wines. Lapierre realized he didn't like the wines he and his neighbors were making, and set about finding ways to improve them.

Lapierre had a friend named Jacques Néauport, a talented, indefatigable wine taster then studying winemaking at the school in Beaune. In 1981, Néauport began an internship with the renowned Beaujolais winemaker and wine scientist Jules Chauvet. Seventy-two at the time and still running his family's wine-dealing business, Chauvet Frères, Chauvet had conducted his own experiments in making wine without added sulfites in the 1950s. Néauport introduced Lapierre to his mentor, Chauvet, who shared his experience with the ambitious young winemaker from Villié-Morgon. Thus began a decade of collaboration, with Néauport playing the role of intermediary in a circle that soon expanded to include many of Lapierre's friends.

Lapierre credited Chauvet and Néauport with providing the scientific framework to confirm his instincts. Together they refined the methodology of sulfite-free vinification in the Beaujolais: mature harvests in small cases, rigorous sorting, no sugar addition, long whole-cluster vatting at cool temperatures, minimal pigéage, and the liberal use of refrigeration for clarification.

While Lapierre practiced organic agriculture in his own vines from 1981 onward, his signature influence was less on farming than on winemaking aesthetics, promoting a respect for purity in vinification. Each July, on Bastille Day, he would throw a lamb roast at his winery, inviting winemakers from throughout France. His charm and hospitality were legendary: he would receive fellow winemakers at his cellar and they would leave hours or days later, worse for wear but convinced of the virtues of additive-free winemaking.

Since his death from melanoma in 2010, Lapierre has passed into the realm of legend. Excellent portraits in English of Lapierre and his work can be found in Ed Behr's 2004 profile in his magazine *The Art of Eating*, and Kermit Lynch's 1988 book *Adventures on the Wine Route*, which also contains the definitive English-language profile of Jules Chauvet.

JULES CHAUVET: THE GODFATHER OF THE NATURAL WINE MOVEMENT

1907–1989

Born into a family of négociants in the village of La Chapelle-de-Guinchay, Jules Chauvet was among the most celebrated wine scientists and wine tasters of his era. In his youth he studied chemistry at the University of Lyon, and his research on respiratory ferments in yeast drew him to the work of 1931 Nobel laureate Otto Warburg, with whom he entered into a correspondence in 1935. Chauvet would intern with Warburg at the Kaiser Wilhelm Institute for Cell Physiology in Berlin for three weeks in 1938, a great honor for a young winemaker from an obscure village in the Beaujolais. The Second World War broke out after Chauvet's return to France. He was conscripted and soon captured and held as a prisoner of war. According to his former assistant and de facto biographer Jacques Néauport, he escaped in time to return to La Chapelle-de-Guinchay for the harvest, and would spend the rest of his life there, running his family business and dividing his time between winemaking, wine dealing, and scientific research.

Chauvet was what is known in perfume circles as a "nose": someone with a preternatural sensitivity to aromas. He was even invited to the French perfumery center of Grasse to attend training in scent science among the renowned perfumers of his day. His tasting notes of great wines, later compiled in such books as *Vins à la Carte*, comprise a sort of exacting scientific poetry. Saying something was "floral" famously wouldn't suffice for Chauvet. He'd cite the exact flower.

Chauvet's experiments with sulfite-free winemaking in the 1950s would form the basis of his later collaboration with Marcel Lapierre and Jacques Néauport. Thanks to the farsighted efforts of Néauport, the Chauvet estate, and the late Parisian publisher Jean-Paul Rocher, many of Chauvet's scholarly works are available in book form. They constitute some of the earliest cautionary notes about the abuse of sulfites in midcentury winemaking.

NATURAL /LEX·I·CON/

THE NÉGOCIANT

The French word *négociant* is usually translated as "wine dealer." There are two types of négociant: the négociant winemaker, who purchases grapes and makes wine; and the "pure" négociant, who purchases tanks of wine to sell as is or blend with other wines.

Jules Chauvet was at once a winemaker, who owned a parcel of gamay vines in the village of La Chapelle-de-Guinchay, and a "pure" négociant, who purchased and resold finished wines. By contrast, the Burgundy winemaker Philippe Pacalet, Chauvet's onetime intern, is a négociant winemaker (see page 201).

JACQUES NÉAUPORT: THE NATURAL WINE SAVANT

1948 to Present

Few are as responsible for the existence of today's natural winemaking community as Jacques Néauport, an eccentric from Ardèche who became the Johnny Appleseed of natural winemaking, traveling widely and encouraging additive-free, unsulfited vinification wherever he went.

While harvesting in the Beaujolais in the fall of 1970, Néauport befriended Marcel Lapierre. Néauport spent the next decade hitchhiking throughout France and tasting wine. His first winemaking experience—and first sulfite-free vinification—occurred in 1975 in the Jura, in the cellar of a neighbor of Pierre Overnoy, whom Néauport also met that year. In 1980, Néauport entered an amateur wine-tasting competition sponsored by *La Revue du Vin de France*, winning the title of Best Amateur of France. Following this success, he enrolled in the Lycée Viticole de Beaune. In the fall of 1981, he first interned for Jules Chauvet, who would become his lifelong inspiration. That same year, Néauport introduced Chauvet to his old friend Marcel Lapierre.

Left to right: Marcel Lapierre, Alain Chapel, and Jacques Néauport at lunch in 1990

Néauport would put Chauvet's ideas into practice at dozens of estates. While working with Lapierre, Néauport helped refine the winemaking styles of Jean Foillard, Guy Breton, Jean-Paul Thévenet, Jean-Claude Chanudet of Domaine J. Chamonard, and Yvon Métras. Néauport consulted with Pierre Overnoy for seventeen vintages, starting with the 1984 vintage. In the 1990s, Néauport began working with another generation of natural winemakers, including François Dutheil de la Rochère of Bandol's Château Sainte-Anne, Pierre Breton in Bourgeuil (see page 172), Thierry Puzelat in Cheverny (see page 150), Michel Favard in Saint-Émilion (see page 307), and Gérald Oustric in Ardèche (see page 236).

The scale of Néauport's reach is astonishing when you consider the practical realities of consulting in natural winemaking. Unlike clients of the famed "flying winemakers" of Bordeaux, Néauport's collaborators had no budget to pay consultants. Néauport was able to gain their trust, and to practice his art of observation in vinification, partly because he never sought to earn a conventional living. During his most active years, he famously slept in tank rooms at times. He walked or hitchhiked everywhere he went, having never learned how to drive.

The unofficial remuneration for Néauport's work with Lapierre was a source of friction between them in later years. In 1999, Néauport published a memoir, *Les Tribulations d'un Amateur de Vins*, in which he mocked Lapierre and other vignerons in their circle, referring to them by pseudonyms. The book precipitated a schism in the natural wine world, with Lapierre and his friends forswearing contact with Néauport.

Today Néauport is semiretired and lives in the mountains of Ardèche with his elderly mother. He still provides counsel to select clients, including several younger winemakers of the Beaujolais.

NATURAL HISTORY

ALAIN CHAPEL (1937–1990)

A close friend of Jules Chauvet, the late three-Michelin-star chef Alain Chapel was an early supporter of natural winemaking at his eponymous restaurant in Mionnay, a suburb of Lyon. Long associated with the "nouvelle cuisine" movement of the era, Chapel emphasized a daring, exotic side of Lyonnaise culinary heritage that was fast disappearing even in his time. (One of his famous dishes involved the entrails of a freshly slaughtered rooster.)

From 1983 until his death in 1990, Chapel tasked Chauvet's erstwhile assistant Jacques Néauport to make wine purchases for the restaurant, so diners in the 1980s enjoyed, alongside the renowned Côte d'Or wines of Domaine Ramonet and the famed Hermitage of Jean-Louis Chave, the Morgon of Marcel Lapierre and the Arbois of Pierre Overnoy.

"When I worked for Alain Chapel, he paid himself less than his two dishwashers. He told me, 'Jacques, you buy from whoever you want, in the quantity you want.'"

—Jacques Néauport

NATURAL HISTORY

SUGAR AND WINE FRAUD

A 1907 cartoon featuring the Languedoc winegrower and political organizer Marcelin Albert, protesting against the sugar addition and illicit wine blending of the era in favor of natural wine

A basic building block of fermentation, sugar has long proved a tempting additive for winemakers. Adding it during fermentation is called chaptalization, after the French chemist Jean-Antoine Chaptal, who articulated the procedure in his 1801 treatise *L'Art de Faire, Gouverner et Perfectionner les Vins*. For as long as some winemakers have added sugar to their wines to increase alcohol content, other winemakers have abstained from the practice and proclaimed their wine to be "natural."

Such was the case during the Languedoc winegrower protests of the early 1900s. Faced with a crisis of poor sales related partly to overproduction, Languedoc winegrowers, led by the winegrower and café owner Marcellin Albert, cast the blame on unscrupulous wholesalers for excessive sugar addition. Their rallying cry? "Vive le vin naturel," or "Long live natural wine."

In French, one still calls sweet wines produced without the addition of sugar *vins doux naturels*. Similarly, sparkling wines produced without the addition of sugar or yeasts have historically been marketed as "vin mousseux naturel," such as the sparkling wines of Gaillac in the southwest.

More Natural Wine Pioneers

Like many great ideas, the rediscovery of the virtues of traditional organic farming and of additive-free, low- or zero-sulfite winemaking occurred simultaneously in many places across the globe. In certain cases, it wasn't a rediscovery at all, but a continuation of how things had always been done. Here are a few other vignerons who arrived independently at what would come to be known as natural winemaking.

The Hacquet Siblings

Precursors to the contemporary natural wine movement, the Hacquet siblings—Joseph, Anne, and Françoise—ran an isolated family farm in Anjou in France's Loire Valley, where, motivated by their father's religious convictions, they toiled in near total obscurity to produce unsulfited and unfiltered wines since 1954. Their Anjou wines became legendary in natural wine circles only within the past decade, thanks to their late-in-life friendship with their natural vigneron neighbor Sébastien Dervieux (see page 161). Joseph Hacquet died in 2004, and the Hacquet sisters retired after the following vintage, ceding a parcel of their vines to Dervieux. Anne and Françoise passed away in 2021.

Pierre Guillot

In the Mâconnais (southern Burgundy), Pierre Guillot, an early proponent of organic farming, was decrying synthetic chemical viticulture and the excesses of modern enology as far back as 1977. Today his grandson Julien Guillot continues the natural tradition at their estate, Clos des Vignes du Maynes (see page 206).

René-Jean Dard and François Ribo

Northern Rhône natives René-Jean Dard and François Ribo met at the Lycée Viticole in Beaune in the early 1980s, where they studied alongside Jacques Néauport. Dard and Ribo carved a different path when they founded their

Left: The Hacquet sisters in the 1930s

Below: Mâconnais vigneron and early organics advocate Pierre Guillot

François Dutheil de la Rochère in an undated family photograph, holding his son Jean-Baptiste

estate together in 1983, eschewing the carbonic maceration and refrigeration of the Lapierre school. The duo's challenging, farsighted work has stayed stubbornly pure for forty years, even as the international luxury wine market has transformed the northern Rhône landscape surrounding them.

François Dutheil de la Rochère

Parisian natural wine buyers of a certain age recall Château Sainte Anne's François Dutheil de la Rochère as having appeared in the early 1990s in the entourage of Marcel Lapierre—and as having died shortly after, in 1996. In fact, Dutheil was a natural wine original in his own right, who had practiced unsulfited vinification on his Bandol reds as far back as the 1970s. Dutheil mentored his young neighbor Yann Rohel, who, after encountering the Lapierre circle during enology studies, would put the two maestros in touch.

Stefano Bellotti

In Italy's Novi Ligure (southern Piedmont), Stefano Bellotti returned to take responsibility for his family's farm in 1977 at the age of eighteen, producing wine without additives as a matter of anarcho-ecological instinct. He died in 2016; today, his daughter Ilaria Bellotti continues his legacy (see page 350).

Tony Coturri

In California, Sonoma winemaker Tony Coturri began producing wine without additives or filtration in 1964. A true original, he was the lone natural radical of California winemaking for his entire career. It was only in 2016 that he traveled to France at the behest of US natural wine importer Joshua Eubank to meet the French natural winemakers who, unbeknownst to him, had been working with similar principles the entire time.

Above right: Stefano Bellotti at his estate, in a photo taken by his daughter Ilaria

Left: François Ribo (*left*) and René-Jean Dard outside their cellar

Below: Tony Coturri in his cellar in Sonoma

How Natural Wine Came to Paris (and the World)

Today you can see natural wine advertised in supermarkets in Los Angeles. It's de rigueur in all the hip restaurants in Copenhagen. Prominent natural wine bars thrive in capitals from London to Tokyo to Melbourne. Natural wine is the subject of think pieces in *Vogue*, *GQ*, and *The New Yorker*. It's fair to say this global takeover came about because, after germinating in relative isolation in the Beaujolais, natural wine took root in Paris, the world's most visited city, the one through which international wine professionals of all stripes pass on their way to Champagne, Bordeaux, or Burgundy.

But this didn't happen overnight. It took the work of successive generations of restaurateurs, chefs, sommeliers, and wine dealers to make natural wine flourish in the City of Light. The most influential establishments appeared in three distinct waves. First there was a handful of natural wine bistros, mostly clustered in the northeast of Paris. Next came a generation of high-end bistros specializing in natural wine. Later, in the late 1990s and early 2000s, a generation of wine shops began serving natural wine alongside simple wine-bar cuisine. The following pages trace the evolution of natural wine in Paris, its window to the world.

"Obviously there was Chauvet, Marcel Lapierre, and Néauport, but it wasn't enough. There needed to be disciples around. And wine drinkers. And so, like many cultural movements, it came from Paris. Natural wine came from Paris and from there went to the whole world."

—Loire vigneron Jean-Pierre Robinot, founder of early natural wine bistro L'Ange Vin

Left to right: François Morel, Jean Foillard, and Jacques Néauport tasting in the Beaujolais, circa 1988 or 1989

François Morel on Discovering *Vins sans soufre*

Author, editor, and restaurateur François Morel is today recognized as the first bistro owner to begin promoting the wines that would later become known as natural wine, at his Belleville wine bar Bistrot-Cave Les Envierges, founded in 1985. After opening Les Envierges, he became part of the team behind the influential French wine review *Le Rouge & Le Blanc,* which he would edit for many years. He is the author of numerous wine books in French, notably *Le Vin au Naturel.*

François Morel in the Parc de Belleville, near the site of his former bistro Les Envierges

"The vignerons are peasants. They have a sense of observation. And they're not imbeciles, at least not in their domaines. They realized that the chemical viticulture that was offered to them was impossible. They didn't have big ideas in their heads, but they knew it would lead to catastrophe."

"The first vins sans soufre I knew were in the entourage of Marcel Lapierre. At the time, there were no interesting wine salons. So Lapierre wanted to do a small salon with his friends in Paris at my bistro. This was throughout the 1980s, six or seven years in a row."

"We drank a lot. We knew it was the sulfur that gave us headaches. Our objective was to manage to drink excessively at night, and be normal the next day. It was this objective that I tried to share in my bistrot."

"In the 1980s, when the first wines that were not yet called natural appeared, we called them vins sans soufre. Which was very simplistic. In reality, the expression 'wine without sulfur' was hiding something more vast. Notably in terms of the work in the vines."

"Obviously, to successfully make wine without sulfur, you had to have a juice that was balanced, which means having vines that were well plowed. These people making vin sans soufre plowed their vines; they worked without herbicides, pesticides. All that was understood. It wasn't just the same shit as normal with less sulfur. It's rigor in the whole path of wine production that makes it so at the end you don't have to put in sulfites."

The First Wave of Natural Wine in Paris: The 1980s

In Paris, the wines that would eventually be referred to as "natural wines" were initially called, imprecisely, *vins sans soufre*, or "wines without sulfur."

The first natural wine destinations in Paris were six small bistros that stocked the wines of Lapierre and his friends.

Les Envierges (now closed)

The first natural wine bistro in Paris was Belleville's Les Envierges, opened in 1985 by François Morel, a former arts editor at Éditions Larousse. Morel hadn't planned to focus on natural wines, since the category didn't exist yet. But the year he opened Les Envierges, a restaurateur friend from the Mâconnais, Leny Chavasson, put him in touch with a young Marcel Lapierre, sparking a revolution.

La Courtille (now closed)

Bernard Pontonnier, former owner of Café de la Nouvelle Mairie on the Left Bank, partnered with his friends Claude Carmarans and François Morel to open the terraced wine bistro La Courtille, down the street from Les Envierges. Writing in the *New York Times* in 1992, food writer Patricia Wells called it "a sparkling bistro-brasserie with a terrace and a view that dominates the city."

L'Echanson (now closed)

The first iteration of the wine bistro L'Echanson was opened in 1985 on rue Daguerre, by Luc Desrousseaux, who would first purchase Jean Foillard's wines in 1986. Since 2008, Desrousseaux has run the wine bistro Le Gibolin in the southern French town of Arles (see page 327).

Le Passavant (now closed)

Just southwest of Belleville, on rue des Goncourt, Bernard Passavant, a former plumber, opened his bistro Le Passavant in 1987. Before his death in 2013, Passavant would have a brief career as a vigneron at Domaine du Pas de l'Escalette.

Le Baratin

In 1988, a young wine clerk named Olivier Camus and his Argentinean girlfriend Raquel Carena, both regulars at Les Envierges, opened their wine bar,

The bar at Bistrot-Cave Les Envierges in 1994

Olivier Camus behind the bar at Le Baratin, circa 1990

Bernard Pontonnier in January 2022, during a wine tasting at Clos du Tue Boeuf in the Loir-et-Cher

Le Baratin, a few streets away on rue Jouye-Rouve. Olivier Camus left Carena and Le Baratin in 2001, and was replaced by the bistro veteran and natural wine aficionado Philippe Pinoteau. Carena and Pinoteau still operate Le Baratin today (see page 403).

L'Ange Vin (now closed)

The first generation of Paris natural wine bistros was completed with the 1988 opening of the first iteration of L'Ange Vin. Plumber turned wine writer Jean-Pierre Robinot, himself a regular of Les Envierges, opened L'Ange Vin on rue Richard Lenoir in the 11th arrondissement. It later moved to rue Montmartre in the 2nd arrondissement before closing in 2003. Robinot purchased vines in the Loire and began a career as a winemaker in 2001 (see page 155).

"It's a story between friends. Friends who saw each other each year, who discussed, who drank a lot, who spent a lot of soirees together."

—Raquel Carena, chef and cofounder of Paris natural wine bistro Le Baratin

---

The Second Wave of Natural Wine in Paris: The 1990s

With the exception of Nicolas Carmarans's Café de la Nouvelle Mairie, the second wave of Paris natural wine destinations was united by the sommelier turned wine dealer Jean-Christophe Piquet-Boisson. A legendarily irascible personality, he would bring Marcel Lapierre's circle of early natural winemakers to the city's nascent *bistronomie* movement, in which young chefs, after training at Michelin-starred restaurants, were forsaking the traditional career paths in gastronomic restaurants to devote their talents to simpler bistro cuisine.

La Régalade (now closed)

Opened in 1991, chef Yves Camdeborde's original La Régalade location in the 14th arrondissement seduced the city's food scene, and turned a generation of Left Bank Parisians on to early natural wines. Camdeborde today runs the luxury hotel and bistro Le Comptoir du Relais, along with several locations of his Avant Comptoir wine bars.

Le Repaire de Cartouche

Opened in 1997 with a loan from Domaine Gramenon winemaker Philippe Laurent, Le Repaire de Cartouche is the bistro of the jolly, towering Normand chef Rodolphe Paquin. Emblazoned on its awnings are the names of the natural wine estates he has never wavered from supporting: Clos du Tue-Boeuf, Domaine Gramenon, and Dard et Ribo.

"Piquet-Boisson converted the whole group: Yves Camdeborde, Rodolphe Paquin, Thierry Breton, Thierry Faucher. They contributed to making natural wine well known, but they also helped make natural winemaking a vocation. Because sometimes buyers give ideas to producers about how to make wine."

—Cheverny vigneron Thierry Puzelat

Left: Rodolph Paquin's Le Repaire de Cartouche in Paris's 11th arrondissement

Below: Wine agent Jean-Christophe Piquet-Boisson (*right*) with Marie Lapierre of Domaine Lapierre in 2002, at chef Thierry Breton's restaurant Chez Michel

The Café de la Nouvelle Mairie

L'Os à Moelle

Chef Thierry Faucher trained under the three-Michelin-star chef Christian Constant before opening, in 1994, his natural wine bistro L'Os à Moelle. Today he maintains L'Os à Moelle in addition to the nearby wine bar La Cave de l'Os à Moelle, and a multilevel restaurant in the Paris-adjacent suburb of Chatillon called Le Barbezingue.

Chez Michel

The wild-eyed Brittany-born chef Thierry Breton once cooked at the Élysée Palace for President François Mitterrand. In 1995 he opened his flagship natural wine bistro, Chez Michel, which offers a scintillating menu of specialties from Brittany, often written in dialect.

Café de la Nouvelle Mairie

In 1994, in an interesting dynastic twist, Claude Carmarans's son Nicolas purchased the Café de la Nouvelle Mairie, once owned by La Courtille restaurateur Bernard Pontonnier, and turned it into one of Paris's premier natural wine destinations. (It would remain so until 2018, when Nicolas's successors, Ben Fourty and Corentin Bucillat, sold the bistro to new ownership outside the friends' circle.) Nicolas later embarked on a career as a vigneron in Aveyron (see page 312).

"Like the industrial chemists who are ruining true gastronomic cuisine, enologists have already destroyed all the winemaking traditions refined by two thousand years of loyal and constant use. The enologist is the avatar of the chemical industry."

—Jean-Christophe Piquet-Boisson, *Siné Hebdo* (October 2008)

The Third Wave of Natural Wine in Paris: The Early 2000s

La Crèmerie in Paris's 6th arrondissement seats just fourteen diners.

The third wave of natural wine in Paris occurred in the early 2000s, led by a generation of wine retailers who ran their shops like wine bars, serving simple food from makeshift kitchens. These new establishments were called *caves à manger*, or "wine shops where you eat." This inexpensive, informal style of wine service struck a chord with a younger generation of wine drinkers, who were discovering what had become, by then, a thriving and diverse French natural wine scene.

Le Verre Volé

Opened in 2000 by Ardèche native Cyril Bordarier, Le Verre Volé helped initiate a revival of its Canal Saint-Martin neighborhood, which, like the natural wines Bordarier promotes, wasn't always as hip and refined as it is today. Bordarier ranks among the most influential wine retailers of his generation. Today the Verre Volé empire also includes a separate wine shop and a grocery in the 11th arrondissement (see page 403).

La Crèmerie

Swiss restaurateur Pierre Jancou (see page 43) opened his jewel box–size wine bar in 2001. From early on, it was distinguished by Jancou's uncompromising taste in natural wine and its heavenly original ceiling of glass-paneled silk heraldry, a registered historic architectural feature. La Crèmerie is now under new ownership; Jancou went on to found and later sell several more influential Paris restaurants, notably Racines and Vivant.

Le Verre Volé's original location beside the Canal Saint-Martin. The restaurant also does brisk business as a wine shop for the throngs of picnicking Parisians who line the banks of the canal in spring and summer.

Le Chapeau Melon (now closed)

Olivier Camus—a cofounder of Le Baratin in 1988—opened his nearby Belleville cave à manger Le Chapeau Melon in 2003 and continued to run it until 2015. It was notorious for Camus's deep wine cellar and his patience-testing handcrafted cuisine.

La Muse Vin (now closed)

Opened in 2003 by Guillaume Dubois and Guillaume Dupré, La Muse Vin occupied the same stretch of road where their forebear Jean-Pierre Robinot's L'Ange Vin once sat. Dupré departed in 2007 to help open the 2nd arrondissement bistro Coinstot Vino. La Muse Vin closed in 2011. Today Dupré runs the 11th arrondissement natural wine bar Goguette.

La Quincave

Homey and wood-paneled, Frédéric Belcamp's La Quincave opened in 2003. Belcamp passed away in early 2022, but the cave à manger he founded and still draws a reliable crowd of Left Bank locals for cheese, charcuterie, and old vintages from pioneering natural winemakers.

NATURAL /LEX·I·CON/

THE CAVE À MANGER

Troyes cave à manger Aux Crieurs de Vin

Cave à manger means "a wine shop that serves food." The idea became popular in Paris and certain other cities in France in the late 1990s and early 2000s, partly because it allowed wine shops to behave like wine bars without the costly bar license. It also promoted a practice that was considered novel at the time: choosing food to go with wine, rather than vice versa. Many of the foundational natural wine destinations of France are caves à manger, like Aux Crieurs de Vin in the Champagne town of Troyes (see page 342) and La Part des Anges in Nice (see page 328).

The concept was soon exported to wherever it's legal to serve food in a wine shop. Many international natural wine destinations today, like Oakland, California's Ordinaire (see page 411) and London's P.Franco (see page 408), take the French cave à manger as a template.

Terroir, Sulfites, and the AOC System

Terroir refers to all the aspects that make a geographical site unique: its soil type, its sun exposure, its average rainfall, its proximity to forests, its native microbial population, its history with a given grape variety or set of grape varieties, within a given winemaking culture.

Don't get hoodwinked by anyone who suggests terroir is an airy French philosophical concept. Terroir is obvious. Examples abound outside the world of wine, too. Everything from Kumamoto oysters to Le Puy lentils to Kobe beef can express the site and the culture where it is cultivated.

Because terroir can determine quality—and therefore price—the notion has long been the subject of contention. France's system of *appellation d'origine controlée* (AOC), created in 1935 and subject to revision ever since, was intended to codify the commercialization of terroir for the sake of preserving it from fraud. But the system's parallel mission is to ensure that winemaking is profitable for the winegrowers of a given region.

Since the mid-1970s, as the cost-cutting and labor-saving effects of synthetic chemical farming and conventional enology grew institutionalized, the two missions of the French appellation system—and systems like it around the world—came increasingly into conflict. Conventional, mass-market wines became yeasted, heavily sulfited, filtered parodies of their appellations, while the natural wines that truly incarnated the historical identity of an appellation were increasingly denied the right to bear its name.

NATURAL HISTORY

"WINEGATE" AND WINE ORIGIN FRAUD

In the twentieth century, the expansion of rail transport made it easy for unscrupulous négociants to import cheap wine and pass it off as more expensive wine. The 1973 to 1974 "Winegate" scandal of Bordeaux is a classic example, in which the influential Cruse wine family and a wine dealer named Pierre Bert were implicated in a scheme to pass off three million bottles of blended Languedoc wine as wine from more expensive Bordeaux appellations.

Journalist Hervé Chabalier, writing in *Le Nouvel Observateur* in 1973, quoted Bert, who made light of the issue: "In Bordeaux, 90 percent of the négociants and 50 percent of the estates are frauds. Two or three hundred thousand hectoliters of Languedoc wine arrive in Bordeaux each year: they do indeed go somewhere."

The négociants and wine estates of the era who employed the phrase "natural wine" were responding to this concern about the origins of a wine. On the label of this 1964 bottle of Château d'Arpayé Fleurie, which the Beaujolais vigneron Marc Delienne found hidden in the wall of the négociant cellar he took over in 2015, is the proud phrase "100 percent natural and from its origin."

A bottle of 1964 Chateau d'Arpayé Fleurie

Jules Chauvet's unsulfited Beaujolais-Villages, which he titled "Pure Origine"

What sulfite addition shares with chemical farming is a capacity to suppress the microbial signature of a vineyard site—part of its geographical identity, its terroir. As Burgundy négociant and natural winemaker Philippe Pacalet wrote of herbicides, pesticides, and sulfites in his introduction to a 2007 edition of Jules Chauvet's *Études Scientifiques et Autres Communications*: "These products act by making a negative selection on [native] yeasts, which no longer permits the delivery of the organoleptic information of the terroir of origin."

"The reputation of French wine was established in the eighteenth to nineteenth century. The research of terroirs happened then. The appellations followed. But the reputations were already acquired. That means they were acquired with vines that were plowed, harvested by hand. There was no liquid sulfur, there were no industrial yeasts."

—Cheverny (Loire) vigneron Jean-Marie Puzelat

In such a way, contemporary natural wine can be considered a response to the failure of appellation systems.

"All of us who began to question things in the 1980s—the bistrot owners, the winemakers—we already understood that the French AOC system was dead," says early natural wine supporter François Morel. "Those who supported the AOCs betrayed them every day. The pesticides, the herbicides, the fertilizers: it comprised a betrayal of the idea of origin."

The system designed to prevent fraud had wound up becoming a kind of fraud in itself. This fraud affects wine on two fronts: wine growing and winemaking.

Natural Wine in the Twenty-First Century

In the wake of the initial waves of natural wine establishments in Paris, the phrase "natural wine" has gone global. It has taken on power in proportion to the scale of the problems it addresses. Today "natural wine" broadly refers to wine made in opposition to systemic agricultural folly and systemic winemaking folly, and the false world of wine aesthetics that results. Any more precise definition remains the subject of debate.

Generally speaking, there are those within the natural wine world who consider organically farmed, nonirrigated, additive-free, unsulfited, unfiltered natural wine to be an ideal to strive toward, while acknowledging that not every wine can be produced that way every year, if a wine is to be pleasurable, and if an estate is to be profitable. (And if a wine is to remain affordable.) In the opposing camp are the vignerons, buyers, and wine drinkers who insist on zero compromise on these matters. For them, a wine must remain 100 percent pure if it is to be worth drinking. A number of neologisms have arisen to refer to the latter aesthetic: "zero-zero" (as in zero sulfites, zero filtration); "nothing added, nothing removed"; "pur jus" (French for "pure juice"), and more.

The contours of this debate arose in the early 2000s, initiated by Loire vigneron Jean-Pierre Robinot, who began an association called Les Péripheriques in the Loire to unite younger vignerons who shared his total opposition to sulfite addition. Robinot's former Parisian colleagues at *Le Rouge & le Blanc* were aghast. But supported by like-minded vignerons like Olivier Cousin (see page 156), Sebastien Dervieux, Patrick Desplats (see page 162), Domaine de Peyra (see page 217), and Pierre Beauger (see page 218), Robinot's ethos captured the imagination of a younger generation of restaurateurs, winemakers, and wine drinkers. Paris restaurateur Pierre Jancou notably took up the cause, as did his protégé, Ewan Lemoigne, who would prove influential in his own right at his Paris restaurants Saturne (since closed) and Clown Bar (Lemoigne has since moved on).

Similar schisms later occurred within natural wine circles in other nations, notably that between the late Friulian vigneron Saša Radikon (who forswore sulfite addition) and his friend Josko Gravner (who insists upon it). The world of natural wine also fractures on the issue of the certification of natural wine, with some high-profile figures insisting on the necessity of certification (Angiolino Maule, Jacques Carroget) and many others on its inevitable futility.

For now these debates are mostly limited to highly developed natural wine markets, like Paris, Copenhagen, Tokyo, and certain cities in the US. Elsewhere in the world, even the pragmatists of natural wine are considered radical.

A summer 2021 pop-up restaurant organized in Paris by the wine bar Chambre Noire in collaboration with Pierre Jancou, during a tasting featuring expat natural vignerons of Burgundy. A successor to Jancou's natural wine radicalism, Chambre Noire serves exclusively unfiltered, unsulfited wines.

Pierre Jancou on Natural Wine Extremism

Throughout the 2000s and 2010s, the well-appointed, radical Paris wine bistros of Swiss restaurateur Pierre Jancou created an influential stylistic template for natural wine service worldwide. Shoebox-size creations like La Crèmerie (2001–2006), Racines (2007–2009), Vivant (2011–2015), and Achille (2016–2018) captured the sublime in traditional Parisian décor, enchanting all comers even as Jancou practiced his trademark extremism in wine selection and ingredient sourcing. Legendarily restless, Jancou has in recent years launched and parted ways with projects in Châtillon-en-Diois, Zurich, and Paris. Most recently he took over the Café des Sports in the Languedoc village of Padern.

Pierre Jancou at his former restaurant in Savoie, the Café des Alpes, which he ran from 2018 to 2020

"I fell into the natural wine movement when I opened La Crèmerie in 2000. I knew of L'Ange Vin, Jean-Pierre Robinot's bistrot, because it was near my first restaurant, La Bocca. I used to go to L'Ange Vin without knowing it was natural wine he served there."

"I don't like what natural wine has become now in Paris. For me natural wine is simple, not chichi, not snobbish. For sommeliers, it's not about just liking these wines. It's about understanding how they're made, and being able to discuss them with the clientele, and transmit a passion about them."

"What I was looking for in a wine was energy and digestibility. For a wine that does me good. Not only good in terms of taste, but for my body and soul. A wine by Patrick Desplats was the first wine that gave me that feeling. That's how I became interested, in an extremist way, in only unsulfited, additive-free wines."

"If I have a little talent, I think it's to give my passion to people, to make them understand. Yesterday an old man from Alsace came into the restaurant. I gave him a Domaine du Mazel Charbonnières from 2000, and now he only wants to drink natural wine."

"I opened Racines in 2007. It was named in homage to Claude Courtois. We were the first place in the world, I think, that had only pure wines on the wine list. We made no compromise. And no compromise in the food. It was interesting."

Olivier de Moor in his aligoté vines before harvest

2. How Grapes Are Grown

Modern Wine Growing

In the Beaujolais, vignerons have a phrase for the poor-quality vineyard sites in the flatlands adjacent to the Saône River: *terroir de maize*, or "corn terroir."

Grapes aren't corn. This is obvious, but it's worth emphasizing. Grapes are quite special in the way they can express the nuances of their farming and terroir origins on our palates. Most of the fundamental problems in modern wine growing derive from the global trend of trying to grow grapes as if they were corn—that is, like a large-scale commodity destined for heavy processing before consumption.

For winegrowers, quality farming is not merely a question of choosing whether to employ synthetic fertilizers, herbicides, insecticides, and fungicides. The issue goes as deep as the choice of planting site and the DNA of the grapevine itself. The modern agro-industrial vineyard has become a complex, highly rationalized system where every stage is informed by a productivist worldview focused on cost cutting, labor savings, and, above all, high, consistent grape yields. This paradigm is worlds apart from wine's origins in mixed-agricultural farming communities, where wine production traditionally complemented the production of grains, fruits, vegetables, cheeses, and livestock.

The story of twentieth-century wine growing is one of increasing specialization, as the naturally fluctuating, symbiotic systems of mixed agriculture were discarded in service of economic rationalization. But the synthetic fertilizers, herbicides, and pesticides widely employed in search of high, consistent agricultural yields have resulted in devastating ecological costs, including a decline of 400 million in French bird populations over the last three decades; insect populations have declined an estimated 80 percent over the same period, according to Vincent Bretagnolle, a CNRS ecologist at the Centre for Biological Studies in Chize.

Synthetic chemical farming is also linked to depleted yeast populations in soils, and the eradication of crucial soil fungi called mycorrhizae, which render mineral nutrients assimilable by plant roots. In such a way, synthetic chemical viticulture tends to yield nutrient-deficient wine musts, thereby creating a need, later in the winemaking process, for enological intervention.

Hazard signs near a vineyard in Napa

MONOCULTURE, POLYCULTURE, AND BIODIVERSITY

Monoculture means the growing or raising of just one thing, and often a lot of it—such as a farm that grows only corn, or a ranch that raises only cattle. It's the dominant method of farming in the contemporary era, because it allows farms to take advantage of highly specialized production systems and equipment.

Polyculture means the growing or raising of many different things together in one place. It was the dominant method of farming in most parts of the world until well into the early twentieth century—think of an autonymous family farm with a diverse vegetable garden, a small fruit orchard, along with cows, sheep, and chickens. Polyculture has the benefit of reducing risk: if it's a bad year for one crop, there are others crops to make up for the shortfall. It also makes it easier to maintain natural, beneficial symbiotic systems within a farm. For example, the manure from a farm's cows can be used to fertilize the farm's fields or vineyards. Sheep can be pastured in vineyards to manage grass cover. Even simple hedges between plots provide a valuable service, housing birds and other fauna that help keep pests under control.

> **"Nature has found only one way to achieve balance. That's diversity. So doing organics or biodynamics or permaculture as a monoculture is impossible."**
>
> —Faugères vigneron Didier Barral

Biodiversity refers to the range of flora and fauna—plant and animal life, right down to microbes and fungi—in a given area. Increased biodiversity is among the chief benefits of polyculture.

Natural winegrowers place a strong emphasis on polyculture and biodiversity. The most progressive natural wine estates are genuine farms, in which a plethora of other crops and animals coexist with—and support—the activity of wine growing. In the Gard, in the Languedoc, for example, vigneron Louis Julian's farm maintains, along with his 22 hectares (54.3 acres) of vines, 300 chickens and 30 hectares (74 acres) of grains.

Geese roaming freely at Cascina degli Ulivi in Piedmont, Italy. In addition to wine, the estate produces fruits, vegetables, cow's-milk cheeses, and breads from heritage grains.

The Conventional Vineyard

Here are nine of the worst excesses of conventional viticulture. Not every conventional winegrower is guilty of all these things, of course. These vineyard features are nonetheless common in much of the world.

1. High-yield vine rootstocks and clones

Most of the world's vineyards are planted with clones—vine cuttings taken from an existing vine, selected for traits beneficial to a winegrower's circumstances. But vineyards planted entirely to the same clones lack genetic diversity, creating mass susceptibility to pests and maladies.

2. Wide vine training, adapted to harvesting machines

Modern vine training systems are designed to allow agricultural machines to pass easily through the rows. Harvesting machines (see page 92) in particular cannot pass through older, denser forms of vine training.

3. Monoculture

Growing only grapes creates diverse liabilities, including erosion and greater susceptibility to pests.

4. Herbicides

Years of herbicide use creates lifeless, almost lunar terrain. In such soils, vines' roots remain superficial, lacking incentive to plunge deeper in search of water and nutrients. Herbicide use also encourages erosion by removing grass that holds soil in place.

5. Insecticides

Insecticides come in two types: contact (ones that wash off) and systemic (ones that enter the plant system). Both are disastrous for vineyard biodiversity and present underexplored health risks.

6. Antifungal agents

Synthetic antifungal agents harm native yeast populations in vineyards. They also harm mycorrhizae, the fungi that permit plant roots to absorb minerals from the soil.

7. Synthetic fertilizers

To boost yields, conventional farmers employ synthetic fertilizers, which tend to exhaust vines and yield grapes of lesser quality.

8. Mechanical pruning

Large-scale industrial grape growers employ machine pruning systems. Vines maintained in such a way see their life spans drastically shortened.

9. Irrigation

Rainfall helps determine yield and harvest quality each year, so conventional winegrowers in dry climates leave nothing to chance, irrigating their vines to deliver the desired water level. The practice leads to superficial root systems and fruit of questionable quality (see page 68).

New plantings in Napa

The Natural Vineyard

Naturally farmed vineyards vary according to region and to the individual approach of the vigneron. But here are nine common features.

1. Massal selection vines

Instead of planting all the same clonal vines, a winegrower practicing massal selection will take cuttings from a variety of vines that have shown positive characteristics. Vineyards planted with massal selection enjoy the benefits of genetic diversity.

2. Sensitive vine training

The common thread among the vineyard-training systems of natural vignerons is an emphasis on vine health and grape quality rather than high yields or cost cutting.

3. Polyculture

Many natural vignerons grow more than just grapes. The aim of polyculture, or cultivating more than just one crop in a given area, is to promote biodiversity.

4. Manual grass management

Natural winegrowers never use herbicides. To manage grass competition, they use a mix of plowing, mowing, hoeing, pasturage, and weeding. Natural winegrowers who practice permaculture (see page 62) strive not to plow or remove grass cover at all.

5. Natural pest management

In an ideal natural vineyard, pests manage themselves. Natural vignerons aim to promote a balance of predators and prey within the vines, for example, by creating hedges to harbor birds (see page 66).

6. Sparing use of copper and sulfur treatments

Natural vignerons employ copper and sulfur treatments (see page 67) as sparingly as possible, to minimize their ecological impact and prevent copper buildup in soil. In rare cases, natural vignerons manage to cultivate vines without such treatments.

7. Minimal—and natural—fertilizers

Natural vignerons seeking to restore soil vitality often employ what is called green manure, cover crops grown and then reintegrated into the soil. Others seek out local horse or cow manure. Some apply compost treatments. Many forswear fertilizers entirely.

8. Sensitive pruning

Natural vignerons prune by hand with manual or electric clippers, taking care to manage crop levels in a qualitative way.

9. Dry farming

Irrigation has little place in traditional natural viticulture.

Young cortese plantings amid fruit trees at Cascina degli Ulivi in Novi Liguri (Piedmont, Italy)

HOW WINE GROWING CHANGED IN THE NINETEENTH AND TWENTIETH CENTURIES

European wine cultures are often depicted as traditionalist and hidebound, compared with new-world wine cultures. Yet the story of European viticulture in the past century is, in fact, one of rapid, jarring adaptation, as age-old regional wine cultures were transformed by pests, war, and technological innovation.

This was the cultural backdrop to the rise of natural wine in France in the 1980s and '90s. To understand today's natural wine scene, it helps to consider what happened to Europe's vineyards just before and throughout the twentieth century.

An undated postcard depicting vineyard workers in the Côte d'Or applying sulfur spray treatments

1863: The American vine louse phylloxera is detected in the vineyards of the southern Rhône, heralding a period of destruction in European viticulture. Efforts to reconstitute the vineyards of Europe by replanting and grafting upon louse-resistant American rootstock continue into the twentieth century. Before phylloxera, viticulture was widespread throughout France and the rest of Europe, even in regions we no longer associate with wine production, like Normandy. Winemaking in these marginal regions never recovered. Even in regions that successfully reconstituted their vineyards, there was a huge loss of biodiversity, as growers neglected to replant traditional blending grapes and instead focused on more famous varieties.

LATE 1800S: As European vineyards are replanted due to phylloxera, vineyard work changes. Formerly vines had been planted in a higgledy-piggledy fashion called *en foule*, which required them to be hoed by hand, since horse plows couldn't navigate the zigzags. Europe's winegrowers replant in clear rows, which permits the use of horses for plowing. Guyot training becomes the norm throughout France, alongside the traditional goblet training.

1905: German chemist Fritz Haber devises a process to convert atmospheric nitrogen into liquid ammonia. His innovation leads to the development of chemical fertilizers, which lead to vastly higher farm yields, along with a corresponding decline in soil health and fruit quality.

1914: World War I leads to diminished harvests throughout Europe, due to a lack of material and labor. Mass rural depopulation begins.

Above left: Women harvesting in the Indre-et-Loire department in 1946

Above center: Horse plowing in Vouvray in 1955

Above right: Pruning in Vouvray in 1955

1935: France passes its landmark AOC system. Intended to curb fraud and safeguard the value of wines' geographic origins, the system initiates further loss of vineyard biodiversity, as winegrowers uproot the minor grape varieties discouraged or forbidden by the new appellations.

1939: World War II devastates rural populations in Europe. It also heralds the age of chemical pesticides and herbicides, most famously the pesticide DDT (invented just before the war) and the herbicide 2,4-D, developed in 1944. The chemical agriculture industry blooms directly in the wake of World War II, repurposing technology formerly used to produce bombs and chemical weapons.

LATE 1950S: Tractors begin to replace horses in vineyards. Row spacing in vineyards grows wider to accommodate larger machines. By the 1980s, very few European vignerons practice horse plowing.

1970S: The first harvesting machines appear. Their use remains banned to this day in many prestigious vineyard zones, but they quickly become normalized in many wine regions, particularly outside Europe.

1974: Agro-industrial conglomerate Monsanto markets Roundup, made up of the systemic herbicide glyphosate. Throughout the 1970s, herbicides, and Roundup in particular, begin conquering viticulture worldwide. Farmers cease plowing and hoeing, and begin spraying herbicides. This leads to a decline in native yeast populations, making native yeast fermentation less reliable. It also leads growers to plant vineyards on terrain impossible to farm without the use of herbicides. Since herbicidal farming requires far less manual labor than traditional farming, the era sees a continued decline in rural populations.

Today the ecological toll of herbicide use is readily apparent. But few natural winegrowers blame their parents for farming this way. Most acknowledge that for farmers persisting throughout the upheavals of the twentieth century, the arrival of herbicides seemed like magic.

Vineyard workers in Luxembourg applying Bordeaux mixture and compost, September 1938

Natural Viticulture

Austrian vigneron Andreas Tscheppe in his terraced sauvignon vineyards in Styria

Natural viticulture is a catchall idea encompassing several different systems of farming. Like natural winemaking, it shouldn't be considered a new phenomenon. It merely comprises a resistance to ascendant intensive methods of growing grapes.

The two most well-known systems of natural viticulture are organic farming and biodynamic farming. Natural winegrowers consider these systems not as ends in themselves, but as a starting point for sensitive, regenerative farming. Natural wine as a category exists today because many vignerons practicing organic and/or biodynamic agriculture acknowledge that these certifications are insufficiently strict regarding winemaking practices.

Both organic farming and biodynamic farming are legally defined systems, subject to certification. As such, they're also subject to loopholes and various forms of exploitation. Some natural vignerons decline certification of their agricultural methods because they resent supporting (and paying for certification from) an insufficiently rigorous system. Other natural vignerons avoid agricultural certification because they prefer not to be associated with organic and/or biodynamic winemakers who do not practice natural winemaking.

Organic Farming

All farming was organic before synthetic nitrogen fertilizers became widely affordable in the early twentieth century. The subsequent decline of traditional agriculture methods, along with innovations in plant breeding and agricultural machinery, gave rise to the notion of organic farming as something distinct from modern farming. The actual phrase "organic farming" was coined in 1940 by the English agriculturalist Walter James in his book *Look to the Land.*

Movements toward certification of organic farming began in France with the founding of the Groupement d'Agriculture Biologique de l'Ouest (GABO) in 1958. The association soon branched into two distinct approaches to organic farming. One was termed the Lemaire-Boucher approach, named for an agribusiness founded by two professors who sought to furnish farmers with organic treatment products. The French consumer-producer association Nature & Progrès, founded in 1964, represented the second approach, which prefigured the notion of natural farming. Nature & Progrès avoids commercial links and emphasizes a wider critique of productivist agriculture.

Organic certification was first legally defined in France in 1980. The first European guidelines for organic agriculture were published in 1991. Ten years later, the United States followed suit with the establishment of its National Organic Standards.

The details of organic certification vary between nations. What they share is an aim to farm without the use of synthetic fertilizers, synthetic pesticides, herbicides, or genetically modified organisms (GMOs).

Traditional farming tools in the Beaujolais, part of a museum in progress at the home of Fleurie vigneron Yvon Métras

In the Wineglass: Organics

Is organic agriculture perceptible in a wine tasting? It depends on the wine tasting.

There are no common traits to the tastes or aromas of "just" certified organic wine, which as a category encompasses far more conventionally vinified wine than naturally vinified wine. It is therefore senseless to announce, as some do, that a wine tastes "organic." People who do this are confusing organic wine (which rarely differs in taste and aroma from conventional wine) with natural wine (which is often very different).

However, if you compare two naturally vinified wines, one from long-term organically farmed vines and another from chemically farmed vines, then, controlling as best you can for all other variables, you often find that the organic natural wine possesses greater depth and nuance.

Biodynamic Farming

Biodynamic farming is the witchy older cousin of what would become known as organic farming. It's a system of agriculture first articulated by the Austrian philosopher Rudolf Steiner in a series of lectures given in 1924 in the then German town of Koberwitz (now the Polish town Kobierzyce). Steiner was responding to farmers' concern about the growing use of chemical fertilizers and their effects on crop quality and farm health. He proposed an alternative form of agriculture based on biodiversity, the lunar calendar, and the holistic idea of the farm as symbiotic entity. He detailed nine preparations to enhance the vitality of a field. The most famous is preparation BD500, a dilution of cow manure that has been buried inside a cow horn for six months.

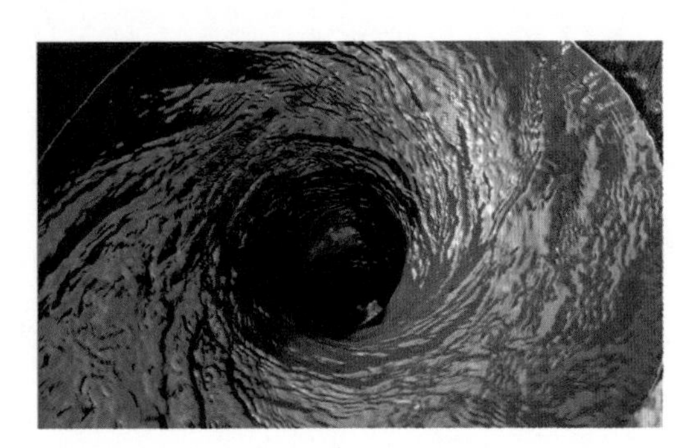

Dynamization of a Maria Thun preparation (a variation on preparation BD500) at Domaine Derain in Burgundy

Steiner died in 1925, but his *Agriculture Course* would have far-reaching influence. In 1938, his student Ehrenfried Pfeiffer, drawing upon the research of Steiner's many followers, published *Bio-Dynamic Farming and Gardening*. The book would influence the English agriculturalist Walter James, who later coined the term "organic farming" in his own 1940 book, *Look to the Land*.

As of 2019, biodynamic farming accounts for roughly 202,000 hectares (499,000 acres) worldwide. Demeter International is the most prominent certification board, with individual Demeter organizations in each nation. In France, Biodyvin also certifies biodynamic wine.

Biodynamic vignerons including Nicolas Joly in Savennières and Mark Angeli in Anjou helped popularize biodynamic viticulture throughout the 1980s and '90s. The natural wine movement arose in the same era, but without significant commercial or ideological overlap with the more high-profile biodynamicists, many of whom continued to rely on conventional vinification techniques like high sulfite doses and filtration.

In the Wineglass: Biodynamics

As with organic wine growing, biodynamic agriculture cannot be directly tasted in a glass of wine. Biodynamic certification does assure a moderate level of purity in winemaking, however, which can make biodynamic wines stand out slightly from their conventional counterparts.

Long-term application of biodynamic agriculture can produce wines that possess greater amplitude and dimension, a more sustained richness and length. In a 2007 *New York Times* piece, wine critic Eric Asimov, a longtime proponent of biodynamic and natural wines, described a sensation of "aliveness" in certain biodynamic wines.

Common Critiques of Organics and Biodynamics

Organic and biodynamic viticulture represent only a tiny fraction of viticulture worldwide. If both are subjects of controversy, it's because consumer awareness of the health and environmental costs of synthetic pesticides, herbicides, and fertilizers poses a threat to estates that employ them. Representatives of such estates tend to employ a familiar set of talking points in efforts to delegitimize organic (and to a lesser extent, biodynamic) agriculture. But large-scale organic and biodynamic agriculture can indeed pose practical and ideological dilemmas, especially when employed in pursuit of excessive yields.

ORGANICS

1. Overuse of copper

For organic winegrowers, the primary defense against fungal attacks is copper sulfate. But applying copper sulfate can cause copper buildup in soil, which is harmful to earthworms and livestock. Some conventional winegrowers argue that employing systemic antifungal agents—whose risks are underexplored—allows them to use less copper.

2. Certification flaws

Organic certification is only as rigorous as the inspectors certifying it. Many organic vignerons report frustration with how easy it is to fool inspectors.

3. Soil compaction and erosion

Most organic practitioners plow to manage grass cover, but the most common modern means of plowing—the tractor—can compact soil, preventing the flow of oxygen and water within it. Excessive or untimely plowing can also harm soil by breaking up biological matter and exposing it to the sun, and can lead to increased erosion (see page 62).

4. Big Organics

Organic agriculture permits certain practices, like irrigation and machine harvesting, that natural vignerons consider unacceptable. Such practices are most common among large-scale organic estates. Some natural vignerons avoid organic certification because they feel it identifies them with those and other excesses of Big Organics.

BIODYNAMICS

1. Lack of scientific proof

Biodynamic preparations rely on homeopathy. Neither biodynamics nor homeopathy has been scientifically proven to have any effect, which can rub die-hard rationalists the wrong way.

2. Complicated history

While Rudolf Steiner's ideas about agriculture have attracted acclaim, his broader belief system, known as anthroposophy, includes some controversial views, such as a developmental hierarchy of races, the belief that vaccination is harmful, and the idea that mistletoe cures cancer. Some vignerons prefer not to be associated with any of his beliefs.

3. Big Biodynamics

While Steiner's original lectures stressed that biodynamic preparations should be sourced as locally as possible, today businesses like BioDynamie Services sell prefabricated biodynamic preparations online. Some vignerons object to such features of Big Biodynamics.

The Natural Vineyard Calendar

Regardless of how they farm, vignerons work according to the life cycle of their vines. It means there are certain very busy periods, like early summer, when grass and plants grow in abundance and the combination of heat and rain can lead to fungal attacks (see page 67). There are also slower periods, like December and January, when all there is to do is prune vines, a task that for some doesn't begin until February or March. To grow grapes is to adapt your lifestyle to the needs of the vines, month to month, week to week, even day to day.

Here's a rough timeline of a year in the life of a French grape grower. Of course, the timeline varies according to region and is entirely reversed in the Southern Hemisphere.

JANUARY TO FEBRUARY

A calm period. Many vignerons make appearances at wine-tasting salons (see page 401), which continue into early April. Pruning often begins during this time (though many conventional vignerons begin earlier).

MARCH TO MAY

An active period. Pruning usually finishes in March. Vignerons focus on planting new vines, debudding, leaf thinning, plowing, copper sulfate treatments, and frost prevention efforts as needed.

JUNE TO JULY

A very busy period. Early summer work may include plowing, sowing cover crops, copper sulfate treatments, and canopy management.

AUGUST

Calm or not, depending on the region. Some vignerons try to take a short vacation before harvest begins, while in other regions, such as Corsica or the Roussillon, harvest occurs in August.

SEPTEMBER TO OCTOBER

A very busy period of harvest and vinification. Many vignerons also turn the soil in autumn.

NOVEMBER TO DECEMBER

As long as vinification has proceeded well, this can be a period of relative calm for a vigneron. There is little to do in the vineyards, because for most, it is too early to prune.

Sébastien Dervieux's gamay at harvesttime in Anjou

Pruning

Pruning consists of cutting away the previous year's vine growth, along with any damaged or diseased wood, and preparing a vine for the present year's growth. In addition to determining a plant's crop levels, winegrowers train its future shape in accordance with their vine-training system and the particular needs of the vineyard.

Pruning generally occurs in winter, from December to April in the Northern Hemisphere. Few vignerons perform the entirety of their pruning themselves; most hire seasonal help. But hiring can be complicated, since this complex and important task must be performed quickly and cheaply.

Beaujolais vigneron Bruno Perraud pruning vines in Vauxrenard

In the Wineglass: Pruning

One of the aims of pruning is to manage a vineyard's crop level, or yield. Yield is expressed as the amount of grapes per unit of vineyard surface. In the Beaujolais and Burgundy, yield is expressed in hectoliter per hectare, whereas in Champagne, it's expressed in kilograms per hectare. In the United States, it is expressed in tons per acre.

One can often get a sense of a vineyard's yield simply by tasting a wine. Vines that are pruned for maximum yields can seem thin, while vines pruned for maniacally low yields can seem intense and overextracted.

There is no such thing as a universal ideal yield, because different grape varieties produce different crop levels according to an infinite number of factors, from soil type to vineyard training system to fertilizer regimen. In general, white grapes yield more than red grapes, and young vines yield more than old vines.

"There's no proportionality in what nature gives. You can make a wine at ten hectoliters per hectare one year, and you can make a wine at forty hectoliters per hectare the following year, and you might find the wine is better at forty hectoliters per hectare than at ten. And it's not four times better."

—Sancerre vigneron Sébastien Riffault

Grass Management

Unless you are a landscaper or a farmer, you probably don't think about grass much, except as something to mow from time to time.

Yet simple grass is among the most pressing, delicate, and vexing issues for farmers worldwide, winegrowers included. Grass is said to compete with vines for soil nutrients. Left unchecked, grass competition can reduce grape yields and lead to grape must deficient in critical fermentation components, chiefly nitrogen. Depending on a vine's training system, grass can also trap humidity at the level of grape bunches, potentially increasing the risk of fungal attack and frost.

"I think we must return precisely to things of the past—which is to say, plowing the vineyards first, and not sprinkling chemicals on the soil, etc. We must plow the vine!"

—Jules Chauvet, *Le Vin en Question*

Until the 1970s, winegrowers traditionally kept grass cover at bay with a combination of plowing, mowing, and hoeing. Herbicides arrived that decade, and within the span of one generation, plowing and hoeing became rare practices among conventional wine estates worldwide.

Today most of us see images of herbicided vineyards and do not even twig the historical peculiarity of the absence of grass on the soil between the vine rows. Our eyes are used to it. But it's fundamentally unnatural.

Today's natural vignerons employ time-honored methods for managing grass cover, such as plowing with tractors, horses, or cable plows. The timing varies by region, but in the Northern Hemisphere, the most important time to plow is from March to July. Many winegrowers also plow after harvest.

In regions adapting to increasingly dry and hot weather in recent years, natural vignerons often reduce their plowing regimen to allow the soil to regenerate and to enjoy the moisture-trapping benefits of grass cover. Certain natural vignerons, like the Andrieu siblings at Clos Fantine in the Languedoc, have ceased plowing entirely, preferring to encourage permanent ground cover (see page 62).

Grass cover on Nicolas Renaud's old grenache vines in Rochefort-du-Gard

Glyphosate (aka Roundup) and Other Herbicides

A heinous example of herbicide treatment on the outskirts of Beaune in Burgundy in April 2021. In this instance the herbicide employed was the brand Basta, which was banned in the EU in 2017.

To plow or not to plow: that was the new choice suddenly offered to the world's winegrowers in the post–World War II period with the introduction of glyphosate. A nonselective systemic herbicide, glyphosate was developed by Monsanto and brought to market under the brand name Roundup in 1974. To a European rural population decimated by two world wars, it was initially hailed as a miracle.

Today the dangers of glyphosate are widely known, but it's impossible to overestimate the herbicide's importance to the global agricultural industry. The global glyphosate market is expected to reach a valuation of $12.54 billion (US) by 2024. In much the same way as the fossil fuel industry has financed scientific studies to question the existence of global warming, the herbicide industry has spent untold sums financing studies that question the human health risks of glyphosate, which include cancer and non-Hodgkin's lymphoma. (In June 2020, Monsanto owner Bayer agreed to pay more than ten billion dollars in to settle thousands of claims linking Roundup to non-Hodgkin's lymphoma—while continuing to market the herbicide.)

The ecological harm of herbicide use is even plainer: soils die. Without grass and weeds keeping the topsoil in place, it blows away in the wind and washes away with the rain. Herbicides also create a costly, self-reinforcing use cycle, as herbicide-resistant weeds have appeared in recent years, requiring ever more herbicides to eradicate.

Many living in agricultural communities have witnessed the health and environmental costs of herbicide use firsthand, losing loved ones to early-onset cancer and watching fields turn to deserts. This is why herbicides are so reviled in the worlds of organic and natural wine.

Plowing

Winegrowers plow their vineyards for different reasons at different stages of the year. It's a delicate task, because they must carefully avoid breaking low branches and roots.

Winegrowers judge when to plow according to the weather and the specific conditions of each vineyard. Often different vineyards require different methods or machines.

Beaujolais vigneron Michel Guignier plowing in Moulin-à-Vent

Tractor

Most vineyards are plowed with tractors. It's the easiest way: all it requires is a tractor and someone to pilot it. But agricultural machinery can be costly to purchase and maintain. It also requires skilled labor, since accidents can be fatal. Professional tractor drivers are in high demand in well-financed regions like Bordeaux and Burgundy.

Beaujolais vigneron Jules Métras operating the winch from the tractor while his vigneron colleague Sylvain Chanudet steers the cable plow below, in the latter's vines on La Madone in Fleurie

Cable Plowing

On vineyards too steep and/or too dense for tractors, cable plows prove handy. A tractor with a winch parks at the top of the vineyard, and a cable is attached to a hand plow at the bottom of a row of vines.

Horse Plowing

Plowing with horses, cows, or mules largely disappeared with the arrival of tractors in the 1960s and '70s, but in recent years many natural vignerons have rediscovered the advantages of horse plowing. Horses do not compact the soil like tractors, do not require gasoline, and can work more nimbly, with fewer broken bushes.

Hoeing

Hoeing is repetitive, backaching labor. Today natural vignerons use hoeing in areas where plows won't pass, as well as to fine-tune the work of a plow, by removing grass cover between vines and at their bases.

Beaujolais vigneron Hervé Ravera plowing in Marchampt with his horse, Reggae Night

Languedoc Vigneron Bernard Bellahsen on Horse Plowing

Bernard Bellahsen (see page 276) is a singular figure in Languedoc winemaking, having successively pioneered organics, biodynamics, horse plowing, and natural winemaking in the region. He began plowing with a horse in 1980, at a time when his neighbors had all abandoned the practice.

Bernard Bellahsen feeding his horse

"Right at the start you're obliged to understand that it's the horse who's strong, not you. And then you come to understand his limits."

"If the earth is so wet that it flows away, after one hundred meters the horse says no. If it's too hard, the plow doesn't enter the earth. It's about balance. You understand that you don't have the right to do just anything. The horse's limited force doesn't allow you to do foolish things."

"With a tractor you can plow excessively, because you have four-wheel drive and you have hydraulics. You can misdirect the earth. Whereas with an animal, whether it's a horse, a cow, or a mule, he's got limited force."

"If the earth wants to let itself be plowed, it's the right time. With a tractor, you wind up doing things because you can. It's a rape of the earth, which disorganizes and destroys the soil structure and the microorganisms."

"I don't care about the past. It's not because it's historical that it's good. There were fools with horses, too."

No-Till Farming

When it comes to healthy grass management in vineyards, you might think: *Plowing good, herbicides evil.* Well, not quite. Plowing is not always a good thing. In fact, among the forward-thinking natural vignerons who practice no-till farming, plowing is considered unnecessary, if not downright harmful. While plowing allows vines to better compete with grass for nutrients, it can also dry out the soil and make it more susceptible to erosion in wind and rain.

"No-till farming" means farming without disrupting the ground cover with plows. In an era of global warming, it's a way to prioritize soil health and moisture retention.

An increasingly widespread systematic variant of no-till farming goes by the unwieldy name of direct seeding mulch-based cropping systems, DMCs for short. Developed on an industrial scale in Brazil in the 1980s and adapted elsewhere since, it consists of planting cover crops (e.g., wheat or fava beans) directly under mulch or residue from a previous crop that has been left on the ground. Winegrowers employing this method use a machine called a Rolofaca to bend existing grass cover flat at specific times of the year, thus precipitating a die-off that creates a mulching effect.

DMC systems are sometimes depicted as an alternative to organic agriculture. There are indeed many large farms practicing DMC systems nonorganically, but there is no inherent contradiction between farming organically and DMCs.

Corine Andrieu of Clos Fantine in Faugères (Languedoc). She and her siblings ceased plowing in 2014, preferring to manage grass cover with a combination of mowing and sheep pasturage.

NATURAL HISTORY

PERMACULTURE

Both no-till farming and direct seeding mulch-based cropping systems are practices aimed at installing permaculture, a regenerative and self-sustaining system of agriculture.

The roots of permaculture are found in a 1943 book by American agricultural agent Edward H. Faulkner titled *Plowman's Folly.* Aghast at the agricultural despoilment that had created America's Dust Bowl, Faulkner advocated the radical idea of no-till farming and warned against soil erosion from excessive plowing. "The 'bigger and better' the plow, the more devastating its effect," he wrote.

Another key permaculture advocate was Japanese agronomist and farmer-philosopher Masanobu Fukuoka, whose 1978 book *The One-Straw Revolution* promoted what he called a "do-nothing" approach to farming. Its first principle? No plowing. "For centuries, farmers have assumed that the plow is essential for growing crops," wrote Fukuoka. "However . . . [t]he earth cultivates itself naturally by means of the penetration of plant roots and the activity of microorganisms, small animals, and earthworms."

More recently, in the 2015 edition of their book *Le Sol, la Terre et les Champs* (*The Soil, the Earth, and the Fields*), acclaimed soil microbiologists Claude and Lydia Bourguignon cite plowing as one of the two principal killers of farm soils. "Plowing," they write, "by exposing the earth to the sun and the rain, is the cause of the disappearance of organic material and the desertification of soil."

Pasturage

You can hoe grass, you can mow it, you can plow it, but in terms of biodiversity, nothing beats pasturage. It's the practice of putting animals like sheep, cows, goats, or horses to pasture amid vines to keep grass growth in check. This throwback to the era of mixed agriculture has proved a valuable aid to natural winegrowers, particularly those concerned with keeping plowing to an absolute minimum. Ideally, it's as simple as the animals eating the grass and nourishing the soil with their droppings.

In practice, pasturage is more complex. It has to be conducted at times of the year when the animals prefer grass to vine buds, leaves, or fruit. Since the animals must be kept elsewhere the rest of the year, many winegrowers borrow herds from other farmers rather than tend their own animals year-round.

Sheep grazing in carignan vines at Clos Fantine

How to Maintain Soil Nutrition

Why maintain soil nutrition?

Healthy soils mean healthier plants, which means healthier—if not necessarily bigger—harvests. And healthy harvests ferment better. There is a direct link between soil nutrition and the presence, in wine musts, of the key elements necessary for a successful fermentation: nitrogen, chiefly, but also various vitamins, minerals, and amino acids.

"You have to always keep asking yourself, what can I do to increase the organic matter? Without the base levels of organic matter, you can spray biodynamic preparations as much as you like, and it's not going to have an impact on the actual farming. But people don't want to talk about compost and organic matter, because working on these things is awkward, time-consuming, and labor intensive."

—Roussillon vigneron Tom Lubbe

Farmers cultivating crops other than grapes employ crop rotation, changing what is sown on a given field regularly to ensure that soil nutrients aren't depleted by long-term demand for the specific nutrient set required for a given crop. As you might imagine, that approach doesn't work for grapevines, which can live for over a hundred years on the same site (which itself might be a valuable terroir that winegrowers are loath to leave fallow for a sufficient period of regeneration, let alone plant with anything other than vines). By necessity, something must be returned to the soil, which gives a crop of grapes every year.

Almost all winegrowers employ fertilizers of some kind. Organic, biodynamic, and natural winegrowers eschew the synthetic fertilizers endemic to intensive agriculture, which they often liken to steroid use in sports. Synthetic fertilizers artificially boost yields, resulting in abundant crops deficient in nutrients and complexity. They also tend to exhaust vine plants, leading to early die-off.

A manure heap beside the vines of Gard vigneron Alain Allier

NATURAL HISTORY

L'AFFAIRE EMMANUEL GIBOULOT

In 2013, Burgundy passed a prefectoral decree requiring winegrowers to use a preventive insecticide against the vine pest flavescence dorée. Flavescence dorée hadn't actually been detected in the region, so Beaune vigneron Emmanuel Giboulot saw no reason to apply an insecticide to his vineyards, which he'd farmed biodynamically for almost two decades.

"It remains a small event in the agricultural world, but it made the system evolve. It awoke a citizens' movement. The authorities were obliged to understand that it's important to listen to what the consumer is saying."

—Emmanuel Giboulot

For failure to comply, Giboulot was threatened with six months in prison and a fine of €30,000. His case went to trial in April 2014. He was found guilty and fined €1,000, but the verdict was later overturned when he won an appeal. Giboulot and his lawyer successfully argued that the decree was invalid because it was not based upon a situation of verifiable urgency.

In the meantime, Giboulot's stand attracted worldwide attention, inspiring articles in the *New York Times* and the *Guardian* as well as the wine media. For wine drinkers, the episode pulled back the curtain on the activities of the Burgundy wine-growing authorities, demonstrating the very real challenges faced by independent winegrowers seeking to farm without pesticides. An online petition in Giboulot's favor garnered more than a million signatures. The message was clear: no one should go to prison for farming responsibly.

Emmanuel Giboulot in his Combe d'Ève parcel of Côte de Beaune

Languedoc Vigneron Jean-François Coutelou on Restoring Hedges

Since 2013, Languedoc vigneron Jean-François Coutelou has sought to address the root cause of biological imbalances in viticulture, by restoring, along his vineyard perimeters, the country hedges that were once common to rural zones. His efforts, sadly, have proved controversial in his native pays d'Oc, where most growers practice intensive, volume-driven chemical viticulture. In 2016 and again in 2021, neighbors have burned Coutelou's hedges, apparently consumed by outmoded fears that the hedges would harbor pests or otherwise reduce yields.

Languedoc vigneron Jeff Coutelou beside one of the many hedges he's planted over the years, intended to harbor the local fauna—and to wall his parcels off from the intensive chemical viticulture surrounding him

"Lately we have invasions of snails, because there are no more natural predators, like game fowl and hedgehogs. As soon as you create an imbalance, you find yourself with this type of problem. Now people put down granules to kill snails, but in doing that, they'll kill the earthworms."

"Around new plantings, I plant trees and hedges to bring in diversity, and create a different ecosystem."

"We plant Mediterranean species that don't need irrigation. Almond trees, field elms, apricot trees, cherry trees. Wayfaring trees, pear trees, rosemary, thorow wax. Coronilla is the first to flower in March. The first for the bees to eat."

"Hedges began to disappear with mechanization, when we went from horses to tractors. With a horse it was possible to turn corners and maneuver around hedges. But winegrowers didn't want to be bothered. The next step was with irrigation and the arrival of mechanical pruning. Winegrowers created spaces that are no longer human: long rows, with no shade, no creeks. The land was adapted to machines."

"You have low, medium, and high flora. It's to have a balance. Up top you have birds, below you have insects, and those types of things, and at the bottom you have small wild animals. You don't have the same wildlife from one part to another."

Fungi

Fungi pose a dilemma for winegrowers. Certain types are crucial to the process of plant growth and fermentation. Other types destroy entire harvests. It's hard to kill the latter types without risking damage to the former.

Conventional winegrowers employ synthetic fungicides to combat mildew, oidium, black rot, and esca. As with insecticides, synthetic fungicides turn up in wine in alarming concentrations.

Natural vignerons refuse to use such products for two key reasons. One, synthetic fungicides, like insecticides, have been linked to increased cancer risk among farmers and agricultural workers.[1] Two, they risk killing off good fungi, like the mycorrhizae plants depend on for nutrients, and the native yeasts vignerons depend on for natural fermentation (see page 74).

To combat fungal attack, natural vignerons rely on copper and sulfur treatments, along with plant-based antifungal preparations, including orange essential oil, nettle infusions, onion decoctions, and milk dilutions.

---

1 *"Prostate cancer and exposure to pesticides in agricultural settings,"* International Journal of Cancer, *2003.*

NATURAL /LEX·I·CON/

COPPER AND SULFUR TREATMENTS

Permitted in both organic and biodynamic farming, copper and sulfur treatments are applied today by almost all winegrowers, natural and conventional, in efforts to prevent outbreaks of mildew and oidium. Unfortunately, excessive and prolonged use can result in copper buildup in soil, as well as groundwater pollution.

Why does the ideology of natural winemaking allow copper and sulfur treatments, while refusing other chemical vineyard interventions? For one thing, these treatments predate synthetic fertilizers (which arrived in the early 1900s), as well as herbicides and pesticides (which came into use after World War II), so it has been a routine part of grape growing since well before the advent of synthetic chemical agriculture.

More important, it's almost inconceivable for most vignerons to produce enough grapes to sustain their estates *without* copper and sulfur treatments. Some speculate that vines need these treatments due to a cumulative weakness in their DNA, related to the transition from European to American rootstock after phylloxera, or to an overreliance on clones (see page 69).

A sack of copper sulfate in a cellar in the Jura

How to Withstand Drought

Thanks to global warming, winegrowers in many historically cool regions like Burgundy, the Loire, and the Jura today struggle with heat waves and drought. The situation is dire in warmer areas like the Rhône, the Languedoc, and the Roussillon, not to mention much of Italy and Spain.

Lack of sufficient rain can create hydric stress in vines, causing growth and ripening cycles to shut down. It also results in a higher proportion of skins to pulp in the grapes at time of harvest, causing a drop in juice yields, along with difficult fermentations.

Good natural winegrowers focus on improving the soil permeability of their vineyards, allowing them to better withstand long periods without rain. Many report better resistance to drought in parcels farmed biodynamically. Beyond these long-terms efforts, however, the only in-the-moment palliative to drought is irrigation, a flash point in the debate about natural wine growing.

Irrigation involves watering vines through systems of pipes threaded throughout the vine-training system. Proponents consider it a lifeline for the vine and a useful tool in sculpting the character of a wine vintage. Critics accuse it of disrupting vine root systems (which stay superficial, having no need to descend to deep water reserves) and eliminating terroir expression and vintage personality.

Irrigation is permitted in both organic and biodynamic viticulture, in both Europe and the United States. It is almost nonexistent in France's natural wine scene, with the exception of newly planted vines (which most agree need some watering).

In the Wineglass: Irrigation

Irrigated vineyards can be perceived in two ways in wine tasting. From a broader perspective, irrigation tends to flatten out differences in vintage personality, by eliminating the variable of rainfall. Irrigation can also yield wines that possess a dilute, simplified fruit, similar to the effects of overcropping. This is most apparent when you compare irrigated natural wines from California beside nonirrigated natural wines from similarly warm areas like southern France, Spain, or Italy. The irrigated wine from California may or may not be higher alcohol and more glycerolic (fatter), but the nonirrigated European wine will almost always leave a greater sense of concentration and direction on the palate.

Sensitive deployment of irrigation is certainly possible, however. At best, it should be imperceptible, leaving both the concentration of the fruit and the personality of the vintage intact.

Irrigation system in the vines at Clos Saron in the Sierra Foothills of California

Jura Vigneron Didier Grappe on Hybrid Grape Varieties

Commonly known as "hybrids," hybrid grape varieties are crossings of European *Vitis vinifera* (the species from which derive all the so-called "noble" wine grape varieties, from pinot noir to syrah to chardonnay) with other species of grapes, such as the American *Vitis riparia*, *Vitis rupestris*, or *Vitis labrusca.*

Hybrids were popular at the turn of the century in Europe for their resistance to the American vine louse phylloxera, as well as to mildew and oidium. If they never truly caught on in commercial wine production in the wine regions of Europe, it's because their rustic flavor profiles rarely offer the fruitful elegance of *Vitis vinifera* grapes.

Didier Grappe sniffs the initial growths on the hybrid vines planted in his rear courtyard garden.

"I like the hybrids that were created one hundred years ago. The new hybrids are good, but the new grape breeders try to put too much vinifera in them, to try to get close to pinot noir, or cabernet. That doesn't interest me at all."

"Nature always needs to cross itself, to create a being that's adapted to dryness, to maladies. The one that isn't adapted won't be able to reproduce. That's how nature works."

"The flesh of hybrids is a bit gelatinous, there's a lot of pectin. You feel like you're eating an eye. That's the big difference with **Vitis vinifera,** ***which is much juicier."***

"The problem is, we've been cloning vines for a thousand years. And genetically, the vine hasn't evolved. There came mildew, black rot, oidium, all that. And the vine doesn't know how to deal with these things, because we've always cloned it."

Vinification in Fleurie. *From left:* Ophélie Dutraive, Justin Dutraive, Sylvain Chanudet, and Jean-Louis Dutraive prepare to devat a tank of gamay.

3. How Natural Wine Is Made

How Grapes Become Wine: The Absolute Basics

Roussillon vigneron Bruno Duchêne stirs the grape cap atop an amphora.

Wine begins, of course, as grapes.

If it's a white wine or a direct-press rosé, a winemaker begins by pressing the grapes. The juice ferments thanks to ambient yeasts passed on from the grape skins and the cellar equipment.

For a red wine (or an orange wine; see page 95), the grapes macerate in a vat for a period of time before being pressed. For these wines, fermentation begins here. Some grapes break—whether from destemming, foot-treading, or simply the fall into the vat and the weight of other grapes piled above them. This releases juice, which begins to ferment. Carbon dioxide is given off, creating an oxygen-free environment in which even unbroken grapes undergo some degree of enzyme-based fermentation within their skins (see page 97).

What is fermentation? It's the activity of yeasts converting sugar into alcohol. Enzymes and bacteria play secondary roles.

After the turbulent first stages of fermentation, the juice—now a wine—is put into another container to age until the winemaker deems it ready for bottling.

In an ideal scenario, a winemaker is like a conductor, directing a symphony of microorganisms in a harmonious performance of natural fermentation. Except the conductor cannot see the players without a microscope and has only a limited idea of how many players there are, and the players are unaware of the existence of the conductor. They come and go onstage as they please, and yet they must be induced to play at the desired tempo; otherwise, it becomes a cacophony. If the performance is a flop, there's no do-over until the following year.

For wine estates, a year's production is at stake. So most winemakers employ additives and corrective procedures at every stage of winemaking, sacrificing a wine's integrity in the name of stability. In refusing these shortcuts, natural winemakers embrace the risks and rewards of natural fermentation, offering wine lovers something unique: a chance to experience the living symphony of the vintage.

The Limits of Modern Winemaking

Underpinning fine wine appreciation is the notion of terroir, or the capacity of a geographic place to transmit its identity in a wine's taste and aroma. Terroir is why even bottom-shelf wine jugs at gas stations claim some vague origin ("France," say, or "Australia") on the label. The more specific the origin, the greater the wine's claim to represent its terroir. But here's some bad news: the natural terroir expression of even the grandest, most sought-after wines is no match for the awesome power of modern winemaking.

Anjou vigneron Richard Leroy holds a bottle of chenin to the light in his cellar.

Modern winemakers have at their disposal colorants, enzymes, micro-oxygenation, fining agents, tangential filtration, aromatic yeasts, powdered tannins, tartaric acid, beet sugar, centrifuges, and more. (More on all these arcane terms later—unfortunately, the diversity of contemporary wine innovation is such that it almost requires a degree in winemaking merely to enumerate what natural wine *isn't.*) These tools give winemakers the ability to adjust everything that defines a wine: its color, clarity, aroma, flavor, acidity, alcohol level, mouthfeel, and so on.

"Our vision of things is so limited. We're taught that there's an alcoholic fermentation and then a malolactic fermentation. But that's not all—there are many other fermentations. It's much more complex than that. Unless you shower it in sulfur."

—Anjou vigneron Richard Leroy

Paradoxically, the more valuable the wine from a given terroir is, the more winemakers feel pressured to employ such invasive techniques to *alter* that wine: to make it conform to consumer expectations, or to secure it from spoilage at all costs.

The result? Terroir is altered beyond recognition. Most wines sold today just taste like modern enology.

One thing modern enology *can't* do is summon a natural expression of terroir. That is the project of natural winemaking. As an art, natural winemaking is at once simpler and more complex than modern winemaking. It is simpler in that it can be achieved without high-tech equipment or additives. Yet natural winemakers must work in close concert with natural fermentation processes that are more complex than science can measure.

Enology

The science of winemaking is called enology. Some enologists work in-house for large wine estates; others work for laboratories analyzing samples and providing counsel for entire communities of winemakers and vignerons. While enology encompasses elements of biology and physics along with a great deal of chemistry, it's important to distinguish it, as a field of science, from something like nuclear physics. We wouldn't have nuclear power if it weren't for nuclear physicists. But wine has successfully been produced since time immemorial without enologists.

That's not to say that enologists are superfluous, or that their influence upon wine is necessarily malign. Some, like Pierre Sanchez and Xavier Couturier of Alsace's DUO Oenologie, actively collaborate with natural vignerons on innovative natural solutions to winemaking problems. Enologists are, however, in the business of ensuring their clients' financial security. So the majority of them overdiagnose wine faults and hawk remedies to winemakers in the form of enological products.

"At enology schools, they teach you how to make industrial wine. They train you for that. If you don't go in already knowing how to make real wine, you're dead."

—Burgundy négociant Philippe Pacalet

NATURAL / LEX·I·CON /

NATIVE YEASTS

Yeasts are the microscopic fungi that permit alcoholic fermentation. Any given vineyard has a native yeast population, as does any given winery. Natural fermentations draw from both of these native yeast populations. Native yeasts comprise part of terroir: they help make something taste like where it comes from.

Native yeasts are sometimes called "wild yeasts," as if they were somehow dangerous or unreliable, when, in fact, almost all wine was fermented on native yeasts until roughly the 1980s.

Natural winemaking effectively begins *before* grapes are brought into a winery, as natural fermentation relies upon cultivating and maintaining healthy native yeast populations through farming and winery maintenance. Synthetic pesticides impoverish native yeast populations in vineyards, while industrial hygiene does the same in a winery. Together, they create conditions that encourage winemakers to employ commercial yeasts.

Commercial yeasts are added for a variety of reasons—sometimes to extract color, or to encourage certain aromas, but mainly to ensure timely, reliable fermentations. Commercial yeasts simplify winemaking, and they do the same to finished wines. A wine fermentation in which one commercial yeast has been prioritized will miss out on the contributions of the diverse relay chain of native yeast families that help bring complexity to naturally fermented wine.

"It's a test of the work in the vines if the grapes hold together in vinification. If it goes balls-up, it means you haven't worked well."

—Roussillon vigneron Tom Lubbe

THE MICROSCOPE GANG

The early natural vignerons who rejected commercial yeast addition weren't shamans who sat around hoping fermentations would start. On the contrary—Beaujolais vigneron Marcel Lapierre and his influential circle of friends were precise technicians, forever seeking to improve their understanding of natural winemaking. In 1994, they established a custom of congregating at the home of Morgon vigneron Jean-Claude Chanudet, where a friend known as l'Oeil de Lynx ("the lynx's eye") would analyze samples of their fermenting wine under a microscope.

An enologist and yeast researcher at the Beaujolais appellation-sponsored (conventional) estate Château de l'Eclair, l'Oeil de Lynx (real name Renée Boisson) was renowned for her understanding of fermentations. She would identify and count the various species of yeast in a given wine sample and warn a vigneron if she identified undesirable yeast or bacterial activity. Now retired, she continues to provide this service to the veteran natural vignerons around Morgon on Wednesday nights during fermentation periods in the autumn.

Marcel Lapierre and L'Oeil de Lynx analyzing fermentation samples at Lapierre's home in Villié-Morgon, 1994

The Natural Winery

While traditional wineries vary by region and local customs, they all share a sensitivity to weather, atmospheric humidity, sun exposure, insulation, drainage, and aeration—the all-important conditions for fermentation and wine aging. Combined, these six aspects create what you might call the terroir of a winery.

1. An enlightened approach to hygiene

Hygiene is of capital importance to all winemaking. But good hygiene is not the same thing as hospital hygiene. Most natural winemakers take care to maintain a healthy microbial biome to ensure healthy fermentation. Natural winemakers use pressure washing and grape distillate, not chemical-based cleaning products, to clean equipment between uses. Things shouldn't be dirty—but they shouldn't be sterile, either.

2. Natural insulation

Fermenting grape juice into wine requires different temperatures at different stages. But as a general rule, yeasts and bacteria don't like abrupt shifts in temperature. Gentle temperature shifts are best, and these are best achieved in well-insulated spaces. This is why so much wine production takes place underground.

3. Natural cold

Different yeasts and bacteria work at different temperatures. Certain yeasts can work at lower temperatures than bacteria. Many traditional wines are fermented underground because it's naturally colder down there. It also minimizes the need for refrigeration and air-conditioning, which consume lots of energy and have a drying effect.

4. Humid cellars

Most wine undergoes a period of aging before release. Cool, humid conditions are best for this, because less wine evaporates and wooden containers (like barrels) don't dry out. This is another reason many wines are aged underground—it's naturally humid down there.

5. Gravity-based layouts

The best winery layouts allow a winemaker to take advantage of gravity for moving wine around. Often wineries are built into hillsides, so incoming harvests can arrive above ground-floor fermentation tanks, which are later emptied into an aging cellar below. In addition to saving labor and energy, moving wine with gravity disturbs its components less than pumping does.

6. Natural aeration

Healthy wineries smell good. Good, natural airflow helps ensure this. It's also a safety issue, since fermenting tanks produce a great deal of CO_2. This is why the first, most turbulent stage of fermentation generally occurs aboveground in well-ventilated areas. Winemakers are free to influence the temperature and airflow simply by opening or closing the winery doors.

Beaujolais vigneron Yvon Métras's cellar in Fleurie

The Conventional Winery

Modern conventional wineries don't vary quite as much throughout the world as traditional wineries. They're a little like fast-food chains, in that they operate according to similar principles everywhere on earth. Features that ensure quality and identity in winemaking are sacrificed for speed and cost efficiency. Here are six hallmarks of contemporary conventional wineries.

1. Industrial hygiene

The institutionalization of yeast inoculation in conventional wineries permitted them to adopt industrial hygiene regimens that destroy native yeast populations. Counterintuitively, the use of chemical disinfectants to clean equipment can cause hygiene hazards, for it creates a biological vacuum ripe for bacterial invasion.

2. The architecture of air-conditioning

Most modern construction is built with air-conditioning in mind. With it comes a wasteful indifference to opportunities to use the landscape itself for natural cooling and insulation.

3. Obsessive temperature control

A meddlesome approach to temperature control is a hallmark of modern conventional wineries. There's nothing wrong with encouraging sluggish fermentations with warmth, or trying to slow down overly vigorous fermentations with cold. But obsessive temperature control uses lots of energy and can result in unnatural temperature maintenance, preventing the biome of a fermenting wine from acclimatizing to a vintage as it is experienced within a cellar.

4. Dry air

Air-conditioning is a fact of life in many wine regions, and natural winemakers use it, too. But overreliance on air-conditioning can cause barrels and wood tanks to dry out.

5. Lots of pumps

Modern conventional wineries typically use large pumps, and many of them. Pumping to devat macerating grapes into a press (see page 98) is unthinkable for most natural winemakers, because it manhandles the grapes and creates excessively leesy juice. But the practice is routine in large-scale conventional wineries.

6. Reliance on fans

Wineries need aeration, particularly during the first, vigorous stages of fermentation, when lots of CO_2 is given off. But fans (along with air-conditioning) create an atmosphere of constant vibration that can perturb yeast activity (while also consuming a lot of energy).

The winery at the cave cooperative Cellier de la Sainte Baume in Saint-Maximin-la-Sainte-Baume in Provence

Sulfites

The use of sulfur in winemaking dates at least as far back as the Roman era, when it was mentioned in works by Pliny, Cato, and Homer. In ancient times, as today, sulfur was employed for its antibacterial and antioxidant properties. At that time, however, its use was limited to the burning of sulfur to clean winemaking containers, and the practice itself seems not to have been widespread. Winemaking of the era relied instead upon the antibacterial properties of the resins used to line amphorae (see page 104). According to Alsace enologist Arnaud Immélé in his book *Les Grands Vins Sans Sulfite*, sulfur burning became commonplace only after the Gauls adapted winemaking to oak barrels, which, unsuited to resin treatment, tended to harbor bacteria. The French agronomist Nicolas Bidet, in his 1759 work *Traité sur la Nature et sur la Culture de la Vigne*, describes winemakers of that era burning sulfur wicks also, during the racking of wine "for fear of it taking on an oxidized taste."

Such examples are often used by natural wine skeptics to imply the impossibility of making wine without sulfite addition. Yet the systematic use of sulfur dioxide (SO_2) in other, more easily dosed forms is more recent. Until well into the twentieth century, the use of liquid solutions of SO_2—the most common form today in Burgundy—was limited to large, industrial wineries, because its production was noxious and cumbersome. Heavy sulfite dosage was then reserved for damaged or rotten harvests and/or for the production of common table wines.

"When I have a wine with a high pH, I put in sulfites. When a wine with a low pH, I don't put any in. The people who systematically add sulfites and the people who systematically don't add sulfites are the same sort of fool, in my opinion."

—Morgon vigneron Jean-Claude Chanudet

An advertisement for the enological additive Sulfo-Hubert (ammonium bisulfite), circa 1920

Artisanal winemaking of the era still placed an emphasis on restraint in sulfite addition. As J. H. Fabre, professor of enology at the Agricultural Institute of Algeria, wrote in his 1920 book *Procédés Modernes de Vinification* (*Modern Vinification Procedures*), "In contemporary vinification in France, we employ barely more than 10–12 grams of SO_2 per hectoliter."

Fast-forward to 1986. By then, standard sulfite doses in vinification ranging from 20 to 200g/hl were described by UC Davis researcher C. S. Ough.[1]

Tablets and powders made from potassium or sodium metabisulfate had arrived in the 1920s,

1 *"Determination of sulfur dioxide in grapes and wines,"* Journal of Association of Official Analytical Chemists, *1986.*

joined later by premade SO_2 dilutions. As the addition of SO_2 in these forms became easier to measure and simpler to apply, its use grew more widespread.

"For half a century," the Burgundy enologist Max Léglise wrote in 1994, "conventional enology has comforted itself in an incessant climb of sulfite dosage."[2]

If, by the 1980s, certain pathbreaking natural vignerons began calling attention to sulfites, it's because sulfite addition in fine winemaking, once limited to burning sulfur at racking, had come to condition the entire winemaking process. Enologists prescribed it upon the harvest, at bottling, and everywhere in between. Vignerons obeyed, eager to exploit the burgeoning wine export market and ensure stability in transport. Excessive sulfite addition quickly became a crutch supporting modern winemaking. It began to define the taste of wine for much of the drinking public.

"I try to learn to make wines without sulfur. If I start to put in a little, I'm no longer learning how to make wines without sulfur."

—Haute-Loire négociant-vigneron Daniel Sage

Today the natural wine scene is divided between winemakers who strive to add an absolute minimum of sulfites in wine production, and those who disavow sulfite addition entirely.

2 *"Les méthodes biologiques appliquées à la vinification & à l'oenologie,"* Vinifications & fermentations, *1994.*

Making Sense of Sulfites

Can you be allergic to sulfites?

Allergic sensitivity to sulfites is a real thing. Serious allergic reactions to sulfites, however, are extremely rare. The most common symptoms of allergic sensitivity to sulfites are asthma-like and include wheezing, coughing, and tightness in the chest.

Do sulfites cause headaches? Or worse hangovers?

There is no scientifically proven link between sulfite levels in wine and headaches or bad hangovers. Research on wine-induced headaches has instead implicated histamines and phenolic flavanoids—compounds more present in red wines, which tend to contain lower sulfite levels than white wines. Sulfites, however, are known to enhance the allergic effects of histamines, meaning there may well be an indirect link.[1]

At any rate, most of us can testify, from personal experience, that highly processed wine makes us feel worse. But is that reaction caused by high sulfite levels, or by the many other additives regularly employed in the production of such wines?

Many common foods contain more sulfites than wine. Does that prove sulfites are harmless?

This is a misleading simplification of how our bodies work. For one thing, no measurable compound we consume arrives in isolation (consider the aforementioned interaction between sulfites and histamines). We don't know whether, or how, sulfites consumed as part of wine affect us differently from sulfites consumed as part of dried apricots or bacon.

1 *Immélé, Arnaud.* Les Grands Vins Sans Sulfite. *Le Tremblay, Vinedia, December 2011.*

How Sulfites Are Added

The most radical natural winemakers add no sulfites whatsoever, at any stage of winemaking. But it is common, in the more pragmatic wing of the natural wine scene, to add tiny amounts of sulfites, often after malolactic fermentation and racking (see page 103) or at assembly before bottling (see page 105). Sulfites can be added to wine in several forms. Here are the most common.

Sulfur wicks

Sulfur wicks are burned inside barrels to prevent and remove bacterial buildup between uses. (Historically, they were also employed to preserve wines during racking, but this use was rendered obsolete in the early twentieth century by other methods that permit more precise dosing.) Overzealous sulfur wick treatment can result in a significant sulfite addition, affecting wines placed in those barrels. Most natural winemakers avoid using sulfur wicks on barrels and other wooden containers more than once per season, and some avoid it altogether. (By comparison, conventional Burgundy estates burn sulfur wicks in their empty barrels up to once a month.)

Aqueous sulfur dioxide

Among natural winemakers who do add sulfites, sulfur dioxide solutions are the most common form of sulfite addition because they're so easy to dose. Common dilutions are 5 or 10 percent.

Top left: Aqueous sulfur dioxide solution

Bottom left: Sulfur wicks in a cellar in the Beaujolais

Right: Sulfur tablets

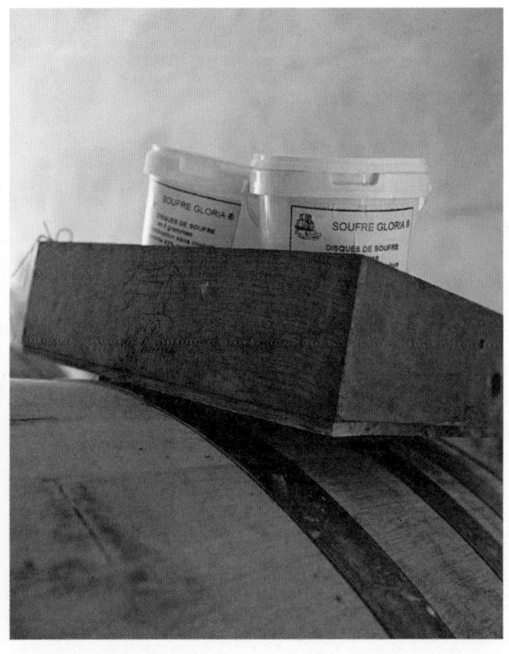

Ammonium bisulfite, potassium bisulfite, or potassium metabisulfite

These come in a powder or tablets. They can be sprinkled directly on the harvest, or added to barrels or tanks. Ammonium bisulfite is rarely employed by natural winemakers because it is not permitted under EU organic wine law.

Sulfur dioxide gas

Pressurized tanks of sulfur dioxide are used to add sulfites to wines. In Muscadet or the Mosel, for example, sulfur dioxide gas is used to stop malolactic fermentation (see page 103). The method is rare to nonexistent in the natural wine scene.

Volcanic sulfur

Many biodynamics practitioners believe that mined volcanic sulfur is different from "manufactured" sulfur produced by petroleum refineries. Though chemically the same element, they are thought to possess different energies. So some vignerons employ volcanic sulfur, burning it and integrating the vapors into wine tanks. The use of volcanic sulfur is technically illegal because its natural state is not chemically pure. Its use is also controversial due to the poor working conditions to which miners are exposed during its extraction.

In the Wineglass: Sulfites

How do you perceive the sulfite levels of a given wine? Excessively sulfited wines announce themselves with an astringence, a bleached-out quality, sometimes causing a tactile tightening sensation on the nose. On the palate, flavors and textures seem simplified and rigidly outlined. On a primal level, the wine becomes unappetizing: you don't feel like another sip.

Funnily enough, the more natural wine you taste, the more excessive sulfite addition tends to jump out at you. Many longtime natural wine tasters report growing increasingly fussy about what sulfite levels they find tolerable. What you once perceived as the normal aromas and tastes of Chablis, might one day scan as the insipid, harsh profile of any oversulfited white wine.

HOW WINEMAKING CHANGED IN THE TWENTIETH CENTURY

Many career wine professionals are skeptical of natural wine, because the notion can seem newfangled or faddish.

It's true that, at least within the high-end wine market, we had less need to specify how natural a wine was until the early 1980s (give or take a lot of illicit blending and chaptalization; see page 40). But by the 1990s, naturally fermented, low-sulfite, additive-free wine had become a rarity, even on the high-end market.

The timeline that follows shows how wine production changed throughout the twentieth century. These are the milestones on the path away from natural methods in conventional winemaking.

The arrival of harvest at an estate in Vouvray in the early twentieth century

1920S: Sulfite addition, long practiced in winemaking via the integration into wine of sulfurous vapors during racking, evolves to encompass innovations that include Campden tablets and aqueous sulfite solutions, both far easier to dose. Initially this yields a net reduction in sulfite addition for many common wines. Sulfite addition in these simpler forms would however creep up throughout the twentieth century, as export markets grew in importance.

A Touraine vigneron measuring sugar density in a wine must in 1951

C. 1919–1939: Electricity comes to rural Europe, leading to changes in many stages of winemaking. Slow, gentle manual presses are increasingly discarded in favor of faster, more thorough motorized versions. Small vertical basket presses—and their positive effects on wine quality—only come back into vogue in the early 2000s (see page 100).

1949: Édouard Leclerc opens France's first supermarket in Finistère, heralding the age of supermarket wine sales and the decline of (better-informed) independent wine retailers. The generalization of wine filtration (see page 109) begins in earnest with the advent of supermarkets, as there are no wine experts around to inform shoppers that deposits in wine bottles are harmless or that mild cloudiness in wine isn't necessarily a flaw.

1956: American trucking tycoon Malcom McClean pioneers the shipping container. Throughout the next four decades, export markets exert increasing influence on European wine regions. Winemakers increasingly revise their practices to prioritize transport stability. Greater demand from export markets also incentivizes winemakers to bring wines to market sooner. All this encourages greater yields, less aging, more sulfite addition, more fining, and more filtration.

Pressing in Champagne, September 1969

1960S: Membrane filtration technology, first invented in Germany in 1916, begins to spread throughout Europe and the United States. Winemakers use the technology to create numerous sweet and off-dry wine styles, sterile filtered to ensure shelf stability. The popularity of these wine styles fades forty years later, but membrane filtration is here to stay (see page 110).

1962: With the advent of freeze-dried wine yeasts, the addition of commercial dry yeasts, previously limited to industrialized, high-volume wineries, catches on in the United States, New Zealand, Australia, and South Africa. Most European vignerons stick with natural fermentation for two more decades.

1978: American lawyer turned wine critic Robert Parker publishes the first issue of what would become *The Wine Advocate*. The late-twentieth-century media climate allows Parker to garner huge influence over the wine purchases of the world's wealthiest economy. Contemporary enology's influence would grow in parallel with Parker's in the following decades, as winemakers the world over modified their wines to suit the tastes of one lawyer from Maryland.

1980S: Having impoverished native yeasts populations with heavy herbicide and pesticide use throughout the 1970s—and having thus rendered natural fermentation unreliable—European winemakers begin to embrace commercial yeast addition.

EARLY 1990S: California winemaker Clark Smith pioneers application of reverse osmosis—a technology initially used to desalinate seawater—to wine, initially in an effort to create alcohol-free wine. Creative use of this technology allows winemakers, over the ensuing decades, to adjust integral wine factors like alcoholic strength, volatile acidity, and more (see page 87).

1991: Madeiran winemaker Patrick DuCournau develops the process of micro-oxygenation (MoX, for short), in which oxygen is injected into fermenting wine, stabilizing color and speeding the polymerization of tannins. Micro-oxygenation is soon adopted worldwide, becoming one of the keys to understanding the "internationalized" wine profiles available on supermarket shelves everywhere (see page 87).

French supermarket tycoon Édouard Leclerc at one of his stores in the 1960s

Common Conventional Wine Additives

So what's in mass-market wine, besides grapes? Conventional, mass-market wine is sold as an expression of its origins: a given place, a given vintage. But the sheer diversity of corrective additives used by conventional wineries gives the lie to this idea. In truth, mass-market conventional wine is a processed confection closer to a soft drink than a genuine agricultural product.

ADDITIVE TYPE	PURPOSE	EXAMPLES
Clarifying Agents	Clarifying wines naturally takes time, which is expensive, and which changes wines in other ways. So conventional winemakers use fining agents to clarify wine quickly.	Calcium alginate, casein, edible gelatin, egg whites, gum arabic, isinglass, milk, plant proteins, polyvinylpolypyrrolidone (PVPP), potassium alginate, potassium caseinate, silicon dioxide, and many others.
Decolorants	There is significant mass-market consumer pressure on certain wine styles to be as pale and limpid as possible.	Activated charcoal is commonly employed in downmarket Champagne and rosé wine styles.
Colorants	Colorants are employed by conventional winemakers seeking to enrich a thin, overcropped harvest, or to enrich the color of a red wine in efforts to appeal to consumers who seek darkly colored reds.	These days the most notorious colorant is Mega Purple, a grape juice concentrate in common use in downmarket California winemaking. It also serves as an enriching agent.
Acidifiers	In very concentrated vintages, when winemakers illicitly add water to their wines, they are then obliged to raise acidity. But more often acidity is adjusted in conventional winemaking to safeguard fermentations from bacterial attack, to which they are more liable at higher pH. Acidity is also adjusted for aesthetic reasons.	Tartaric acid is used to boost acidity in musts with high pH.

ADDITIVE TYPE	PURPOSE	EXAMPLES
Deacidifiers	In cold, rainy vintages from cool climate regions, winemakers sometimes use additives to lower acidity, chiefly for aesthetic purposes.	Lactic acid bacteria, neutral potassium tartrate, potassium bicarbonate.
Enriching Agents	Enriching agents are used for aesthetic purposes (to improve "mouthfeel," etc.). The Champagne method also calls for the addition of an enriching agent—sugar, in some form or another—to create the bubbles.	Sugar and oak tannins can be added to wines to enrich them. So can concentrated grape must. Tanninization is common in bad Burgundy, and adding concentrated grape must is the norm in all Champagne, even among the vignerons who produce Champagne following otherwise natural principles.
Fermentation Aids	Tanks deficient in yeast nutrients (such as nitrogen) struggle to ferment and risk spoilage. So winemakers routinely add fermentation aids to encourage a speedy, clean fermentation. What determines whether a harvest has sufficient native yeast nutrients? The quality of farming.	Ammonium bisulphite, ammonium sulfate, ammonium sulphite, commercial dry yeasts, diammonium phosphate, thiamine hydrochloride, yeast cell walls, and betaglucanase, pectolytics, and urease enzymes.
Preservatives	Winemakers use a range of preservatives to prevent oxidation and discourage bacterial activity during the winemaking process.	Allyl isothiocyanate, ascorbic acid, CO_2, dimethyl dicarbonate (DMDC, aka Velcorin), lysozyme, potassium bisulphite, potassium metabisulphite/disulfite, potassium sorbate, sorbic acid, sulfur dioxide.

Eight Common Conventional Wine Manipulation Techniques

Natural winemaking isn't simply about not *adding* anything to wine; it's also about not intervening in high-tech ways to fundamentally transform a wine. All winemaking involves human intervention and transformation, of course. But the outsize transformative power of the modern techniques below tends to standardize wines. That's why natural winemakers shun them.

Top: A flash détente apparatus at the Mont Tauch Cave Cooperative in the Languedoc

Bottom: A centrifugation unit at Bodegas Frutos Villar in Castilla y León

1. Thermovinification

Thermovinification just means heating a wine to kill off yeast and bacterial activity—a technique pioneered by Louis Pasteur. It also concentrates color and yields uniform stewed cassis aromas. The technique came into vogue in the Beaujolais and (briefly) in Burgundy in the 1980s, but quality-minded winemakers soon abandoned it.

2. Flash détente

You could consider this the nuclear version of thermovinification. Flash détente consists of superheating a harvest (to 71° to 88°C/160° to 190°F) and pumping it into a vacuum chamber, which explodes the grapes in their skins. Significant water content disappears as steam, concentrating the resulting grape must. Certain flavor and aroma compounds like smoke taint and underripeness also disappear in the steam, leaving winemakers free to decide whether to add them back later, and in what proportion. The method imparts a characteristic fruitiness and low tannicity to wines, but in other ways functions like the Transmogrifier in Bill Watterson's *Calvin and Hobbes*: it turns anything into anything. Because flash détente systems are large and costly to install, their use remains limited to enormous industrial wine estates.

3. Centrifuging

Centrifuges are used at high-volume conventional estates to immediately separate press juice from lees. This almost instantly clarifies the juice, and allows winemakers to immediately dose it with other additives, including sulfites. The practice is most common in high-volume production of rosé and white wines, where an artificially pale color is often prized.

A reverse osmosis machine at Château Paloumey in Bordeaux

4. Reverse osmosis

Reverse osmosis employs filtration technology to perform a sort of wine surgery. First used to concentrate grape must in cool vintages by removing water content, it has since become a way to reduce alcohol content in wines. The process involves removing a water-and-alcohol solution from the wine, then distilling the solution and returning the water to the wine. Reverse osmosis is also used to lower volatile acidity in wines.

5. Excessive micro-oxygenation

Micro-oxygenation involves the controlled introduction of tiny amounts of oxygen to tanks of wine through a porous sparging stone. It has multiple uses, from nourishing yeasts to preventing stuck fermentation to stabilizing color to softening tannins. Used sparingly, it can be a helpful tool in a winemaker's tool kit, but clumsy or indiscriminate use of micro-oxygenation has a standardizing effect on wine, sanding away its nuance and polishing its mouthfeel to a new-car sheen.

6. Excessive filtration

Filtration is a complex subject. Some styles of filtration, like diatomaceous earth and plate filtration, are relatively common for white wines and rosés in the more conservative wing of the natural wine scene (see page 109). But tangential or cross-flow filtration is a more heavy-handed technique rarely used by natural winemakers. Excessively filtered wines become thinned out, lacking in texture and body.

7. Cryomaceration

Cryomaceration refers to a broad set of practices of rapid cooling of grapes—sometimes to freezing point or below—before maceration or pressing. The technique breaks down cell walls within the grapes, increasing extraction. While many natural winemakers chill their harvest before vatting (see page 114), none go as far as freezing temperatures, which fundamentally alters the harvest.

8. Dialysis

Dialysis in winemaking is a variation of reverse osmosis. But where reverse osmosis consists of diffusion across a semipermeable membrane, dialysis is a separation of the molecules themselves. In winemaking, the technique can be used to reduce sulfite levels, adjust pH and tartaric acid levels, create tartrate stability, and more. As with any powerful technology, it tends to result in standardized wines, custom-built to appeal to the broadest swath of consumers.

How a Natural Wine Is Made

Natural winemaking techniques vary from estate to estate and region to region. But here's a general idea of how a still natural wine is made. (The process for macerated rosés and orange wines is much the same as for light reds; see page 95 for more on orange wines.)

1 Manual harvest *(ideally in small cases)*

2 Refrigeration of harvest *(optional)*

3 Destemming or foulage *(optional)*

4 Vatting & maceration *(for reds, orange wines, and some rosés)*

5 Pigéage & pumpovers *(optional; for reds, orange wines, and some rosés)*

6 Devatting *(for reds, orange wines, and some rosés)*

7 Pressing

8 Lees settling *(optional)*

9 Racking & élevage

10 Racking & assembly

11 Bottling

12 Bottle aging before release *(optional)*

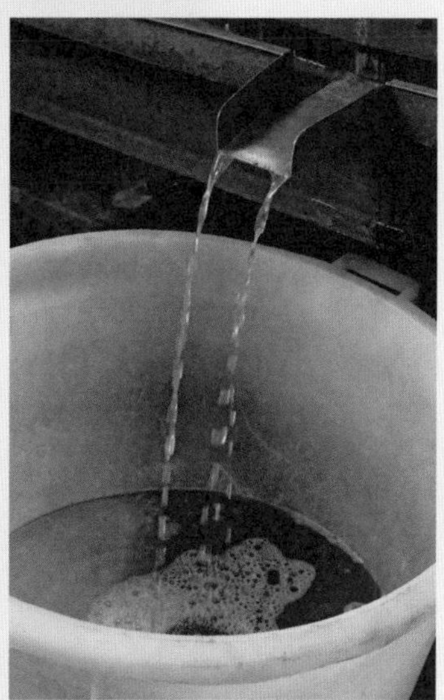

Anjou vigneron Sébastien Dervieux (aka Babass) assembling and pressing his chenin

Making Natural Wine, Step 1: Harvest

An estate's harvest practices affect every subsequent stage of winemaking. If you think of winemaking like cooking, harvest is how you handle the ingredients. As any chef will tell you, meat, vegetables, and other ingredients will cook quite differently depending on their initial state. You can't make a steak out of ground beef.

When it comes to grapes, you have to be gentle. That's why natural wines are *always* hand-harvested. (This is such a rule that it's assumed in natural wine circles, and rarely stated on the wine label.) Harvesting machines (see page 92)—even the fancy, high-tech ones—simply break too many grapes. As soon as a grape is separated from a bunch and its skin ruptures, the juice begins to oxidize and becomes the site of bacterial activity. This is why estates that practice machine harvesting need to use more additives, including sulfites, often applying them to the grapes immediately upon harvesting.

Small cases

The best natural wine estates harvest in small cases, often around 25 kilograms (55 pounds). That way the grapes don't get crushed by their own weight.

Morning harvest

The warmer the wine region, the more crucial it is that the harvest be conducted in the cool hours of the morning. Why? Because hot grapes can lead to temperature spikes in fermentation, which can stress yeast populations, causing stuck fermentations.

Sorting in the vines

Expensive sorting tables are nice, but most natural winemakers simply instruct their teams to sort grapes in the vines. This has the advantage of preventing rotten bunches from coming into contact with other, healthy bunches in harvest buckets.

Saint-Aubin vigneron Julien Altaber during harvest 2019 in the En Remilly vineyard

How to Join a Harvest Team

Local customs—and labor laws—vary wildly between wine regions. But it's not impossible to join the exhausting fun at a natural winery at harvesttime. Here are a few rules of thumb to point you in the right direction.

1. In Europe, you'll need a work visa.
Most wine regions are rigidly overseen by their local labor boards, so you'll need the right documentation to join a harvest team. It's simply not possible to volunteer, unless the estate knows you well enough to bend the rules for you. If you can't get a work visa, try looking for estates that accept volunteers as part of the WWOOF (World Wide Opportunities on Organic Farms) program.

2. Ambience varies wildly.
Some estates are happy to lodge and feed their harvest team. Others are less generous and less festive. It's worth doing a little research to get a feel for the ambience at a given estate.

3. Try your luck.
Most estates firm up their harvest team by the July before the harvest. If you're dead set on working at a certain estate, send them an email and call to follow up. It's sometimes worth approaching their importer for an introduction.

4. Language is not really an issue.
Cutting grapes is not rocket science. It's simple, repetitive work, so there's no need for perfect fluency in the local language.

5. Wear long pants and practical footwear.
Unless it's an exceptionally hot region, it's worth wearing long pants to avoid scratching your legs on grass and weeds. (Remember: There are no herbicides used at natural estates!) Practical footwear is a must.

Sommelier and import specialist Jing Song harvesting in Saint-Aubin in Burgundy

In the Wineglass: Machine-Harvested Grapes

There's no telltale characteristic that defines the profile of a wine made from machine-harvested fruit. Machine-harvested wines tend to be monodimensional—but this is due to a host of factors. Machines bring in messy, broken harvests, which winemakers often treat with heavy sulfite addition, as well as synthetic antioxidants like glutathione, used systematically in conventional Champagne production. The way harvesting machines manhandle grape bunches has led some to jokingly call them "field destemmers," and indeed, harvests brought in by machine are unsuitable for carbonic, whole-bunch maceration (see page 97), which requires intact bunches. In such a way, harvesting machines effectively limit the purity and stylistic range of the wines made with the fruit they bring in. Estates that practice both manual and machine harvesting never bring in their best fruit by machine.

"There's not one person who was at the table when the appellations were declared in the 1930s who would have thought you can pick grapes with a machine. They couldn't have thought of that. The appellations should have changed when people started to harvest with machines."

—Cheverny vigneron Jean-Marie Puzelat

A harvesting machine in the Mâconnais

Destemming vs. Whole-Cluster Maceration

Before vatting grapes, winemakers decide whether to remove the stems. When they leave the harvest intact, it's called whole-cluster fermentation.

Wine literature tends to frame the decision about destemming as a purely stylistic choice. Stems, we are often told, bring green tannins and astringency. Or structure, depending on whom you ask. Indeed, it really depends on the grape variety, the region, the vintage, and the winemaker's specific aims with a given lot of grapes.

"Destemming developed in the 1960s. It accompanied the industrialization of cellars. If you don't destem, you can't have tanks that are too big, because you'll have to do pigéage."

—Burgundy négociant Philippe Pacalet

Ardèche vigneron Gilles Azzoni destemming merlot. Destemming by hand is also common among natural vignerons in the Jura region.

Systematic destemming, however, is a hallmark of industrialized winemaking. Stems are the last thing to ripen, so if you've harvested too early, you destem. If you've harvested humongous grape yields and want to fit more volume in a tank, you destem. If you want a liquid—rather than partly solid—fermentation milieu in which sulfite solution, commercial yeasts, and other additives will diffuse evenly, you destem. If you want to make life easier by pumping the harvest, you destem. If you want to cut labor costs by reducing the hours spent strenuously foot-treading wines during harvest and fermentation periods, you destem. You get the idea.

By eliminating the stems, you remove whatever it is they might contribute to the wine, be it structure or greenness. But you also remove the influence of intracellular fermentation, which otherwise occurs in unbroken grapes in an anaerobic environment, and which endows an agreeable fruitiness to wine. Stems tend also to absorb a small percentage of alcohol, which can yield lighter, more graceful wines.

A destemmer at Château de Lacarelle in the Beaujolais

Making Natural Wine, Step 2: Vatting

Grapes destined for red wines, orange wines, and some rosés see vatting before pressing. (Some white wines see a very short vatting, too.) It just involves putting them in a container of some kind for a certain period of time. It's not quite as simple as just chucking the grapes in a bin, however. It all depends on the condition of the harvest and the desired fermentation ambience.

Temperature of the harvest

Hot grapes in the tank can cause chaotic fermentations. Some natural winemakers prefer to refrigerate their grapes for twelve to twenty-four hours before vatting to lower their temperature.

Working with gravity

Pumping fresh grapes into tanks mauls them, which limits winemaking options afterward. Winemakers concerned with the integrity of their harvest vat use gravity. Sometimes wineries are built so grapes arrive above maceration containers. Other winemakers use conveyor belts or forklifts to vat their grapes.

Using a starter

Some winemakers ensure a good start to natural fermentation by using a starter ferment, known as a *pied-de-cuve* in French. Before the real harvest begins, they'll select a small quantity of grapes and crush them in a small container, sometimes gently warming it to initiate fermentation. After verifying that the starter ferment is healthy, the winemaker will add it to the bottom of their first tank, where it serves to kick-start fermentation.

Foulage

Foulage is a light treading of the grapes that occurs before fermentation. It's performed to release juice and to encourage fermentation to begin.

Fleurie vigneron Yvon Métras vatting whole-cluster gamay by gravity using a forklift

A cellarhand raking whole-cluster grapes into a cement tank at the cellar of Yvon Métras in Fleurie

Ardèche négociant Antonin Azzoni (*right*) and a friend performing foulage before vatting grapes

NATURAL /LEX·I·CON/

ORANGE WINE

"Orange wine" is the popular term for skin-macerated white wines, which are also known as amber wines or skin-contact white wines. These wines have roared back into vogue since being introduced to Western wine markets by Friulian vignerons Josko Gravner, Saša Radikon, and others in the early 2000s (see page 351). Certain wine pedants make a distinction between long-macerated orange wines, which are actually orange in color, and short-macerated whites, which can resemble white wines in most ways. It's simpler to call them all orange wines and accept that orange comes in a variety of hues: some darker, some lighter.

Contrary to popular belief, orange wine is not a new wine style, nor are all orange wines *necessarily* natural wines. Orange winemaking, in fact, is older than direct-press white winemaking. It is the traditional white-grape wine style in many parts of the Republic of Georgia, where the earliest known traces of wine production have been documented (see page 358). Allowing white grapes to sit in a tank (or other container) on their skins for a period before pressing softens them, making them far easier to press manually—an important consideration before the advent of motorized wine presses.

It's true that most natural wine lists contain many orange wines, and that many orange wines are also natural wines. But orange wine is a style that can be commodified and adulterated like any other. If at the moment you see more of it in the natural wine scene, that's because natural wine drinkers tend to be more stylistically open-minded than conventional wine drinkers. It's also because natural winemakers have realized that skin-macerated whites ferment more reliably than direct-press whites, benefiting from time spent in contact with native yeasts on the grape skins.

"Skin maceration is the origin of vinification. It's electricity that allowed for the invention of direct-press whites. If you had to press everything with a manual press, believe me, you'd opt to do skin macerations all by yourself."

—Alsace vigneron Jean-Marc Dreyer

Languedoc vigneron Jean-François Coutelou's OW, a maceration of muscat d'Alexandrie, at Paris natural wine bar Goguette

Making Natural Wine, Step 3: Maceration

A macerating tank of poulsard at Domaine des Bodines in the Jura

So you've got the grapes in the tank, perhaps destemmed, perhaps not. They're macerating, absorbing color, tannins, and flavors from their skins. What now? The manual procedures winemakers employ during maceration depend on what kind of wine they wish to make with the grapes at hand.

Healthy, low-yield grapes often benefit from longer macerations, while damaged and/or overcropped grapes call for shorter macerations to prevent the extraction of unripe tannins or unpleasant flavors. Winemakers often practice shorter macerations on grapes from younger vines as well, because they anticipate less potential for complexity from them. Rosés usually see very short macerations, if any.

If you want a dense, structured wine, you put in work on the macerating grape mass with a mixture of punchdowns (*pigéage*) and pumpovers.

If you want a lighter, silkier, fruitier wine, you leave the grape clusters whole and make sure to seal the tank from oxygen contact, allowing the intact grapes to ferment within their skins (see opposite). Often juice is created at the bottom of the tank due to the crushing weight of the grape mass. This begins to ferment, too.

Several times per day, you observe the progress and intensity of fermentation by taking the tank's temperature and measuring the sugar density of the juice in the tank.

CO_2 seal

Some natural winemakers use canisters of CO_2 to expel air from their whole-cluster maceration tanks, creating an oxygen-free environment. Other natural winemakers consider this an unnatural contrivance, and prefer to pipe in naturally created CO_2 from other fermenting tanks.

Foot-treading and punchdowns (aka *pigéage*)

Pigéage is the practice of stamping on a grape mass to release juice and moisten the "cap" of grapes atop a tank during fermentation. Why use feet? Human weight exerts a gentle pressure on the grape mass, disturbing it less than most mechanical means of pigéage. Using bare feet also allows the person performing pigéage to remain sensitive to heterogeneous temperatures amid the grape mass, a warning sign that fermentation is proceeding unevenly. As maceration continues, it becomes important to keep the grape cap moist to prevent off aromas from developing as it dries out.

Pumpovers

A pumpover is when juice from the bottom of the tank is pumped over the cap of grapes at the top of a tank. It serves to homogenize the grape mass, moisten the grape cap, and aerate the juice, which can help manage reduction and prevent the formation of volatile sulfur compounds (see page 376). The same effect can be achieved on a smaller scale without pumps by drawing off a few buckets of juice from the bottom of the tank and pouring them over the top.

Carbonic Maceration

Carbonic maceration: Sounds technical, right? Carbonic maceration is actually very simple. In its "classic" form, it involves saturating a tank of whole-cluster grapes with CO_2, creating an oxygen-free environment, which causes an enzyme-based fermentation within the grapes themselves. The grapes ferment to between 1 and 3 percent alcohol. Then the grapes are pressed off, and the wine continues a normal fermentation in a liquid state. The process is known to lower acidity and yield fruitier, paler, less tannic wines.

The vinification style was first elucidated by the enologist Michel Flanzy in 1935; a decade later, the natural wine progenitor Jules Chauvet (see page 27) made it the subject of further study. It's wrong to say that carbonic maceration was "invented," however, as the enzymatic fermentation that it emphasizes is a natural part of almost all winemaking (Louis Pasteur noted it back in 1872). Whenever grapes are vatted whole-cluster and not immediately broken up, an anaerobic environment soon forms as the juice at the bottom of a tank begins fermenting and gives off CO_2. The fermentation dynamic that results is often called, imprecisely, "semi-carbonic."

Fleurie vigneron Justin Dutraive (*top center*) preparing to devat a carbonic-macerated tank, aided by (*clockwise from Dutraive*) Jean-Louis Dutraive, Sylvain Chanudet, Yvon Métras, and Ophélie Dutraive.

"Some people are persuaded that carbonic is a method that flattens terroir expressions. I find it doesn't. If you can't see the differences, you don't know how to taste it. That's the error."

—Philippe Pinoteau, proprietor of longtime Paris natural wine bistro Le Baratin

---

Pure carbonic maceration requires that no juice at all be created at the bottom of a tank, so it's only possible in small containers where the grapes will not be crushed by their own weight. Wines made with pure carbonic maceration are known for a keen, piercing fruitiness.

In the Wineglass: Carbonic Maceration

Broadly speaking, wines that undergo a strict or pure carbonic maceration do share a certain profile. They reveal glimmery fruit, soft tannins, and an unmistakeable candied quality, alongside whatever terroir nuances they might express.

Longer carbonic macerations can yield more persistent, profound wines. (But clumsy or overlong carbonic macerations can yield acetate aromas and a limp, jellied fruit.)

Making Natural Wine, Step 4: Devatting and Pressing

If you're making a white wine or a direct-press rosé, you'll skip right to pressing, avoiding maceration. But if you're making a macerated rosé, an orange wine, or a red wine, you have to get the grapes out of the maceration tank and into the press. This is called devatting.

Then there is the press itself: the stage when pressure is applied to the grape mass to extract the juice, leaving behind a dry(ish) mass of pressed skins, seeds, and stems called marc. Wine presses are a highly technical subject of underestimated importance. The type of press and how it is used determine the conditions for what you might call the birth of the wine must, i.e., the unfermented juice. The must is cloudy, sweet, and rich with lees. How it is handled and how much oxygen it sees in this phase can greatly affect the profile of a finished wine.

Conventional winemaking doctrine advises adding sulfites at this stage, and more conservative natural winemakers sometimes add a microdose here. (It's common among those working in a natural way in Champagne, for example.)

Free-run vs. press juice

The juice that flows freely from a maceration tank at devatting is called free-run juice. It differs from the juice that's later produced by pressing the devatted grape mass—but just how different depends on the type of maceration effected beforehand. From tanks that have seen heavy pigéage or destemming, the free-run juice is often considered tastier than the press juice, which extracts coarser elements. But from tanks that have seen carbonic maceration, the press juice is usually finer than the minimal free-run juice, which hasn't seen enzymatic fermentation.

Manual devatting

Just as with vatting, it's good to avoid pumping at devatting, too, even if the unpressed grape mass has been softened by maceration and pigéage. Why? Pumping presses the grapes prematurely and yields a juice suffused with gross lees.

A slow press

Wine presses come in many styles (see page 100). A regular feature of quality winemaking, however, is a slow press. It allows the fresh juice to filter over the grape mass, enriching itself with native yeasts from the grape skins all the while. Almost all winemakers claim to enact a slow press. But how slow? Some consider three hours to be a slow press. Others press over the course of twelve to twenty-four hours.

A hard press? Or no press?

Some natural winemakers, like Alsace's Bruno Schueller or the Frères Soulier in the Gard, work with a very thorough, hard press, extracting the maximum from their marc. Others, like those working with traditional qvevri methods in the Republic of Georgia, conduct no press at all after maceration. There's no right or wrong way; it depends on the many other factors involved in a winemaker's methodology.

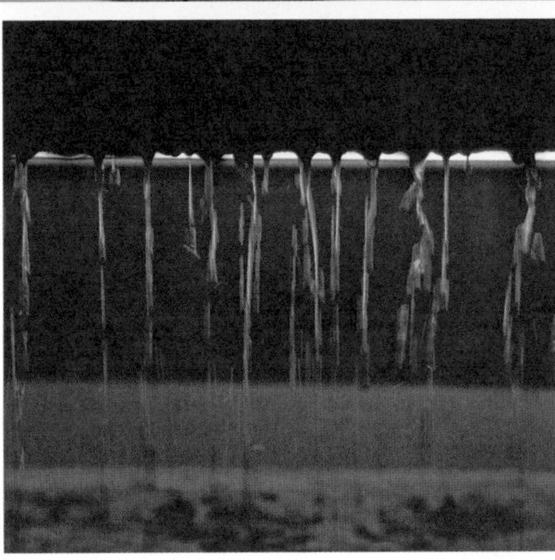

Roussillon vigneron Edouard Laffitte devats a tank of carbonic-macerated syrah.

Wine Presses

Wine presses fall into two broad categories: batch presses, which must be unloaded and reloaded between batches, and continuous presses, which are built to handle large volumes by continuously ejecting grape marc as they press. Only batch presses are used by natural winemakers, as continuous presses are said to extract grape matter more harshly. Here are five wine presses in common use within the natural wine world.

1. Pneumatic or bladder press

Pneumatic presses work by inflating a rubber sack inside the press chamber, which presses the grape marc toward slits on the chamber exterior.

2. Horizontal screw press

In this type of press, a screw system applies pressure from both sides of a grape marc.

3. Vertical or basket press (motorized)

Modern motorized vertical presses come in many variations. They're often considered the most quality-oriented type of press, because they allow the juice to flow slowly through the marc.

4. Vertical or basket press (manual)

Some natural winemakers use manual vertical presses. They are hard work. But they turn a pressing session into a hard-drinking, hard-exercising, daylong party.

5. Feet

The world's first wine press, feet are still used to press grapes by small-scale natural winemakers in certain parts of the world, like the Republic of Georgia. (Wines made in traditional southern Italian *palmenti* also see a foot pressing.)

Top: A horizontal screw press at Domaine David-Beaupère in the Beaujolais

Middle: A vertical press at the cellar of Marc Soyard at Domaine de la Cras in Burgundy

Bottom: A pneumatic press at Domaine Bonnet Cotton in the Beaujolais

Making Natural Wine, Step 5: Clarification and Racking

After pressing, most winemakers assemble the pressed juice into a tank to allow the solids—called the *gross lees*—to settle. Known as *débourbage* in French, it's a simple process: most natural winemakers just pump the juice into a tank and wait a night or two for the autumn chill to cause lees to settle. The juice is then transferred—or "racked," in winemaking parlance—to another container, where fermentation begins. (This stage doesn't occur in traditional qvevri winemaking. Instead, a natural clarification occurs after fermentation, when solids fall to the bottom of the qvevri.)

For natural winemakers, the only intervention (if any) during débourbage is to reduce the temperature in the tank to encourage a more thorough clarification, should they desire it.

Just how much lees a winemaker will keep varies depending on their individual style, the grape variety, and the condition of the harvest. Some perform no débourbage at all. Healthy lees act as antioxidants, so they protect a wine from oxidation. But they can also send it into reduced states of varying desirability (see Reduction, page 376).

For conventional winemakers, débourbage can include a flurry of manipulations, including fining, filtration, sulfite addition, and the addition of clarifying aids, such as pectinase.

Morgon vigneron Guy Breton (*right*) uses a refrigeration device inserted into a ceramic-lined chamber in his winery floor to clarify some free-run juice.

Making Natural Wine, Step 6: Liquid Fermentation and Cellar Aging

Winemakers often wait until fermentation has begun in tank before transferring wines to smaller containers, such as oak barrels, small *foudres*, or amphorae, where fermentation later finishes. Alongside alcoholic fermentation (in which yeasts transform sugars into alcohol), another fermentation occurs, either in parallel or afterward: malolactic fermentation, in which lactic acid bacteria transform malic acid into lactic acid (see opposite).

After fermentation, wine undergoes other changes as it ages. Its color stabilizes. It also settles, becoming clearer. This process is widely known by its French term, *élevage*.

How long does a wine age? In what kind of container? The answers tell you something about what made a wine the way it is. But it's important to recognize that for natural winemakers, the processes of fermentation and aging take as long as a given wine needs, so the answers to these questions change every year. Natural winemakers follow their own recipes like good chefs—very loosely.

Conventional winemakers, by contrast, do what they can to routinize and simplify the fermentation and aging process, including (but not limited to) adding commercial yeasts, adding yeast nutrients, and inoculating the wine with lactic acid bacteria.

Anjou vigneron Olivier Cousin (*right*) putting his cabernet franc into demi-muids

Temperature and humidity

The temperature sets the rhythm of fermentation. Natural winemakers, faced with slow or stuck fermentations, sometimes intervene to raise the temperature of a fermentation with space heaters, heating panels, or even by wrapping the entire tank or barrel in makeshift jackets for insulation. Humidity plays a secondary role; wines lose more liquid to evaporation in drier spaces.

Stirring lees (aka *bâtonnage*)

This is when a winemaker stirs a wine in the course of aging, or, less often, fermentation, in order to put fine lees back into suspension. The practice favors extraction in both white and red wines; it also encourages the wine to remain in a reductive state, which can help prevent oxidation.

Topping up

Standard winemaking practice in both conventional and natural winemaking is to ensure that aging containers remain "topped up," preventing contact with oxygen in the air. This is achieved by regularly refilling aging containers, which lose a small amount of wine to evaporation throughout the aging period. Certain winemakers make a point of *not* topping up their aging containers, preferring to age their wines in an oxidative milieu.

Adding sulfites

Avoiding sulfite addition is, of course, the rule in natural winemaking. Many more conservative natural winemakers nonetheless add microdoses of sulfites during the aging period, most often after malolactic fermentation (see opposite).

MALOLACTIC FERMENTATION

Often shortened to "malo" in industry parlance, malolactic fermentation is a bacterial (as opposed to yeast-based) fermentation that occurs spontaneously—and finishes—in natural wines in all but the coldest wine regions. (Lactic acid bacteria have difficulty with extreme cold.) During malolactic fermentation, bacteria transform a wine's sharp, green-appley malic acid into rounder lactic acid.

Sometimes malolactic fermentation happens right after alcoholic fermentation. More often there is some overlap between the beginning of malo and the end of alcoholic fermentation. This creates a tense moment for natural winemakers, who wish to avoid a situation in which lactic acid bacteria begin eating a wine's residual sugars. When this happens, it can cause a spike in volatile acidity (see page 376) and turn a wine to vinegar.

Malolactic fermentation is like puberty for wines. Wines are ugly during malo—cloudy to the eye and prickly on the tongue. It's a difficult stage, but a necessary one for wines to reach adulthood. They gain complexity and depth, losing some juvenile malic acidity in the process.

Many conventional winemakers, particularly in warm climates, choose to prevent malolactic fermentation in some or all of a wine in order to preserve more of its malic acidity. This is called "blocking malo." In wine media, it's usually discussed as an aesthetic choice, like how much lemon juice to add to a soup. But blocking malo invariably requires heavy sulfite addition and filtration, without which those lactic acid bacteria remain present and ready to act. This is why blocking malo is anathema to natural winemaking.

Beaujolais vigneron Julie Balagny uses a malolactic chromatography test kit to determine the progress of malolactic fermentation on some wine samples.

In the Wineglass: Blocked Malolactic Fermentation

Blocked malolactic fermentation reveals itself in a finished wine in several ways. High sulfite dosing can contribute a hardness to a wine's palate. The filtration removes its texture, leaving it with a squeaky sheen on the tongue, particularly when young. The wine's acid, meanwhile, is high and thin, a falsetto of acidity. Blocked malolactic fermentation doesn't necessarily mean a wine can't be complex (see most "classic" German riesling), but it does often diminish complexity. The technique—and its resulting superficial wine profile—is endemic to cheapo white and rosé winemaking worldwide. Blocking malolactic fermentation is also common in the vinification of base wines for conventional Champagne.

AMPHORAE: QVEVRI, TINAJAS, DOLIA

An amphora is an earthenware vessel. Specifically, it refers to a smaller vessel with handles that was historically used to transport wine and other liquids. But the term has come into colloquial use to refer to several types of traditional wine fermentation vessels, including (but not limited to) Georgian *qvevri*, Spanish *tinajas*, and Roman *dolia*.

Qvevri are traditional Georgian earthenware winemaking vessels, which are buried up to their necks in the earth, sometimes inside wineries, sometimes outside. Qvevri volumes vary, but around 800 liters is common. They've grown in use throughout Europe in recent years, particularly since Cher valley vigneron Thierry Puzelat (see page 150) brought some back from Georgia and gave them to vigneron friends throughout France in the mid-2010s. Traditional qvevri are lined with beeswax, which must be carefully maintained between uses. Qvevri differ from other sorts of amphorae in that they are fired at higher temperatures, which makes them less porous.

Tinajas are smaller, slimmer clay vessels, native to Castilla–La Mancha in Spain. Traditionally, they are freestanding rather than buried in the earth, though burial is also practiced by some contemporary estates. Vignerons who employ tinajas include Paul Estève and Chrystelle Vareille of Domaine des Miquettes in the northern Rhône (see page 242), Elisabetta Foradori in Italy's Trentino valley, and Giusto Occhipinti of Azienda Agricola COS in Sicily.

Dolia (singular: *Dolium*) are native to Rome and other parts of Europe, including southern France. They are larger and wider-mouthed than tinajas, making them especially suitable for initial fermentation. They have been used traditionally in both freestanding and buried forms. Vignerons who use dolia for fermentations and aging today include Bertrand Gautherot of Vouette & Sorbée in Champagne (see page 337) and Dominique Hauvette of Domaine Hauvette in Provence (see page 324).

Qvevri at the production site of Zaliko Bodjadze

Making Natural Wine, Step 7: Racking and Assembly

Some wines are bottled the year they are harvested. Others age for many years before bottling. Either way, at some point the wine is usually removed from its aging container and transferred to a new (often larger) container before bottling. This homogenizes individual lots of wine that may have diverged from each other during the aging period. It also removes the wine from contact with lees that have settled during the aging period. Some natural winemakers add sulfites at this stage.

This moment of transfer, called "racking" in industry parlance, is of the utmost importance. Inattentive racking can cause too much oxygen to reach wines, resulting in oxidation. It can also cause settled lees to get mixed back up, which can send wines into states of fierce reduction. (It's even possible to make both errors at once; see pages 376–377.)

It's ideal to rack using gravity, but relatively few winemakers' cellars are built in ways that allow them to gravity rack from aging containers. Instead, most use pumps—simple devices that in practice require great attention and skill to operate well. There's a whole jujitsu involved in attaching and detaching pipes and operating pumps.

Suffice it to say, name a great winemaker, and you've named a racking expert.

Roussillon vigneron Edouard Laffitte employs a special extension on his forklift to permit racking by gravity.

VIN NOUVEAU (AKA VIN PRIMEUR)

Capturing the heady, fruity immediacy of freshly finished fermentations is the idea behind *vin nouveau*, also known as *vin primeur*, or "new wine." Such wines don't see extended aging in tank or barrel; they're bottled in the autumn, as soon as fermentation finishes.

Many historical wine cultures have traditions of releasing new wines, the most famous being Beaujolais Nouveau. France's Beaujolais region first saw global success marketing its vin nouveau in the 1970s and '80s. Unfortunately, the combination of global market demand and the challenges inherent in making primeur so quickly each autumn soon incentivized the region's winemakers to use every cynical agricultural and winemaking shortcut available to them. They'd crop too high, harvest too early, add sugar, add commercial yeasts to finish the fermentation quickly, and heavily filter and sulfite the finished wine, in order to ship everything to importers by late October. Such practices gave not just the Beaujolais region, but the whole style of vin nouveau a bad reputation in conventional wine circles for decades.

But Beaujolais isn't a bad word in the natural wine world, and neither is vin nouveau. Many natural winemakers from around the world make nouveau wine, from Marcel Richaud in Cairanne in the southern Rhône (see page 231) to Thierry Puzelat in the Cher in the Loire to Sonoma's Martha Stoumen. Releasing primeurs is a way to celebrate the birth of a new vintage. If you want to read into things, it's also a gesture that insists that before wine became a heavily traded futures commodity, it was an inexpensive agricultural product, to be enjoyed in the moment.

Roussillon vigneron Jean-François Nicq bottling his primeur Octobre on-site on the day of the release of Beaujolais Nouveau at Paris wine shop Caves Augé, November 2013

Making Natural Wine, Step 8: Bottling

Bottling might seem like a prosaic step in the winemaking process. The wine's done, right? Well, not quite. The conditioning procedures that precede the act of bottling—which include fining, filtration, degassing, and sulfite addition—can be a transformative moment at conventional wine estates. Sometimes you can taste a wine from barrel that is brimming with energy and life, only for a winemaker to remove all that in the process of bottling.

Beaujolais vigneron Julie Balagny (*right*) bottling her Fleurie

---

"A lot of wines are damaged at bottling. Until bottling, they're good. Then there's an enologist who comes in and filters, and degasses poorly, and adds gum arabic. That's the moment when good wines are damaged."

—Morgon vigneron Jean Foillard

Natural winemakers come in all stripes when it comes to bottling practices. Some, like Frédéric Cossard in Burgundy (see page 202) or Frank Cornelissen on Sicily's Mount Etna (see page 351), eagerly embrace innovations in contemporary wine-bottling technology in order to finely calibrate their wines. Others, like Patrick Desplats in Anjou (see page 162), shun modern devices, preferring to bottle small lots by hand with primitive equipment.

Most natural winemakers fall between these two extremes, employing contemporary bottling lines, but taking care to avoid the bottling interventions that tend to standardize wine. The latter include fining, filtration (see page 109), excessive degassing (see page 108), and high sulfite doses.

When to Bottle?

Winemakers bottle wines when they feel that further aging is unnecessary for the type of wine they wish to make. But winemakers also bottle wines when bills are due and they need wine to sell. It's always a compromise. This helps explain why procedures like fining and filtration are so widespread: they help winemakers bring wines to market sooner, at the expense of a wine's integrity.

NATURAL /LEX·I·CON/

DEGASSING

The process of fermentation leaves wines naturally embedded with CO_2. Degassing is when winemakers remove some of the integrated CO_2 from a tank of wine a week or two before bottling through a process called sparging. A device that resembles a microphone is fed on a line into a tank. The device passes a fine bead of nitrogen or argon through the wine in the tank, which releases entrapped CO_2. A mousse forms on the surface of the wine in the tank during the process.

Why remove CO_2 from a wine? Because the level of entrapped CO_2 in a wine determines its rhythm on the palate. Wines with a lot of CO_2 can be spritzy, acid, and saline; many consumers dislike the sensation, often confusing it with that of a wine that has refermented in the bottle. By contrast, wines with very little CO_2 can seem heavy and lugubrious, seeming to move like pancake batter on the palate.

Natural winemakers leave more CO_2 in their wines than conventional winemakers, because it acts as an antioxidant, which is particularly helpful when avoiding sulfite addition. Also, natural wine appreciation circles tend to have more tolerance for spritziness and are more likely to consider it a pleasant contribution to a wine's personality. So what are the downsides of leaving a lot of CO_2 in a wine? In certain cases, it does its antioxidant job too well, sending wines into states of stubborn reduction.

> **"You shouldn't degas. We bottle at 1,150 to 1,200 milligrams of CO_2. There are people who that shocks. But CO_2 is an antioxidant."**
>
> —Burgundy vigneron-négociant Frédéric Cossard

NATURAL /LEX·I·CON/

FINING

Fining is when a substance that bonds to particulate matter is passed through a wine, so that the particles become large enough to settle or be filtered out. In the past, ox blood and egg whites were often used to fine wines. (Ox blood was banned from use in 1997 due to fears of mad cow disease.) More common fining agents today include bentonite (a type of clay), isinglass (derived from fish bladders), casein, polyvinylpolypyrrolidone (PVPP), and gelatin. In conventional wine production, fining often occurs at several stages, not just before bottling.

Natural winemakers do not practice fining because it fundamentally changes wines, removing good particles along with the bad.

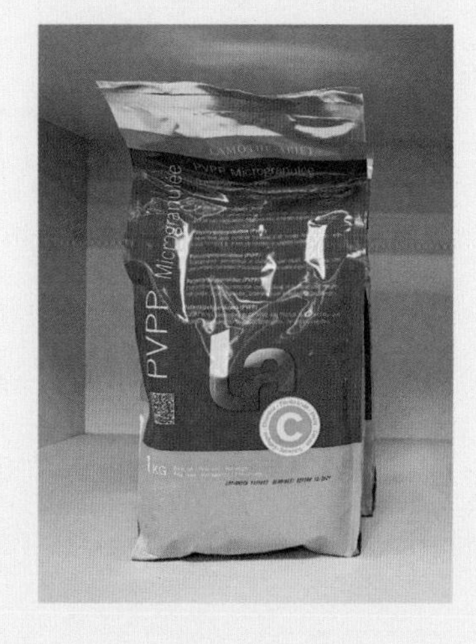

A pouch of the synthetic fining agent PVPP

FILTRATION

Natural winemakers avoid filtration. But—and this is a big "but"—it all depends on what you mean by "filtration."

Gaillac vigneron Bernard Plageoles (see page 309) filters his sparkling wines through custom-woven canvas sheets, an extremely light, historical method of filtration. Alsace vigneron Jean-Pierre Frick (see page 249) plate filters most of his white and orange wines before bottling, passing the juice through a series of cellulose sheets—which he subsequently dries in the sun and burns to heat his house.

These are rough styles of filtration that retain most of a wine's microbial life. By contrast, large-scale conventional wineries routinely employ cross-flow membrane filtration, a comparatively expensive, high-tech setup that eliminates all microbes, along with phenolic elements that contribute to color, texture, and body, to ensure total microbiological stability.

"The more you tighten the filtration, the more stays in the filter—and at a certain moment you have nothing left in the wine."

–Sancerre vigneron Sébastien Riffault

The problem is, because we don't have the right terminology to distinguish between them, we tend to call all these things "filtration," which leads to a lot of confusion on the subject. All forms of filtration perturb wines in some way (in part due to the extra pumping involved). And truly great wines are almost invariably unfiltered. But there is a lot of gray area between an ultra-filtered, biologically dead, mass-market wine and a truly unfiltered natural wine.

Top left: Diatomaceous earth filtration setup at conventional négociant Henry Fessy in the Beaujolais

Top right: Canvas used for traditional, very coarse filtration of sparkling wine at Domaine Plageoles in Gaillac

Bottom left: Plate filtration system at Domaine Pierre Frick in Alsace

NATURAL /LEX·I·CON/

PIQUETTE

Accompanying the global rise of natural wine, in recent years, has been a renewed interest in piquette, a thin, spritzy, wine-adjacent beverage produced by passing water (often mixed with sugar) over pressed, already fermented grape marc and fermenting the mixture again. (It's a similar principle to re-using a teabag.) Piquette is as old as wine itself. If you haven't heard of it before, it's probably because, for much of history, it was considered beneath the purview of wine appreciation. It was the beverage of Roman slaves, and, later, French miners and factory workers. Today, outside the EU (where it technically cannot be sold, and where most consumers still turn up their noses at it), well-made piquette from natural winemakers has proved an unlikely hit with health-conscious drinkers, enchanted as much by its refreshingly low alcoholic degree as by the cute sonics of its name.

Meanwhile, for natural winemakers—ranging from Romain des Grottes in the Beaujolais (see page 143) to Julien Guillot in the Mâconnais (see page 206)—piquette production is a way of valorizing every last drop of their harvest, a gesture of kinship and support for traditional paysan thrift.

Roussillon vigneron Loïc Roure (*right*) and his cellarhand loading destemmed grape marc into a tank for an experimental piquette

Natural Sweet Wines

Sweet wines in both the US and the EU are allowed to have significantly higher sulfite levels than their dry red and white counterparts. Why is this?

It's related to the inherent instability of residual sugar. Fermentation, you see, is a natural process. It'll happen of its own accord if you do nothing to stop it. (Think of what happens if you leave a jug of fresh apple cider open in the fridge for a month.) Because of this, conventional producers of sweet wines use filtration and high sulfite doses to ensure no active yeasts or bacteria are left in a wine. This has been the norm since the 1960s, when atomic membrane filter technology became widespread. It offered winemakers a quick way to achieve microbial stability in sweet wines. (It also gave rise to a glut of cheap, sugary wine brands that still scar the memories of several generations of wine drinkers.)

If today even natural winemakers resort to some form of filtration for sweet wines, it's because the truly natural method is expensive, risky, and time consuming. It consists of aging a sweet wine until it becomes biologically stable, a process that can take many years. Even then, bottling without sulfites or filtration means there's always a risk of fermentation restarting in the bottle. When that happens, the best-case scenario is that the wine becomes a little spritzy and pushes its cork out a millimeter or two. In the worst-case scenario, the bottle explodes.

This is why truly natural, unfiltered sweet wines are rather rare, and worth treasuring if you can find them. For a start, check out the opulent, long-aged natural sweet wines of Anjou vignerons like Jérôme Saurigny, Didier Chaffardon (see page 160), or Philippe Delmée, or the declassified Vin de Paille of Jura vigneron Julien Labet (see page 188).

NATURAL ROSÉS

Wines to Know

To find natural rosé, we have to hearken back to a time before refrigeration and atomic membrane filtration made artificially pale Provence-style rosés possible. What was rosé like then? It was a very light red wine—the kind of thing you feel like drinking if you work outdoors in a hot, predominantly red-grape region like the southern Rhône or Provence. Natural rosés are made either by directly pressing red grapes or by letting red grapes macerate for a short period of time, from a few hours to a few days.

Natural rosés tend to be darker and more vinous than conventional rosés. But they're no sweeter than conventional rosés—quite the opposite, in fact. Color intensity has nothing to do with sweetness. And since the best natural rosés are unfiltered, they must be fermented to dryness; otherwise, they would risk undergoing further yeast or bacterial activity in the bottle.

Here are three benchmark natural rosés to get to know.

Domaine l'Anglore | Vin de France–Chemin de la Brune

Once a parcel cuvée, Chemin de la Brune has in recent years been reconceived as an assemblage, uniting all the Pfifferlings' direct-press wines. As such, its varietal composition changes each year. (The 2020 is carignan, syrah, mourvèdre, and grenache.) Used-barrel aging is a constant—as is the wines' fleshy, cantaloupe-toned grace. It is notable for being the estate's only rosé that more or less resembles a rosé.

Domaine la Paonnerie | Coteaux d'Ancenis–Le Rosé d'Ancenis

From their organic, schist-soil gamay vines west of Angers and east of Nantes, Agnès and Jacques Carroget make an unfiltered rosé the color of onion skin, redolent of white cherry. The grapes spend twelve hours macerating before pressing.

Les Foulards Rouges | Vin de France–La Soif du Mal Rosé

A barrel-aged direct press of syrah and carignan grown on granitic soils in the Roussillon, Jean-François Nicq's Soif du Mal rosé possesses a buoyant, lip-smacking black-grape character, belying both its pale, bluish color and its malign name, which translates to "a wicked thirst."

The Pétillant Naturel

Pétillant naturel—or *pét-nat* for short—refers to a sparkling wine whose effervescence comes from the fermentation of its own yeasts and sugars in the bottle. It's the rare French neologism that has crossed the Atlantic and entered mainstream Anglophone wine discourse in recent years.

The term *pétillant naturel* was coined by the late Vouvray and Jasnières winemaker Christian Chaussard (see page 154), who experimented with the wine style in the 1990s alongside cohorts like Pascal Potaire (later of Les Capriades; see page 153) and Thierry Puzelat. But the wine style itself is about as old as the glass wine bottle, with traditional versions having long been produced in the southern French regions of Gaillac and Limoux, where they're known as *méthode ancestrale* ("ancestral method") wines, as well as in parts of Italy and Spain. If you bottle partially fermented grape juice without sulfites or filtration, it will keep fermenting in the bottle, yielding bubbles. Chaussard's neologism is best understood as a distinction that he made between his own pétillant naturel, produced without adding yeast or sugar, and the various crémants of the Loire (including Vouvray), which are generally produced with added yeast and sugar.

Pétillants naturels are typically lower pressure and less bubbly than sparkling wines produced via the Champagne method (also referred to as the *méthode traditionnelle*, or traditional method). They can be red, white, orange, or rosé. They can exhibit a yeasty character, thanks to their fermentation on lees. Most are disgorged to expel those lees, but some aren't. The majority are intended to be simple, refreshing apéritif wines; something to kick-start the palate.

Roussillon vigneron Jean-Louis Tribouley examines the deposit in a bottle of pétillant-naturel rosé.

PÉTILLANTS NATURELS

Wines to Know

Pétillants naturels hail from all over the wine world. As a wine style, pétillant naturel has significant overlap with the *méthode ancestrale*, or *col fondo* in Italy. But we call something a pétillant naturel when it's fermented on its own yeasts with its own sugars and is *also* a natural wine. (It is perfectly possible to make conventional méthode ancestrale or col fondo wines.)

Pétillants naturels are full-flavored, fun, and purer than any Champagne, at a fraction of the price. What's not to love?

Château Lafitte | Vin de France–Funambule

Based on the gros manseng grape of Jurançon, Antoine Arraou's Funambule *pét-nat* is a southwestern rocket trailing vapors of almond and finger lime, more complex than most in its genre.

Patrick Bouju | Vin de France–Festejar

Patrick Bouju's Festejar contrasts the joyous, redcurrant fruit of gamay d'Auvergne with a subtle savory aspect attributable to his volcanic basalt soils.

Les Capriades | Vin de France–Piège à . . . Rosé

Off-dry, deep pink, and scrumptious, Piège à . . . is like a rebuke to the self-serious, ultra-dry Champagne trends of the day. An unsulfited blend of côt, gamay, and cabernet franc, it's the ultimate apéritif, a sunburst of wholesome pomegranate sappiness.

Six Flash Points in Natural Winemaking

Naturalness in winemaking is in perpetual debate among winemakers, importers, wine retailers, négociant houses, wine critics, and, of course, wine lovers. So it should come as no surprise that even the natural vignerons at the heart of the discussion are rarely in total agreement with one another. Certain issues are subjects of perennial controversy: natural to some, unacceptable to others. Here are the major sticking points that have frustrated many who've attempted to impose a strict charter on natural winemaking. (These topics are limited to the subject of vinification—that is, what happens in the winery. For flash points in natural viticulture, see page 55.)

A tank of CO_2 in a winery in the Gard (France)

1. Commercial CO_2 and argon

The late Morgon vigneron Marcel Lapierre and his cohorts were the first to begin promoting the "wine without sulfites" ideal, which evolved into what we know as natural wine today. But their cold-carbonic winemaking method involves sealing maceration tanks with commercial CO_2, which some natural winemakers consider an artificial intervention.

Similarly, some winemakers, including Lapierre's peer Guy Breton (see page 128), use canisters of argon instead of nitrogen to degas wines before bottling, believing that argon yields superior results. Some natural winemakers object to this on the grounds that argon is a rare gas, and somewhat expensive. Others object to degassing with any commercial gas canisters at all, nitrogen included.

2. Refrigeration and temperature control

How much temperature control is too much? Many natural winemakers use mobile refrigeration units to cool their harvests before vatting. Some object to this as an energy-wasting intervention. Nowadays, most natural winemakers agree that using dry ice (CO_2 in solid form) to cool fermentation tanks is an overly aggressive winemaking intervention, but some natural winemakers have experimented with the practice. Some natural winemakers also consider excessive or prolonged temperature control a form of unnatural meddling.

3. Négociant work

In Anglophone culture, we tend to applaud the can-do spirit of businesses that diversify to expand production. Not so in the world of natural wine, where many natural vignerons are leery of their peers who make négociant wines (wines from purchased grapes). To some vignerons working in the paysan tradition, making négociant wine constitutes an admission that it is no longer possible to make a living as a "pure" vigneron.

This attitude is changing, as climate shifts cause grape losses that oblige more and more vignerons—even the radical purists—to establish négociant winemaking businesses merely to stay afloat.

4. Yeast nutrients

It takes years of thoughtful farming to obtain a healthy wine must, naturally replete with all the nutrients native yeasts need to ferment a wine to dryness. Nitrogen, one of the critical components for yeast activity, is sometimes lacking in wine musts. Natural winemakers faced with nutrient-deficient musts have to decide whether to compromise their principles by adding yeast nutrients. Unlike with sulfites, wine critics rarely ask about yeast nutrients, and in tasting it is extraordinarily difficult to discern whether such additives have been used.

5. Filtration

Filtration is such a handy tool to winemakers, in so many diverse situations, that very few natural winemakers will disavow it completely. Most know that they, too, will have use for it now and then. Only the very bravest and most dogmatic artisanal natural winemakers never employ it.

"The problem is that lysozymes, in stopping the lactic [acid] bacteria from functioning, they stop all the other bacteria from functioning. We can't reasonably think that it's only yeasts that work during a fermentation, or that only yeasts are responsible for a wine's gustatory characteristics. Even if the enologists tell us otherwise."

—Roussillon vigneron Jean-Louis Tribouley

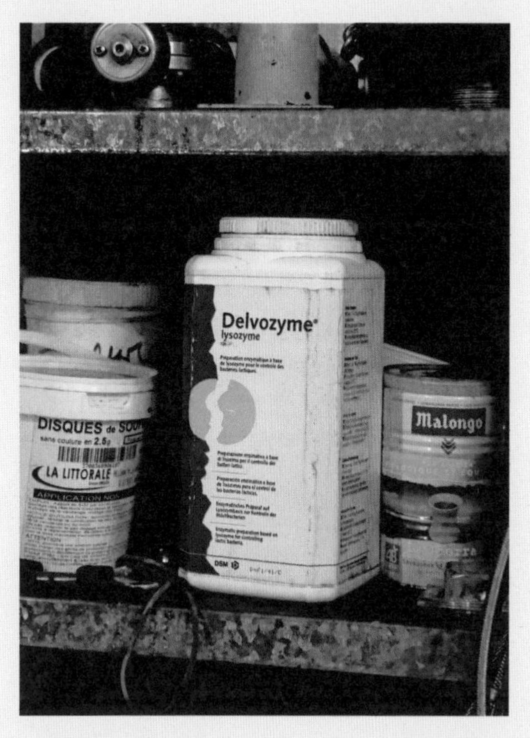

Lysozyme on the shelf of a vigneron-négociant in the Loire

6. Lysozyme

Lysozyme, an enzyme derived from egg whites, is sometimes used in food and wine production for its bactericidal properties. Adding lysozyme is permitted in US organic wine, but illegal in EU organic wine. It degrades the cell walls of lactic acid bacteria and is sometimes used to delay malolactic fermentation or stave off lactic spoilage in red wines. Its use is taboo in natural winemaking, but that doesn't stop many otherwise natural winemakers from using it.

Natural Wine Certification

The subject of natural winemaking is as complex as winemaking itself. Defining what is "natural" means defining what constitutes an unacceptable transformation of wine. This, in turn, requires taking aesthetic stands about wine quality that are often rather subjective. The issue is rendered all the more intractable by the near impossibility of ever truly verifying a wine's purity through its entire journey from vineyard to wineglass. Many common winemaking interventions, from enzyme additions to commercial lactic acid bacteria, leave no trace in a finished wine. They can only be detected by analyzing the fermenting wine must—or, with something less than legal certainty, by the palate of an experienced wine taster.

The practical impossibility of certifying a wine's naturalness has not stopped many from trying. As with the "organic" label in the 1980s and '90s, there's a lot of money and market influence at stake in defining natural wine. Here are the most prominent natural wine certification organizations active today. Some are stricter than others, and none is the last word in natural wine—for now.

L'ASSOCIATION DES VINS NATURELS

Founded: 2005

Founded by: Marcel Lapierre

Current president: Sancerre vigneron Sébastien Riffault (see page 171)

Vigneron members as of 2020: 30

Sulfite limits: Less than 10mg/L at analysis for red wines, 20mg/L for white wines. Vignerons adding zero sulfites to at least 80 percent of their production are permitted to use the green AVN logo sticker on zero-added-sulfites wines.

Requires: New members must be nominated by existing members.

Permits: Only the purest possible natural winemaking.

Forbids: Any additives whatsoever. Micro-oxygenation, must concentration, alcohol reduction, thermovinification. Cleaning products other than water and steam. Négociant work.

Verification: Vignerons are required to submit lists of wines, accompanied by analyses and proof of organic certification each year. Analyses are subject to verification at random.

The verdict: The AVN, as it's known in France, has become a cautionary tale about the difficulties of managing winemaker associations. As the movement for natural winemaking picked up steam in France in the 2000s, Marcel Lapierre and subsequent president Christian Chaussard were criticized for being too permissive with entry to the association. Under Sébastien Riffault, the AVN has instead adopted the strictest possible label charter—and has experienced an accompanying loss of membership.

SYNDICAT DE DÉFENSE DES VINS NATURELS

Founded: 2019

Founded by: Loire vigneron Jacques Carroget, with lawyer Eric Morain and wine journalist Antonin Iommi-Amunategui

Current president: Loire vigneron Jacques Carroget (see page 164)

Vigneron members as of 2020: 109. The Syndicat also welcomes the support of journalists and consumers.

Sulfite limits: Less than 20mg/L for the more rigorous of the Syndicat's two labels, which amounts to a declaration of zero added sulfites. The Syndicat also offers a label bearing the declaration "Less than 30mg/L total sulfites" to estates that add sulfites at bottling.

Requires: Organic certification for grapes.

Permits: Irrigation. Chemical cleaning products in the winery.

Forbids: "Brutal and traumatising techniques," including reverse osmosis, thermovinification, fining, and filtration. Any additives besides sulfites at bottling.

Verification: Certification conducted by existing French organic certification inspectors. Roughly 3 percent of wines bearing the Syndicat's label are laboratory tested per year.

The verdict: It's too early to say whether the Syndicat de Défense des Vins Naturels will have a significant market impact. But the organization demonstrated ambition and administrative savvy by gaining approval for its "Vin Méthode Nature" labels in France. Its relatively flexible charter has proved welcoming to many more recently converted natural wine estates.

NATURE & PROGRÈS

Founded: 1964

Founded by: Government official André Louis, Languedoc vigneron Mattéo Tavera, and agrobiologist André Birre

Current president: Éliane Anglaret

Vigneron members as of 2020: 59

Sulfite limits: Doses are permitted up to 70mg/L for reds, 90mg/L for whites and rosés, and 210mg/L for sweet wines. But total free sulfites at final analysis must be no more than 10mg/L for reds and 15mg/L for whites or rosés.

Requires: Organic certification for grapes.

Permits: "Finishing yeasts" and other commercial yeasts, as long as their use is not systematic, and as long as they are not genetically modified. Tannin addition upon the harvest if overseen by an enologist. Flash pasteurization. Micro-oxygenation. Casein, bentonite, and organic egg whites for clarification. Centrifuges. Basically all forms of filtration. Acidification under certain conditions, as well as deacidification with potassium bicarbonate under certain conditions.

Forbids: Any synthetic chemicals in agriculture or winemaking. Thermovinification. Flash détente. Dialysis for tartrite stabilization.

Verification: Nature & Progrès functions as a participatory guarantee system, in which members are obliged to take active participation in the establishment and verification of production standards.

The verdict: Nature & Progrès is a vast association, covering all realms of agriculture. Its charter governing wine production is admirably strict in some ways and weirdly loose in others. It is no longer a reliable indicator of the naturalness of a wine estate; more often it indicates the estate's commitment to an ideal of a peasant or locavore economy.

LES VINS S.A.I.N.S.

Founded: 2012

Founded by: Mâconnais vignerons Catherine and Gilles Vergé (see page 207) and Loire vigneron Jean-Pierre Robinot (see page 147).

Current president: Catherine Vergé

Vigneron members as of 2020: 13

Sulfite limits: Only trace elements, none deriving from any sulfite addition.

Requires: "Clean agriculture"—no synthetic chemicals. Manual harvest.

Permits: The charter is quite vague. But the organization consists exclusively of the Vergés' radical friends, all very small-scale vignerons. In practice, no one cheats.

Forbids: Any additives, including sulfites. Négociant work. The charter must apply to 100 percent of a member vigneron's production (meaning no less-natural cuvées).

Verification: None, just a statement of honor from each vigneron. As a group, it is "participatory, friendly, and convivial."

The verdict: It's a well-intentioned but marginal group of radicals, with little influence on vignerons outside their immediate social circle. The participating vignerons are as natural as can be, however.

TRIPLE "A"

Founded: 2002

Founded by: Italian natural wine and spirits wholesalers Luca Gargano and Fabio Luglio

Current president: There is no president, as such. It is a commercial grouping linked to the Vellier import and distribution company.

Vigneron members as of 2020: 86

Sulfite limits: Can be added "only in minimal quantities and at time of bottling."

Requires: The manifesto implies that it requires massal selection vines, cultivated without synthetic chemicals. But in practice, it is not necessary to be certified organic.

Permits: Sulfite addition at bottling (see above). In practice, anything is permitted, as long as the finished product passes muster with the Vellier team.

Forbids: Nothing is explicitly forbidden; the organization is based on a manifesto, not a charter.

Verification: No verification is conducted; the organization is based on a system of trust between its founders and the estates they distribute.

The verdict: Founder Luca Gargano anticipated the movement toward natural winemaking in a way that was nothing short of visionary. His organization, for better or for worse, is more concerned with promoting natural wine as an ideal—alongside Gargano's own commercial interests—than with defining it or enforcing any definition among its members.

VINIVERI

Founded: 2003

Founded by: Veneto vignaiolo Angiolino Maule, Friuli vignaiolo Stanko Radikon (see page 354), Tuscany vignaiolo Fabrizio Niccolaini (see page 353), and Umbria vignaiolo Giampiero Bea

Current president: Giampiero Bea (see page 350)

Vigneron members as of 2020: 131

Sulfite limits: 80mg/L for dry wines and 100mg/L for sweet wines.

Requires: Cultivation of indigenous grape varieties. All new vine plantings must be mass-selected.

Permits: Organic vineyard treatments.

Forbids: Herbicides. All systemic vineyard treatments. Mechanical harvesting. Chemical fertilizers. GMO vines. Yeast inoculation. Yeast nutrient additives. Enzyme inoculation, bacteria inoculation. Must concentration. Temperature-controlled fermentation. Clarification and filtration.

Verification: No verification is conducted.

The verdict: ViniVeri united a pioneering generation of passionate, quality-oriented, mostly natural Italian vignaioli, at a time when quality Italian wine was benefiting from a commercial renaissance in high-end Italian restaurants. The group later fractured over issues of enforcement. The members who were more ideologically committed to enforcing a collective definition of natural wine left to join Angiolino Maule's VinNatur.

VINNATUR

Founded: 2006

Founded by: Veneto vignaiolo Angiolino Maule

Current president: Angiolino Maule (see page 353)

Vigneron members as of 2020: 180

Sulfite limits: 50mg/L for whites, rosés, or sparkling wines. 30mg/L for reds.

Requires: Organic certification.

Permits: Up to 10 percent of purchased grapes as part of an estate's production, or 40 percent in case of natural disaster. Irrigation. Filtration of no less than 5 microns for whites and rosés and 10 microns for reds. Micro-oxygenation.

Forbids: Synthetic herbicides, fungicides, and insecticides in viticulture. Mechanical harvesting. Any vinification additive other than sulfites. Yeast inoculation. Fining agents. Enzyme inoculation.

Verification: Since 2007 the association has analyzed at least one of each member estate's wines per year. Since 2016, the association has put in place a verification system in which specially trained inspectors visit 40 percent of the member estates one or two times.

The verdict: Angiolino Maule's VinNatur is the most well-run and administratively ambitious natural wine association. More radical than its Italian forebears yet less radical than its French counterparts, it seeks to negotiate a middle path for commercially conscious natural winemakers. Its quest to ensure adherence to its charter through novel enforcement schemes, however, seems quixotic.

PART II

A Pantheon of Natural Wine

FRENCH WINE REGIONS DISCUSSED IN CHAPTERS 4–16

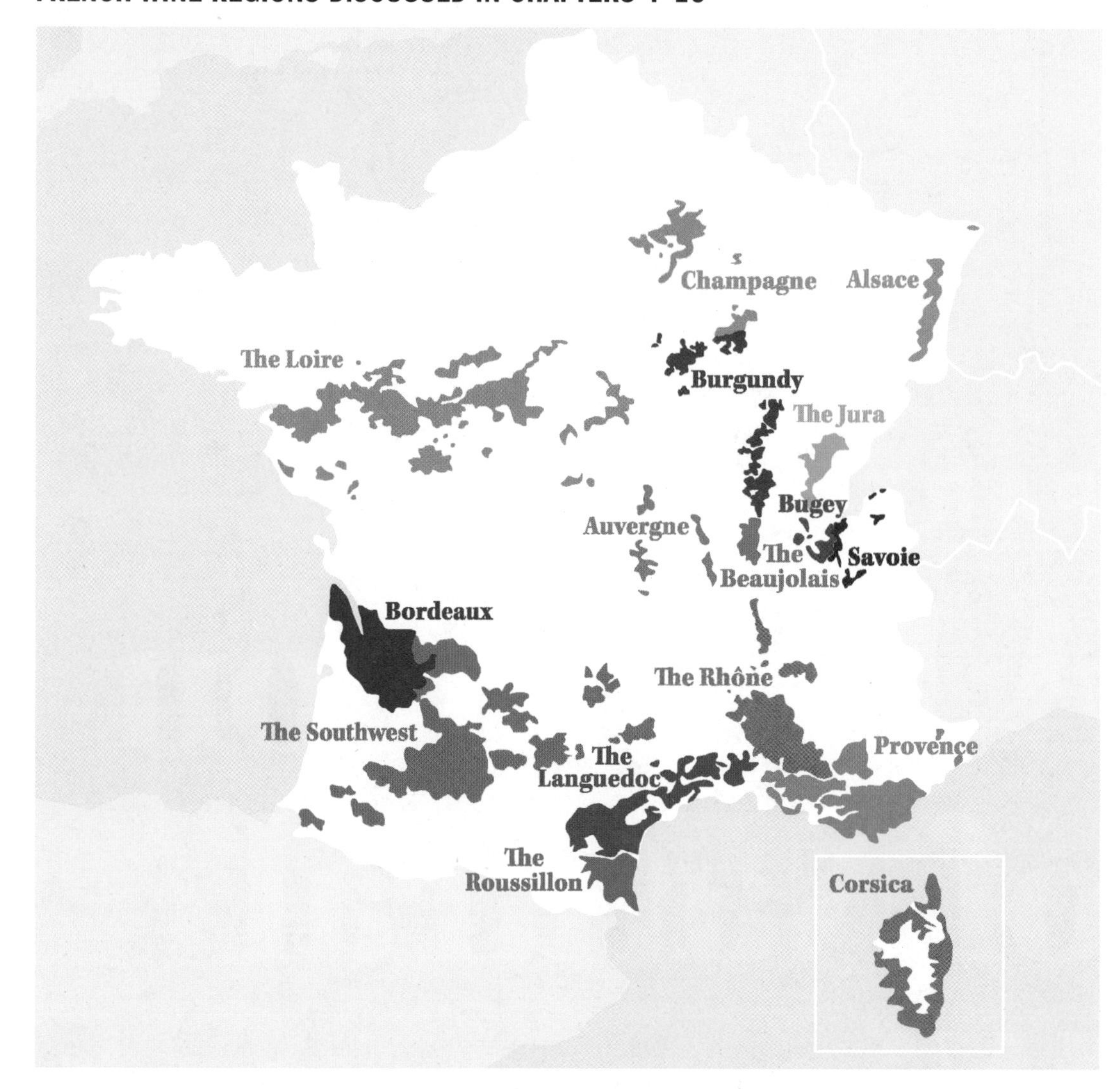

NATIONS OF EUROPE AND THE CAUCASUS DISCUSSED IN CHAPTER 17

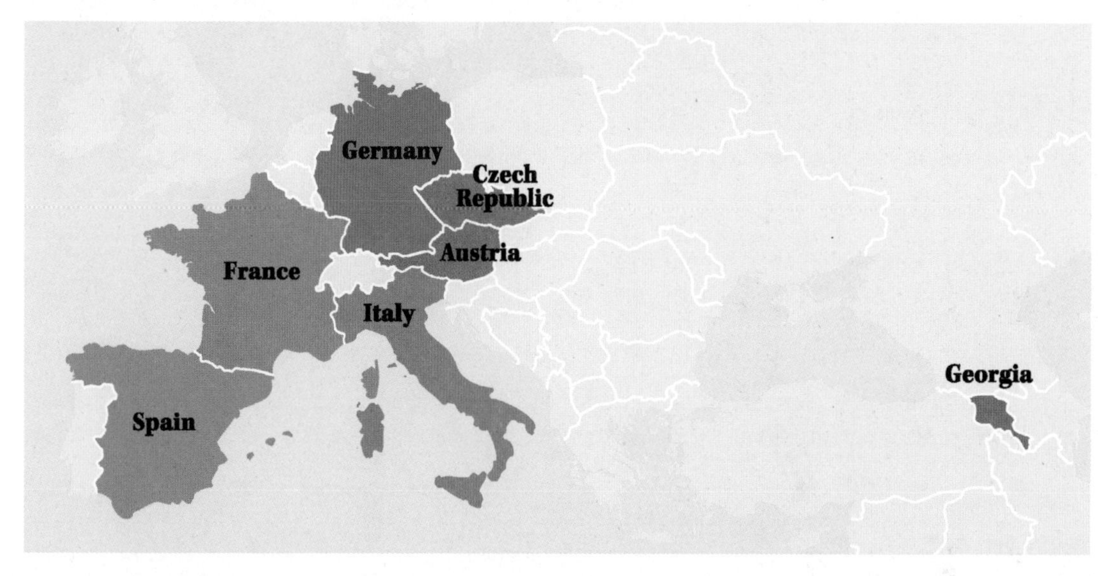

The world of natural wine is a community, made up of individuals united by common beliefs about wine growing and winemaking. This section illustrates what you could call the pantheon of natural wine: the most influential natural winemakers throughout France, Europe, and the Caucasus.

In this pantheon of natural wine, France predominates. (For more on this, see page 23.) So I've organized the French regions into individual chapters. The more nascent or fragmented natural wine cultures in Europe and the Caucasus are grouped into one broad chapter.

Each regional chapter opens with an overview of several benchmark wines from the region's most celebrated natural winemakers. These aren't necessarily the "greatest" or most expensive wines from each winemaker. They represent a winemaker's signature cuvée: the wine that best illustrates their personal style.

I have not cited vintages, since such information goes out of date very quickly.

With a little research, and a lot of tasting, a few decades of natural wine history goes by fast. Consider the Wines to Know an introduction to all the great bottles to come.

Note: Each estate profile contains basic figures about its founding date, its size, its agricultural certification, which grapes it works with, and which appellations it produces. Also included is information on whether an estate conducts négociant grape purchases, and whether it has recourse to sulfite addition or filtration in winemaking. It is, however, in the nature of this information to change from one year to the next. The figures cited in the profiles should be considered provisional: they represent a snapshot of the workings of a wine estate.

Pages 120–121: Loire vigneron Jean-Pierre Robinot disgorging a bottle of pétillant naturel outside one of his cellars in Chahaignes

Harvest in Chiroubles, 2016

4.
The Beaujolais

Today the rolling, fairy-tale landscape of the Beaujolais boasts a broad and dynamic community of natural winemakers, befitting its history as the cradle of natural wine. Why here, though? Why this particular stretch of granite hills north of Lyon?

Partly it is a story of personalities. There was the leadership of Morgon vigneron Marcel Lapierre from the 1980s until his death in 2010. There was the groundbreaking research of the Beaujolais wine scientist Jules Chauvet and the influence of his eccentric assistant Jacques Néauport. There was the early support of the three-Michelin-star Lyonnais chef Alain Chapel. (For more on these figures and the Beaujolais influence on natural wine, see pages 24–29.)

But cultural and agricultural factors also made the Beaujolais a ripe place for natural wine. The Beaujolais Nouveau marketing phenomenon in the 1970s and '80s coincided with the advent of herbicides and contemporary enology, which the region's vignerons embraced with a particular gusto. Herbicides replaced plowing on its steep, densely planted, thin-soiled hills. Commercial yeasts, sugar, and heavy sulfite addition were deployed to make the most of overcropped harvests destined for cheap Beaujolais Nouveau on the export market.

The Beaujolais's natural wine pioneers were thus witness to an early, exaggerated display of the same trends that were transforming winemaking worldwide. And they took an influential stand against them.

Visitors to the region today are often struck by the geographic proximity that bound Lapierre and his friends. The cellars of Jean Foillard, Guy Breton, and Jean-Paul Thévenet are *walking distance* from that of Lapierre. This still comes in handy, given the region's deserved reputation for voluminous drinking. Historically a land of peasants and sharecroppers, the Beaujolais is known as much for its generosity as its humility.

Total vineyard surface:
14,200ha (35,089 acres)[1]

Rate of organic agriculture:
16 percent[2]

Key grapes:
Gamay, chardonnay

1 Inter Beaujolais
2 Agence Bio 2020 figures for Rhône department. Does not include several northerly Beaujolais appellations within the Saône-et-Loire department.

Wines to Know

The Beaujolais's reputation as a source of cherry-fruited, easy-drinking reds is ennobled by its great natural wine estates, whose wines show there's zero conflict between innate glug-ability and complexity. There is a stylistic range among the cool-carbonic masters of the region, with the wines of Jean Foillard and Jean-Paul Thévenet typically showing more force and extraction than the glowier wines of Guy Breton or Domaine Lapierre. In recent years, keener, brighter, more radically unsulfited work from the region has been popularized by the likes of Romain des Grottes and Hervé Ravera (see page 143).

Domaine Jean Foillard | Morgon–Côte du Py

A remnant of an ancient volcano, the Côte du Py is a bulging hillside of magmatic schist beneath the hamlet of Morgon, the historical center of the appellation. Its present-day renown is due to Jean Foillard, whose singular Morgon, Côte du Py, is the most universally acclaimed wine of the region, a suave masterpiece of stone fruit, boasting the typical aromatic structure of its terroir, known as *pierre bleue* (blue stone) within the region. The wine sees long, cool-carbonic maceration in cement tanks before aging in used oak barrels. Historically, Foillard's accomplices in the final blending decisions each year were the journalist and early natural wine bistro proprietor François Morel (see page 33) and Le Baratin co-owner Philippe Pinoteau (see page 403).

Domaine Lapierre | Morgon

What defines the profile of this genre-defining wine? A delirious drinkability, thanks to long, cool-carbonic maceration in wooden vats, from which the free-run juice (see page 98) is often removed for use in other wines. The succulent press juice that leaves the Lapierres' custom vertical presses is aged for a season in used oak barrels. Domaine Lapierre produces no less than three versions of this wine: one filtered and sulfited, another sulfited but not filtered, and a third with neither sulfites nor filtration. The latter two are worth seeking out for their classic flavors of iron and kirsch (cherry liqueur), hallmarks of great granite-soiled Morgon.

Yvon Métras | Fleurie–Vieilles Vignes

Yvon Métras's Fleurie–Vieilles Vignes benefits from impeccable vineyard work on two renowned terroirs, La Madone and Grille-Midi. The wine strikes a stylistic synthesis between the more consistent, cool-carbonic winemaking practiced by Métras's peers in the Lapierre band, and the edgier, zero-added-sulfites work of radicals like Christian Ducroux (see page 138) and Michel Guignier (see page 139). Fermented in concrete tanks and aged in used oak barrels, the wine shows delicate wilted-rose notes amid its dark-cherry fruit. Warm vintages show a kinship with the great syrah wines of Saint-Joseph.

GUY BRETON

VILLIÉ-MORGON

Known as much for his personal charm as for his temper, Guy Breton is a winemaker's winemaker, a monomaniacal craftsman whose wines are typified by a dancing lightness and crisp acidity. More commonly referred to as "Le P'tit Max"—a reference to the first name of his father, a blacksmith—Breton was among the first acolytes of Marcel Lapierre in the 1980s, and worked closely with him and Jacques Néauport well into the 1990s. Breton, who formerly sold heating fuel for a living, began making wine in 1988 with 2 hectares (4.94 acres) of vines from his mother's family in Morgon, which he soon supplemented with grape purchases to produce Beaujolais Nouveau and Beaujolais-Villages.

Breton practices a strict style of carbonic maceration, removing free-run juice from fermenting tanks in Morgon or Régnié and using it for his primeur or his Beaujolais-Villages. He is adept at the use of cold in natural winemaking: he vats around 5°C (41°F), and has no qualms with employing CO_2 in powdered dry ice form to further chill tanks if necessary. He'll also chill his wines directly after pressing to clarify them, and occasionally during barrel aging of his Morgons. Seek out his potent Morgon–Le P'tit Max, produced from his own hundred-plus-year-old high-sited vines in the village of Saint-Joseph.

"When the vigneron working my father's vines died, Lapierre told me, 'You've got to take the vines!' He gave me a press and four tanks and told me to go for it."

THE WINE TO TRY

Beaujolais Primeur–Cuvée Fanchon

Guy Breton's Beaujolais Nouveau, named for his youngest daughter, is perennially in a class of its own: vivid and diaphanous, flitting across the palate with the spontaneity of birdsong. It is a minor wonder of the wine world, owing to the sleepless dedication and miniaturist's precision of its creator.

Since: 1988

Winemaker: Guy Breton

Vineyard surface: 4ha (9.88 acres)

Viticulture: Practicing organic

Négociant work: Yes

Grapes: Gamay

Appellations: Morgon, Régnié, Chiroubles, and more

Filtration/Fining: None

Sulfites: Dosage of about 10mg/L at assemblage before bottling, and another 10mg at bottling

DOMAINE J. CHAMONARD

Jean-Claude, Geneviève, and Jeanne Chanudet

VILLIÉ-MORGON

Embodying the grit and ingenuity of the Beaujolais paysan, Villié-Morgon's Jean-Claude Chanudet is among the most enterprising of the late Marcel Lapierre's acolytes. Widely known as "Le Chat," the stern, soft-spoken Chanudet began making wine in 1990, after inheriting the estate of his wife Geneviève's father, the venerated vigneron Joseph Chamonard. (The estate has been in Geneviève's family since the nineteenth century.) Today Chanudet is recognized as much for his winemaking talent as for his practice of bottle aging portions of wine for long periods—often more than a decade—before release. Chanudet can afford this thanks to his other businesses, notably his work in partnership with Marie Lapierre at the value-driven natural Beaujolais estate Château Cambon. (In addition, Chanudet runs a business based on a machine he invented to unbottle and reassemble lots of wine.) During the fall fermentation period, the Chanudets' home plays host to Renée Boisson's weekly microscope sessions, during which the celebrated winemakers of their circle gather to talk shop and analyze wine samples.

"There are two very important things for a winemaker: he must not fall into habits in winemaking, and he must have a very strong mentality to resist the pressure of modern enology."

—Jean-Claude Chanudet

THE WINE TO TRY

Morgon

From vines in the Corcelette site, Domaine Chamonard's Morgon is a monument to the age-worthiness of great gamay, thanks to Chanudet's habit of retaining back vintages for later release. Bottle aging accentuates notes of tea, mulling spices, and, in good vintages, a buoyant, grenache-like fruit.

Jean-Claude Chanudet in his cellar in Les Marcelins

Since: 1990

Winemakers: Jean-Claude and Jeanne Chanudet

Vineyard surface: 6ha (14.8 acres)

Viticulture: Practicing organic

Négociant work: None

Grapes: Gamay

Appellations: Morgon, Fleurie

Filtration/Fining: None

Sulfites: Usually 10–20mg/L at bottling

DOMAINE LAPIERRE

Mathieu and Camille Lapierre

VILLIÉ-MORGON

From left: Camille, Marie, and Mathieu Lapierre, at the entrance to the cellar at Les Chênes in Villié-Morgon

Marcel Lapierre's influence on natural wine is almost incalculable. Lapierre's friends and former employees span the globe, from the Loire Valley (Olivier Lemasson, Catherine and Pierre Breton, Thierry Puzelat) to Auvergne (Jean Maupertuis) to Ardèche (Gérald Oustric) to Chile (Louis-Antoine Luyt) to Rioja (Olivier Rivière). It's only fitting that this historic estate, which, during Marcel's lifetime, was the front line for many of wine's ideological battles, should remain controversial today under the stewardship of his son Mathieu and eldest daughter, Camille. In Marcel's era, the wines were seen as too radical for the mainstream. Today Mathieu and Camille find themselves in the peculiar position of being considered too mainstream for later generations of radical natural winemakers—the same ones who profit from the international natural wine market Marcel and his friends created. In this, the Lapierres are in a similar position to their cousins, the Beaujolais négociant Christophe Pacalet and the Burgundy négociant Philippe Pacalet, whose

---

Since: Marcel Lapierre took over from his father in 1973. The estate has been in the Lapierre family since 1904. Mathieu took over winemaking in 2010; Camille joined in 2014.

Winemakers: Mathieu and Camille Lapierre

Vineyard surface: 15ha (37 acres). The Lapierres purchase the fruit from another 15ha each year.

Viticulture: Estate grapes are organic. Purchased grapes are not necessarily organic.

Négociant work: Yes

Grapes: Gamay

Appellations: Morgon, Julièнas; Vin de France (Raisins Gaulois)

Filtration/Fining: None, except for one bottling of Morgon

Sulfites: Certain cuvées are unsulfited. Others see 30mg/L at bottling.

work could be considered the supporting legs of the Lapierre throne.

Domaine Lapierre purchases grapes, and not always organic ones; only the very limited Cuvée Camille and Cuvée Marcel Lapierre derive entirely from organic estate fruit. Mathieu, who inherited his father's contrarian inclinations, is prone to declaring what journalists least expect from him: for example, that he prefers larger sulfite doses than his father's peers, or that gamay is a "commonplace, rustic" grape, or that "carbonic maceration can gum up the perception of terroir." It's as if he seeks to prevent his father's followers from believing too readily in natural wine sloganeering. Like their father, Mathieu and Camille abhor chemical and high-tech innovation in vinification, but remain more agnostic about more traditional correctives like chaptalization (rare), water addition (also rare), or sulfite addition during vinification (only when absolutely necessary).

> **"Natural wine is a freedom to do what we want—a freedom to not all do the same things."**
>
> —Mathieu Lapierre

Are these winemaking correctives natural? Mathieu suggests that the ideology of natural wine is less a set of rules than a way of avoiding rules.

The Lapierres continue to practice what has come to be called "Lapierre-style" vinification: long, cool-carbonic maceration, refrigeration of the harvest before vatting, and saturation of maceration tanks with CO_2. Visitors to their cellar at Les Chênes marvel at the famous ceiling, painted to resemble the cosmos by the painter-sommelier-cellarhand Denis Pesnot. No less impressive are the estate's two custom stainless-steel vertical presses, an old idea of Marcel's realized by his children after his death.

THE WINE TO TRY

See Wines to Know: The Beaujolais, *page 127.*

TASTING DESTINATIONS
FESTIVAL DEZING

A three-day, all-ages throwdown of live music, natural wine, clowns, cuisine, theater performance, and hot-air balloons, Camille Lapierre's Festival Dezing has carried on the spirit of her late father's famous lamb roasts (see page 25) since 2015. Held in late August just before harvest, the festival attracts a Who's Who of Beaujolais winemakers and their families, along with fellow natural winemakers from farther afield. Its name is an imaginary word that combines *zinc*, referring to the material traditional French bar tops are made of, and *dézinguer*, slang for "to massacre."

DOMAINE JEAN FOILLARD

Jean and Agnès Foillard

VILLIÉ-MORGON

Jean and Agnès Foillard took over Jean's family estate in 1981. That year, Jean joined his friend and neighbor Marcel Lapierre and the itinerant winemaker Jacques Néauport in low-sulfite vinification on native yeasts and in bottling without filtration. Among the original Lapierre circle, Foillard is distinguished by his significant vineyard holdings on Morgon's Côte du Py, the volcanic hill composed of magmatic schist that yields gamay of exceptional power and structure.

Foillard arrived at his present cool-carbonic winemaking style in 1989, when he began refrigerating the harvest before vatting. Today his iconic Morgon–Côte du Py is the most lauded wine of the Beaujolais, acclaimed in natural and conventional wine circles alike.

"Wine is not a merchandise like the others, right? There's a culture behind it. There are two manners of doing it, slow and fast. If you do thermovinification, you'll have it even faster, but you won't want to drink it."

—Jean Foillard

Jean Foillard in his backyard in Villié-Morgon

The Foillards produce an old-vine, selection-level Côte du Py called 3.14, along with a Fleurie, two other Morgon *climats*, Les Charmes and Cuvée Corcelette, a Beaujolais-Villages from purchased grapes for the US, a Beaujolais primeur, and a generic Morgon. Like Domaine Lapierre, Domaine Jean Foillard blends organically farmed estate grapes with purchased fruit of varying viticutural methods, meaning no wines are certified organic. (Only 3.14 and, since 2018, the Fleurie consist solely of estate fruit.) For the Beaujolais primeur and the generic Morgon, Foillard occasionally purchases finished wine to blend with his own. These practices, coupled with the Foillards' success, have made the estate a minor flash point within the natural wine community. Jean himself is a natural wine classicist, with a maestro's impatience for natural winemakers who neglect to perfect their craft.

THE WINE TO TRY

See Wines to Know: The Beaujolais, *page 127.*

Since: 1981

Winemaker: Jean Foillard

Vineyard surface: 15ha (37 acres). Foillard purchases the equivalent of another 15ha.

Viticulture: Practicing organic

Négociant work: Yes, in both grapes and finished wine

Grapes: Gamay

Appellations: Morgon, Fleurie, Beaujolais-Villages, Beaujolais primeur

Filtration/Fining: None

Sulfites: Generally 20mg/L at assemblage, before bottling

DOMAINE JEAN-PAUL THÉVENET

Jean-Paul, Annick, and Charly Thévenet

VILLIÉ-MORGON

Among the early followers of Marcel Lapierre, Jean-Paul Thévenet has always stood out for not sticking out. Familiarly known as "Polpo," he's a genial paysan, gifted with a frank and straightforward charm, not known for the cutting sarcasm of his peers. His warm simplicity was long evident in his wines, a slim range comprising just his Morgon–Vieilles Vignes, a lesser Morgon Tradition, and a sweet sparkling gamay for the British market. It was the first that made his renown, a marvel of unfiltered, low-sulfite, old-vine Morgon from the Douby site. Unlike his natural Morgon peers, Thévenet never expanded into négociant work, so his wine is noteworthy for the uniform rigor of its farming.

"We make the wines of an older time. Fifty years ago, people put grapes in the tank and they waited. When it was good, it was the same fermentation we do now."

—Jean-Paul Thévenet

Thévenet practices cool-carbonic fermentation in concrete tanks, vatting at slightly higher temperatures than Guy Breton and Yvon Métras. Free-run juice is not removed from the final Morgon–Vieilles Vignes (as it is chez Breton or Lapierre). This yields a richer wine, closer to the historical archetype of Morgon as a muscular, Beaujolais cru. In 2018, Thévenet restructured his estate in partnership with his son, Charly Thévenet, a noted natural vigneron in nearby Régnié since 2007. Charly shares his father's aesthetics, preferring slightly longer oak élevage to early spring bottlings. In 2022, the Thévenets' debuted a new cuvée from a recent acquisition of 1.3 hectares (3.2 acres) of organically-farmed Morgon in the lieu-dit La Roche Pilée.

Charly Thévenet is also a founder and organizer of the Festival Dezing (see page 131) with his childhood friend Camille Lapierre.

THE WINE TO TRY

Morgon–Vieilles Vignes

Jean-Paul Thévenet's old-vine Morgon is a model of baritone grace, offering a sort of alternate-history glimpse of Lapierre-style winemaking upon well-farmed Morgon without the influence of négociant grape purchases.

Jean-Paul Thévenet tamping down grapes in his press

Since: 1985

Winemaker: Charly Thévenet (since 2018)

Vineyard surface: 7ha (17.2 acres)

Viticulture: Organic

Négociant work: None

Grapes: Gamay

Appellations: Morgon

Filtration/Fining: None

Sulfites: Generally 20–25mg/L at bottling

YVON MÉTRAS

FLEURIE

A grizzled, blue-eyed bear of a man, Yvon Métras stands apart from his peers in the early Lapierre-Néauport circle. He's the only one from Fleurie, for one thing. For another, he chose to forge a separate commercial path, declining to work with the influential American importer Kermit Lynch and instead exporting more to Europe and Japan. It hardly seems an accident that Métras (now joined at the estate by his son Jules) became the most radical of his peers, ceasing sulfite addition on most cuvées. With the exceptions of an early foray into Mâcon-Villages and a brief experiment purchasing southern Beaujolais fruit for two cuvées in 2017 and 2018, Métras has avoided becoming a négociant, preferring to stick to his own vineyards of Fleurie, Moulin-à-Vent, and Beaujolais.

"In this winery, the wine works. We don't know why. But you can't seek to know. You just have to do everything for it to continue."

His Fleurie vineyards are in two of the appellation's greatest sites, La Madone and Grille-Midi. His "basic" Beaujolais is located in an altitudinous, productive site called Cercillon. Father and son conduct monthlong cool-carbonic maceration, refrigerating the harvest and scrupulously gassing their cement tanks. The Métras use a motorized vertical press, pressing over the course of a day. Wines age in used oak barrels, partly in a cellar at Yvon's home in the forest in nearby Vauxrenard.

Métras's influence on neighboring vignerons rivals that of Foillard or Lapierre. He mentored the future Fleurie star Julie Balagny when she arrived in the region. Younger natural winemakers from Yann Bertrand to Pierre Cotton to Paul-Henri Thillardon cite him as an inspiration, as do friends farther afield, like the duo behind Domaine Les Bottes Rouges in the Jura.

THE WINE TO TRY

See Wines to Know: The Beaujolais, *page 127.*

Since: 1994

Winemakers: Yvon and Jules Métras

Vineyard surface: 7ha (17.3 acres)

Viticulture: Organic

Négociant work: Briefly between 2017 and 2018, and once in 1996

Grapes: Gamay

Appellations: Fleurie, Moulin-à-Vent, Beaujolais

Filtration/Fining: None

Sulfites: Often none; otherwise, a microdose of 10mg/L at bottling

DOMAINE DE LA GRAND'COUR

Jean-Louis Dutraive

FLEURIE

No Beaujolais estate has seen its star rise as meteorically in recent years as Fleurie's Domaine de la Grand'Cour, where winemaker Jean-Louis Dutraive has assumed a community leadership role akin to that of the late Marcel Lapierre. His estate has become the social center of Fleurie, where winemakers young and old congregate in the evening to enjoy Dutraive's wit and hospitality along with his vivid gamay. The estate was founded in 1969 by Dutraive's father, Jean, who moved the family from Charentay, where they had raised cows and cereal along with Brouilly grapes. Jean-Louis Dutraive took over in 1989, stopped chemical farming in the early 2000s, and obtained organic certification in 2009.

"We're relaxed during harvest, because in vinification we don't have much work. There's observation. But there's no pumping, no yeasts, no chaptalization, no chilling down, nothing else."

By all accounts, Dutraive wasn't close with Marcel Lapierre during Lapierre's lifetime. Only since 2012, thanks to the influencs of his Fleurie neighbor Yvon Métras and Dutraive's cousin Rémi Dufaitre, has Dutraive employed Lapierre-style, cool-carbonic vinification on all wines. (Previously the wines saw semi-carbonic vinification.) The results of this new approach, on Dutraive's superbly farmed sites, are bombshells of grainy, soulful Fleurie that have attained cult status among a new generation of natural wine fans.

Hailstorms and mildew have cost Dutraive dearly in recent years, wiping out the entire vintage in 2016. In response, he and his children (Ophélie, Justin, and Luca, all of whom have joined him at the estate) began producing négociant wines in 2016.

THE WINE TO TRY

Fleurie–Clos de la Grand'Cour

Representing the heart of Dutraive's estate, aged half in barrel and half in foudre, Clos de la Grand'Cour is a keen, cherry-toned, pink-granite gamay with a finely spiced complexity born of decades of organic farming.

Since: 1969

Winemaker: Jean-Louis Dutraive

Vineyard surface: 11.5ha (28.4 acres)

Viticulture: Certified organic

Négociant work: Yes, mostly under the "Famille Dutraive" label

Grapes: Gamay; occasional négociant work with cinsault, carignan, chardonnay

Appellations: Estate wines are Fleurie and Brouilly. Négociant Chénas and Saint-Amour.

Filtration/Fining: None

Sulfites: Occasionally none, but more often 10mg/L before bottling

GEORGES DESCOMBES

VILLIÉ-MORGON

In the Morgon hillside hamlet of Vermont (not to be confused with the US state of the same name), Georges Descombes has created an empire to rival that of his mentor, Marcel Lapierre. An imposing, robust man whose hairstyles have grown increasingly wizardy in recent years, Descombes is known by the amusing nickname "La Noune," short for *nounours*, or "teddy bear." The son of a winemaker, Descombes worked on a bottling line alongside his neighbor Jean Foillard in the early 1980s.

Upon taking the reins of his family domaine in 1993, Descombes established organic agricultural practices and natural vinification, working tirelessly to promote the wines in regular deliveries to Paris bistros. Today Descombes is the patriarch of a winemaking clan that includes his stepson Damien Coquelet and his son Kewin Descombes. All three produce négociant wines in addition to estate wines, which adds up to a dizzying array of low-sulfite, mostly organic gamays. Collectively, they could be said to embody the most pragmatic, nonideological, salt-of-the-earth wing of natural winemaking in the Beaujolais. Descombes's own estate wines are muscular and ambitious, more extractive and aged longer in oak than those of his peers in the cool-carbonic style. The best marry this power to a marked florality.

Since: 1988

Winemaker: Georges Descombes

Vineyard surface: 15.5ha (38.3 acres)

Viticulture: Practicing organic

Négociant work: Yes

Grapes: Gamay, chardonnay

Appellations: Morgon, Régnié, Chiroubles, and more

Filtration/Fining: None on estate wines. Négociant wines sometimes see filtration.

Sulfites: Dosage of 20–30mg/L at bottling

JEAN-CLAUDE LAPALU

SAINT-ÉTIENNE-LA-VARENNE

Hailing from a wine-growing family at the southern border of the Beaujolais crus, Jean-Claude Lapalu notably *didn't* arrive at natural winemaking through the band of natural winemakers farther north in Villié-Morgon. Lapalu's epiphany came in 1998, when a retailer in Provence introduced him to the natural wines of Provençal vigneron Henri Milan. Lapalu, who had been growing grapes for the local cave cooperative since 1982, began a conversion to greater purity in farming and winemaking.

Lapalu's handsome, rich Brouilly and Côte de Brouilly wines soon saw success in the Paris natural wine market. He found friends and allies in the Lapierre gang when he finally met them in the early 2000s. Lapalu obtained organic certification for his estate in 2010. Lapalu's wines throughout the 2000s were muscular in style, thanks to foot-stomping and pumpovers; in the past decade, he's adopted a lighter touch in vinification. His masterful cru wines are supplemented with a bevy of experimental cuvées, like the amphora-aged Alma Mater and the late-harvest Rang du Merle.

Today's thriving natural wine scene in the southern Beaujolais crus owes a lot to Lapalu's mentorship. Many younger natural Beaujolais winemakers spent time as cellarhands chez Lapalu, from Raphael Champier to Jérôme Balmet to Cyrille Vuillod.

Since: 1996

Winemaker: Jean-Claude Lapalu

Vineyard surface: 12ha (29.6 acres)

Viticulture: Organic

Négociant work: Yes

Grapes: Gamay, chardonnay, pinot noir

Appellations: Brouilly, Côte de Brouilly, Beaujolais-Villages

Filtration/Fining: None

Sulfites: Certain cuvées see zero sulfites, but most see dosage of around 20mg/L at bottling.

TASTING DESTINATIONS
BEAUJOLAIS NATURAL WINE SALONS

These Beaujolais wine-tasting salons are the Glastonburys of gamay. While some are nominally professional tastings, in practice they're open to anyone with a passionate interest in wine.

Bien Boire en Beaujolais

When: Early April

Who: At La Beaujoloise tasting, the original natural wine pioneers of the Beaujolais. At Biojolaise, several succeeding generations of organic and natural winemakers from the region.

For: Wine professionals

Where: In recent years the tasting has been held across three chateaux in the region: Château de Pizay, Château des Ravatys, and Château de Corcelles.

The premier wine salon of the Beaujolais, Bien Boire en Beaujolais ("Drink Well in the Beaujolais") unites no less than five distinct wine tastings, spread out over three sites. Of these tastings, only two are of real interest for natural wine lovers: La Beaujoloise, founded by Marcel Lapierre and his friends, and the more recent La Biojolaise, which is limited to organic-certified estates and mostly features a more recent generation of natural winemakers.

Biojoleynes

When: Mid-April

Who: Philippe Jambon and his sulfite-free cohorts

For: The local public, as well as wine professionals

Where: The northern Beaujolais village of Leynes

Created in 2009 by Philippe Jambon (see page 140), neighboring vigneron Pierre Boyat, and their friend Yann Desgoulle, Biojoleynes brings twenty to thirty certified organic winemakers from the Beaujolais together in a communal event hall in Leynes, a ten-minute drive away from the TGV station at Mâcon-Loché. Attendees include Vauxrenard's Michel Guignier, along with many lesser-known sulfite-free winemakers from the south of the Beaujolais.

Bojalien

When: Early April

Who: Romain des Grottes and his extended friend circle of charming zero-sulfite radicals

For: Wine professionals

Where: Often at Domaine des Grottes in Saint-Etienne-des-Oullières, but the location shifts.

Bojalien is a satellite salon of the Bien Boire en Beaujolais tasting and is held on the same Monday. Founded in 2015 by erstwhile Fleurie winemaker-restaurateur Denis Baldin, the salon has become increasingly relevant under the direction of biodynamic natural vigneron Romain des Grottes, who hosts two dozen sulfite-free winemakers from throughout France beneath a tent at his estate in Saint-Étienne-des-Ouillères. Attendants are encouraged to wear a green paint fingerprint on their foreheads as a sign of astral solidarity.

Beaujolais vigneron Romain des Grottes on piano at the Bojalien salon

CHRISTIAN DUCROUX

LANTIGNIÉ

Like the moon, the story of Beaujolais natural winemaking has a hidden face. It is that of austere Lantignié vigneron Christian Ducroux. A precursor of the radically pure wing of natural winemaking, Ducroux obtained organic certification as early as 1980, and converted to biodynamics in 1985, forsaking commercial yeasts and sulfites in vinification soon after. Built like a beanpole and thrifty with words, he evinces a monklike devotion to wine growing, insisting he is "not a winemaker at all." Since 2013, his wines have been produced without appellations.

"We have to collect what is given to us. If you have to seek something with force, it's not the same."

Ducroux emphasizes biodiversity and microbial health in his thin granite soils, where fruit trees, flowers, and oats grow amid the vines. His vineyard work has influenced many fellow zero-sulfite winemakers of the region, including Vauxrenard's Michel Guignier and Marchampt's Hervé Ravera.

During harvest, Ducroux collects grapes from each of his parcels in a given day, with the result that all his wines derive from the same blend of terroirs. He conducts seven- to ten-day vattings of his gamay, which is pressed over twenty-four hours with a vertical press dating from 1873. His fermentations finish in large barrels. Unlike the Lapierre band, Ducroux does not use outside CO_2 or refrigerate his harvest. In his surprisingly complex wines, you can taste the fruit of his sensitive gardening; the juice is dense, almost opaque in its youth, growing increasingly luminous and violetty with age. Since 2018, he has produced Cuvée Thibault, an extremely rare white wine from pinot gris and riesling.

THE WINE TO TRY

Vin de France–Exspectatia

Since 2018, Exspectatia has been made from the middle of the very long press of Ducroux's blended parcels. Its wholesome dark fruit is the signature expression of his holistic approach to natural winemaking. It demands bottle aging.

Since: 1970

Winemaker: Christian Ducroux

Vineyard surface: 4ha (9.88 acres)

Viticulture: Certified organic and biodynamic

Négociant work: None

Grapes: Gamay, riesling, pinot gris, and more

Appellations: All wines are produced as Vin de France. Until 2013 he made a Régnié.

Filtration/Fining: None

Sulfites: None added

MICHEL GUIGNIER

VAUXRENARD

Michel Guignier of Vauxrenard is one of the most exacting winegrowers in the Beaujolais, a seasoned vigneron who found a late-career mentor in Lantignié's Christian Ducroux, and followed him into organics in 2000, biodynamics in 2001, and sulfite-free vinification in 2007.

Guignier took over his family domaine in the late 1970s. He farmed and vinified conventionally until the turn of the century. As he switched to more labor-intensive agriculture in the 2000s, Guignier reduced his vineyard holdings, most recently in 2014 when he ceded control of his rented parcel of Fleurie in Les Labourons. Among the Beaujolais crus, there remains only Moulin-à-Vent in his vineyards; the rest are Beaujolais or Beaujolais-Villages. Nowadays, like Ducroux, he bottles everything as Vin de France.

"How can we speak of terroir, when most people's vineyards are dead? I changed my agriculture in 2000, and I see the difference between my wines from before and my wines today."

Today Guignier practices semi-carbonic vinification, with no refrigeration or outside CO_2. His impeccably farmed gamays are capricious, keenly acid, and often very rewarding. A true believer in the benefits of biodiversity, Guignier also raises cattle and grows cereals.

THE WINE TO TRY

Vin de France–La R'vole

Formerly known as Festivitas, Guignier's unfiltered, unsulfited primeur is ripe and granular, bright as a spark, and rare as a truffle.

Since: 1976

Winemaker: Michel Guignier

Vineyard surface: 4.5ha (11.1 acres)

Viticulture: Certified organic and biodynamic

Négociant work: None

Grapes: Gamay

Appellations: Previously Fleurie, Moulin-à-Vent, Beaujolais-Villages; today all wines are Vin de France.

Filtration/Fining: None

Sulfites: None added

PHILIPPE JAMBON

Philippe and Catherine Jambon

CHASSELAS

Originally from the vineless part of the northern Beaujolais, Philippe Jambon began his career as a sommelier, working at the three-Michelin-star Restaurant Girardet near Lausanne. A fateful visit to the nearby restaurant of the chef Pascal Santailler led Jambon to discover the early natural wines of Marcel Lapierre and others. That's when he decided to quit being a sommelier.

In 1997, he purchased vines in Leynes, not far from his native village of Ouroux. He would become one of the visionary natural vignerons of his generation, emphasizing extremely low yields and forswearing all sulfite addition from the 2000 vintage on. He seeks maximum ripeness, despite the cold weather and hail risks of his area, and prizes vineyard sites that enjoy isolation from nearby chemical agriculture. In practice, this means Jambon often makes very little wine. This dovetails with his tastes in winemaking, as he often practices extremely long macerations (up to a year!) and extremely long élevage. Sometimes his wines bear names of individual vineyard sites in Ganivets, Baltailles, or Balmont. More often his wine names change according to the vagaries of the vintage. Jambon supplements his slim wine production by releasing wines from natural winemaker friends as part of his "Une Tranche de . . ." négociant wine series.

"There's more gap between zero and a little bit than between a little bit and even more. If a wine is a little filtered, it's filtered. If it's a little sulfited, it's sulfited. If it's zero, it's zero!"

—Philippe Jambon

In all his wines, Jambon seeks a challenging, borderline masochistic style all his own. Most winemakers avoid volatile acidity at all costs; Jambon considers it a crucial element of the architecture of a great wine.

Philippe Jambon in his vines

THE WINE TO TRY

Vin de France–Les Baltailles

Comprising a mere half hectare of very old vines on a granite plateau, Les Baltailles is only rarely made into its own cuvée, a black, grainy masterpiece of Leynes terroir.

Since: 1997

Winemaker: Philippe Jambon

Vineyard surface: 3.5ha (8.6 acres)

Viticulture: Organic

Négociant work: Yes, under the label "Une Tranche de . . ."

Grapes: Gamay, chardonnay, various others as a négociant

Appellations: All estate wines are produced as Vin de France.

Filtration/Fining: None

Sulfites: None added

Natural Wine Dining

In the United States, the wines of the Beaujolais have often been marketed in tandem with Thanksgiving, as a low-cost option for easy drinking with family and friends. And indeed, the wines tend to marry well with poultry of all kinds. But the Beaujolais has many of its own regional specialty foods, from the fresh and aged goat cheeses of the western mountains to grattons, a variation on pork rinds, to andouillette à la fraise de veau, a pork sausage stuffed with the meaty membrane known as "calf's ruffle."

While still a far cry from its culinary heyday in the 1970s and '80s, the Beaujolais has seen a resurgence of fine bistros in recent years, often backed by local vignerons. Many of the best offer a tantalizing glimpse of the rich, bygone glory of Lyonnais cuisine.

Restaurant Éphémère in Vauxrenard

Le Port de By

GRIÈGES

An idyllic riverside bistro run with charm and aplomb by Cécile Ducroux, a cousin of Lantignié's Christian Ducroux and a longtime habitué of the region's natural wine scene (she used to run a bistro in Lyon). Don't miss the true Saône fish fry, a rare speciality.

Épicerie

SAINT-ÉTIENNE-DES-OULLIÈRES

Opened in 2019 by local natural vigneron Rémi Dufaitre, Épicerie is a wine bar, table d'hôte, and gourmet grocer, a cornucopia of good taste and a welcome addition to the Beaujolais dining scene.

Auberge du Col du Truges

SAINT-JOSEPH

For over thirty years, chef Jean-Jacques Soudeil has prepared hearty Beaujolais specialties like quenelles and coq au vin at his lunchtime restaurant in the hilltop town at the border between Chiroubles and Morgon. Marcel Lapierre organized the after-party lunch for his lamb roasts here for years, and the small flute bottles in which Lapierre sold his early unsulfited wines are still used to hold flowers on the tables.

Restaurant Éphémère

VAUXRENARD

Every summer since 2018, Yvon Métras's backyard has hosted an elegant pop-up restaurant by his girlfriend, Gusta van Walsem, a Dutch corporate human resources specialist who partners with a different young chef each season. The constant? A slim, select natural wine list sourced from among their vigneron friends.

Legends in the Making

Julie Balagny on her porch, July 2015

The pioneering natural winemakers of the Beaujolais changed the wine world with their work in the 1980s and '90s. But the Beaujolais today remains a thriving and close-knit hotbed of natural winemaking. Here are eight gifted vignerons to watch from the region's current generation.

Julie Balagny

ROMANÈCHE-THORINS

Fomerly the winemaker at biodynamic estate Terre des Chardons near Nîmes, Julie Balagny established her own estate in the Beaujolais in 2009, working with her then companion, Yvon Métras. Since then she has established a ramshackle farm in the hamlet of Moulin-à-Vent, where she produces magisterial, pure, unsulfited gamays that have captivated the natural wine world.

Yann Bertrand

FLEURIE

A childhood friend and neighbor of Jules Métras—and a recent protégé of Jacques Néauport—Yann Bertrand took the reins of his family estate in 2012, when he was just twenty-two years old. He quickly established a name for himself as a winemaking prodigy, taking his family's Morgon and Fleurie wines to new heights of craft and ambition, and expanding into organic négociant work in Saint-Amour and Juliènas.

Domaine Chapel

LANTIGNIÉ

David Chapel and Michele Chapel-Smith moved to the Beaujolais in 2015. For Chapel-Smith, former sommelier of Brooklyn Fare, it was a bold leap into a new life as a vigneron in France's most hard-drinking wine region. For Chapel, the son of famed Lyonnais chef Alain Chapel (see page 29), it was a homecoming, for he had grown up close family friends with the Lapierre family. Chapel worked for Domaine Lapierre that first year, and the couple produced their first wine, a Julienas from purchased fruit, in 2016. They've since taken on vines in Chiroubles, Fleurie, and Lantignié, from which latter they produce a masterful Beaujolais-Villages. For now all wines age in fiberglass tank. Just over a half-decade into their careers as vignerons, the Chapels have emerged as keen disciples of the Lapierre school, carrying careful cool-carbonic vinification into a new generation.

Pierre Cotton

ODENAS

Blessed with serious talent and sterling Côte de Brouilly terroir, Pierre Cotton made his first wine in 2014. He took over winemaking for the entirety of his parents' estate in 2016. Notably, his wines attain all the depth and complexity of his forebears in the Lapierre gang, even though he employs few of the same methods: he uses no refrigeration, often with a *délestage* (one big pumpover) later in fermentation.

Romain des Grottes

SAINT-ÉTIENNE-DES-OUILLÈRES

Since taking up viticulture at his family's château in 2003, Romain des Grottes has distinguished himself as one of the region's most radical farming and winemaking visionaries. It's fair to say that he reconceived the local winemaking style from A to Z, adapting it inextricably to his own unusual biodynamic farming methods, which involve plowing only rarely on vines of very low plantation densities. His brief, slender gamay wines hit like minor epiphanies.

Laurence and Rémi Dufaitre

SAINT-ÉTIENNE-DES-OULLIÈRES

An irascible knucklehead with a gift for natural vinification, Rémi Dufaitre and his then wife, Laurence, began bottling their own wines in 2010. Dufaitre had the sense to choose the right mentors: his older cousin Jean-Louis Dutraive, Jean Foillard, and Jean-Claude Lapalu all consider him a protégé. Recent vintages have seen Dufaitre emerge from their shadows with striking innovations, including cinsault and grenache from his own plantings, as well as a bottling of gamay co-fermented with chardonnay.

Hervé Ravera

MARCHAMPT

Once a nurse, now an aesthetic successor to Christian Ducroux and Michel Guignier, Hervé Ravera founded his small estate in the high granite slopes of Marchampt in 2007. He soon ceased all sulfite addition and filtration on his wines, all produced as Vin de France, and adopted biodynamic methods and horse plowing in his vines. Ravera's wines stand out among the work of the region's radical, zero-sulfite vanguard for their lightness and grace.

Domaine Thillardon

CHÉNAS

Brothers Paul-Henri and Charly Thillardon moved from their native Frontenas in the southern Beaujolais to establish their organic estate in the overlooked cru of Chénas in 2008. They came under the influence of natural winemakers like Jean-Louis Dutraive and Yvon Métras; by 2015, they had eliminated all destemming and filtration and adopted cool-carbonic maceration. Successive vintages have seen Paul-Henri emerge as a world-class winemaking talent, and the brothers have ceaselessly reinvested to improve farming methods and winemaking equipment.

Hervé Ravera with his plow horse

Loire vigneron Sylvie Augereau's old-vine chenin overlooking the Loire River

5. The Loire

France's Loire Valley, named for the country's longest river, is known for a cornucopia of wine styles. Its whites—both still and sparkling—run the gamut from sweet to dry; its reds offer everything from crisp refreshment to earthy, baritone depths. Where did such diversity in grapes and wine come from?

Well, what we tend to call "the Loire" actually comprises at least seven distinct subregions, each with its own historical wine culture.

There are the famed sauvignons (and less famed pinot noirs) of the Cher and Nièvre departments; the flinty menu pineau and romorantin of the Loir-et-Cher; the electric chenin and perfumed pineau d'Aunis of the Sarthe in the north; the buoyant grolleau and delicate chenin of Anjou; the structured cabernet francs and chenins of the Saumurois; the tense, opulent chenin and grainy cabernets of Touraine; and, lastly, the abundant, refreshing melon de Bourgogne wines of the Loire-Atlantique. Referring to all of these cultures collectively is a misleading habit of the wine media.

If the Loire as a whole is home to a thriving and diverse community of natural winemakers, they're far from evenly spread among its subregions. Most Loire appellations remain downright antagonistic when it comes to natural winemaking, particularly Sancerre, Pouilly-Fumé, and Muscadet, where just a handful of natural vignerons struggle against shortsighted local wine authorities. Other Loire areas, like Anjou and the Loir-et-Cher, are home to some of the most influential natural vignerons in France and scores of their acolytes.

Some of these differences among the subregions are due to cultural factors. But individual vignerons played an outsize role. The Loire does not have one towering giant of natural winemaking à la Marcel Lapierre in the Beaujolais; instead, there are a myriad of influential natural winemakers. Natural vignerons like Olivier Cousin, Christian Chaussard, Claude Courtois, and Thierry and Jean-Marie Puzelat initiated a sea change in the region's winemaking in the 1990s, while pathbreaking, intellectual vignerons like Mark Angeli and Nicolas Joly made the Loire famous for organic and biodynamic agriculture.

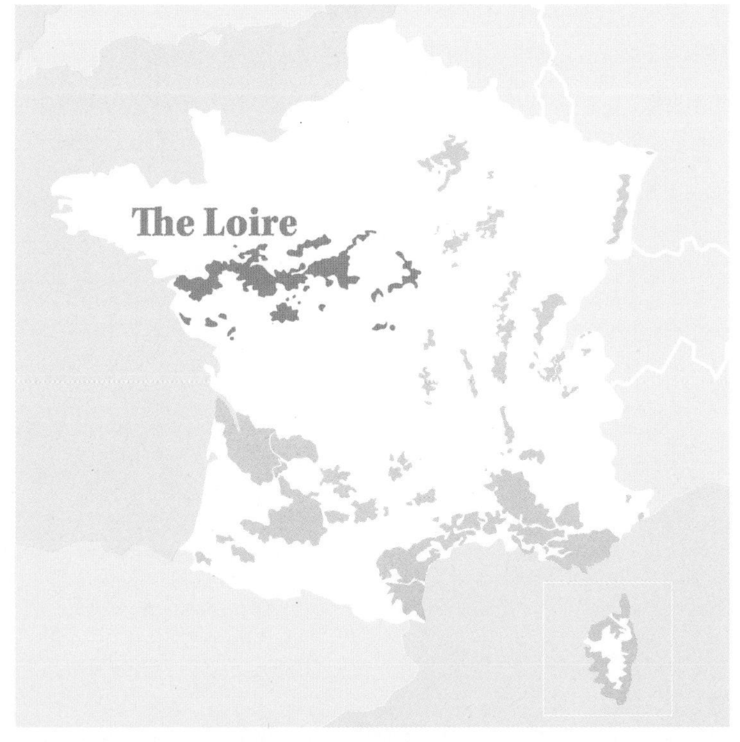

Total vineyard surface:
35,000ha (86,487 acres)[1]

Rate of organic agriculture:
12 percent[1]

Key grapes:
Chenin, sauvignon, melon de Bourgogne, menu pineau, romorantin, cabernet franc, cabernet sauvignon, pinot noir, gamay, grolleau, côt

1 Bio Pays de la Loire

THE LOIRE

Wines to Know

The precise white wines of certain qualitative but very conservative Sancerre, Vouvray, and Montlouis estates are dear to the hearts of wine lovers around the world. By comparison, unfiltered natural white wines from these areas are deeper, more textured, and ultimately more nourishing. Where natural reds are concerned, the Loire's slim range extends from the brisk gamays and pinot noirs of Cheverny to the perfumed pineau d'aunis of Anjou and the Sarthe to the more concentrated and broody cabernet francs from Bourgeuil and Saumur.

ANJOU

Richard Leroy | Vin de France–Les Noëls de Montbenault

Richard Leroy's masterful dry chenins—unsulfited and unfiltered since 2011—have become almost impossible to get hold of. They represent the high-water mark among natural wines from the local schist terroir in the Layon. Les Noëls de Montbenault shows a concentrated, nervy white fruit, flecked with lemon zest, along with the towering minerality common to Leroy's wines.

CHER AND NIÈVRE

Alexandre Bain | Vin de France–Pierre Précieuse

It's difficult to choose favorites among Alexandre Bain's seven parcel bottlings of declassified Pouilly-Fumé sauvignon. All strike a rewarding stylistic middle ground between the historical classics of ripe, well-farmed sauvignon (François Côtat, Didier Dagueneau) and the more lush, oxidative, sulfilte-free style of Bain's neighbor Sébastien Riffault. The steel-aged Pierre Précieuse, from Portlandian limestone soils, is a good place to start: a nervy, muscular sauvignon, ripe but never fat, with notes of citrus confit.

LOIR-ET-CHER

Claude Courtois | Vin de France–Racines

Claude Courtois's renowned Racines red is a blend of dozens of impeccably farmed grape varieties that cannot be named because many of them are of questionable legality in his region. Some come from as far afield as Italy and Spain. Does the wine's svelte, earthy profile—vaguely reminiscent of Auxey-Duresses—derive from the flinty, poor soil? From the proportion of exotic young vines, still adapting to their new surroundings? From the long old-oak aging? Deciphering Racines is one of the natural wine world's great puzzles.

LOIRE-ATLANTIQUE

Marc Pesnot | Vin de France–Chapeau Melon

Unique in Marc Pesnot's oeuvre, Chapeau Melon is an unsulfited, unfiltered, undisgorged pétillant naturel of old-vine melon de Bourgogne. (If you find a bottle, open it near a sink with glasses at the ready, because it might be volcanic.) It's yeasty and densely layered, yet never loses the hallmark acid and stoniness of Muscadet. Old vintages lose their sparkle and become something greater, showing captivating almond and honey flavors.

(continued)

THE LOIRE

Wines to Know *(continued)*

THE SARTHE

Jean-Pierre Robinot | Vin de France–Les Années Folles

Jean-Pierre Robinot's wines are as impressionistic as their labels, which bear his jazzy long-exposure photo art. They can be luminescent if you catch them on the right day. If you don't feel like rolling the dice, try Les Années Folles, a spiffing pétillant naturel rosé of pineau d'aunis (and sometimes a smidge of chenin) that marries the bristling, high-acid immediacy of his northerly red fruit with a refined, yeasty splendor.

THE SAUMUROIS

Clos Rougeard | Saumur-Champigny–Le Clos

The wines of Clos Rougeard have become collector's items, but these profound, immortal wines are hard to overvalue. Le Clos, this estate's equivalent of an entry-level wine, is a cabernet franc from fifty- to seventy-year-old vines that sees macerations of over a month before aging for two years in oak barrels that are at least four years old. In lighter vintages, it can dazzle with winter spice aromas at only a few years of age.

TOURAINE

Catherine and Pierre Breton | Bourgeuil–Clos Sénéchal

What with the ever-expanding range of estate and négociant wines produced by the Breton family, it can be easy to lose sight of the traditional Bourgeuil wines that made them famous. The foudre-aged parcel bottling Clos Sénéchal, like its barrel-aged brother Les Perrières, benefits from long bottle aging. But it's more approachable in youth, showing the Bretons' consummate craftsmanship in its deep, earthy flavors of cassis and shiitake.

LES CAILLOUX DU PARADIS

Claude and Etienne Courtois

SOINGS-EN-SOLOGNE (LOIR-ET-CHER)

Claude Courtois labeling a bottle

The vineyards of Les Cailloux du Paradis had been abandoned for seven years and farmed chemically for the previous two decades. Its clay-flint soils were rocky and thin, prone to drying to an impenetrable crust between rain showers. It was here that Claude Courtois saw paradise, when he purchased it in 1991, after a previous career in Provence at Domaine Saint Cyriaque.

Thirty years later, as he cedes responsibility to his son Etienne, Claude Courtois's vision has been realized in an estate hailed as a model for freethinking viticulture and renowned for its unique wines.

Back in Provence, Courtois had scandalized fellow organic winemakers by leaving alternate rows grassed-over, rather than plowing everywhere. In Sologne he went further, cultivating fruit trees, woods, and prairie alongside his vines. Most famously, he engaged in a long-term act of rebellion against the local appellation authorities by resurrecting the forgotten gascon grape variety and planting scores of other varieties, many unauthorized. The results of this ampelographical free-for-all are in the white and red Racines cuvées, which Courtois still vinifies. Since 2007, his son Etienne has produced a range of sought-after monovarietal wines. His other son, Julien, makes a range of equally coveted wines at an adjacent estate.

"To habituate a plant somewhere, it takes at least fifteen years. The ancients knew that after three generations, the beast was native. So we take nearly fifteen years with each grape variety to adapt it to the sector."

—Claude Courtois

THE WINE TO TRY

See Wines to Know: The Loire, *page 147.*

Since: 1992

Winemakers: Claude and Etienne Courtois

Vineyard surface: Claude farms 1ha (2.5 acres) as of 2019. Another 5ha (12.3 acres) are farmed by his son Etienne.

Viticulture: Organic

Négociant work: None

Grapes: An innumerable quantity, including pinot noir, romorantin, menu pineau, and others

Appellations: Vin de France

Filtration/Fining: None

Sulfites: Usually none; occasionally a microdose at racking or bottling

CLOS DU TUE-BOEUF

Thierry, Jean-Marie, and Zoé Puzelat

LES MONTILS (LOIR-ET-CHER)

Puzelat brothers Thierry and Jean-Marie joined forces at their family estate, Clos du Tue-Boeuf, in 1994. By the time Jean-Marie retired in 2018, they had built their family farm in the Blois backwoods into the most influential natural wine estate of its generation.

Thierry attended winemaking school in Bordeaux and Mâcon, against his will. It was while working in Bandol in 1991 that he encountered the young enologist Yann Rohel, a classmate of Philippe Pacalet's in Dijon. Rohel invited Thierry to attend one of Marcel Lapierre's renowned Bastille Day lamb roasts. Thierry fell for natural wine and soon communicated the passion to his brother.

Thierry (*left*) and Zoé Puzelat in their cellar

Upon uniting at Clos du Tue-Boeuf, the Puzelats set about converting to organic farming and producing natural wines. For the first three vintages, Jacques Néauport provided counsel, which the Puzelats paid for in cases of wine.

Today an emphasis on the overlooked Loir-et-Cher grape varieties côt, pineau d'aunis, menu pineau, and pinot gris typifies the Clos du Tue-Boeuf production, which also includes négociant wines purchased from organic growers in Touraine, Cheverny, and the Beaujolais.

The top Clos du Tue-Boeuf wines are the unfiltered, low-sulfite parcel cuvées, which see at least partial aging in used oak barrels. Red grapes are refrigerated before vatting. Since 2012, Clos du Tue-Boeuf has begun experimenting with several cuvées aged in qvevri each vintage, notably a bristling, characterful sauvignon. Whites, once direct-pressed, since 2019 see a day or two of skin maceration before pressing.

A new chapter at the domaine began in 2018, when Jean-Marie retired and Thierry was joined full-time by his daughter Zoé, an enthusiastic traveler herself, who spends part of each year in the Republic of Georgia.

THE WINE TO TRY

Vin de France–Le Brin de Chèvre

A tense, splendorous white from a clay-flint-soiled parcel of old-vine menu pineau, Le Brin de Chèvre is historically Clos du Tue-Boeuf's most capricious yet most rewarding wine. Some vintages have been off-dry, and all the more sumptuous for it.

Since: 1994

Winemakers: Thierry and Zoé Puzelat. Jean-Marie Puzelat also vinified until 2018.

Vineyard surface: 14ha (34.6 acres)

Viticulture: Organic

Négociant work: Yes, since 2003. Generally the Puzelats purchase the equivalent of another 4–5ha (10–12 acres).

Grapes: Côt, gamay, pinot noir, and more

Appellations: Cheverny, Touraine, Vin de France

Filtration/Fining: Early-release, steel-aged wines see light filtration.

Sulfites: Sometimes a small dose at bottling

CHRISTIAN VENIER

CANDÉ-SUR-BEUVRON (LOIR-ET-CHER)

Half carefree paysan, half questing aesthete, the shaggy-haired Cheverny vigneron Christian Venier was initiated to natural winemaking in the mid-nineties by his cousins, the Puzelat brothers (see opposite). At the time, Venier did seasonal vineyard work and traveled the world as a sheep shearer. When he decided to become a vigneron like his father, Venier attended viticultural school at Amboise, where he studied under Christian Chaussard alongside fellow students Agnès and René Mosse.

Today Venier takes pride in having remained exclusively a vigneron rather than establishing a négociant business like many of his peers. He refuses organic certification, because he feels those farming correctly shouldn't have to pay extra to prove it. He's nonetheless a respected vigneron whose acumen is evident in a range of soulful, saline whites and brisk, wholesome reds. For reds, he prefers wood tanks for maceration; his whites, all direct-pressed, ferment in a variety of steel and fiberglass tanks. Venier uses barrel aging with restraint, preferring to blend barrel-aged wine with an equal quantity of tank-aged wine. The exact grape assemblages of his wines can be difficult to follow; the whites can vary greatly in proportions from year to year.

Since: 1999

Winemaker: Christian Venier

Vineyard surface: 7ha (17.3 acres)

Viticulture: Practicing organic

Négociant work: None

Grapes: Chardonnay, sauvignon, pinot noir, and more

Appellations: Cheverny

Filtration/Fining: Whites and rosés are sometimes plate filtered.

Sulfites: Generally 10–20mg/L at bottling

HERVÉ VILLEMADE

Hervé and Isabelle Villemade

CELLETTES

Based in Cellettes, near the town of Blois, Hervé Villemade is a salt-of-the-earth vigneron whose modest nature belies his huge drive and ambition. Villemade took charge of his family estate in 1995 and made his first sulfite-free vinification in 1997, around the same time that he developed a skin allergy to sulfur. He began practicing low-sulfite vinification and converted his estate to organics in 1999.

Pragmatism defines Villemade's approach in both farming and winemaking. In 2002, after several years of small harvests due to frost, he set up a négociant winemaking business in partnership with his neighbor Olivier Lemasson, a Marcel Lapierre acolyte who would continue the business after splitting with Villemade in 2006. By 2008, Villemade had started a new négociant business on his own; its labels are defined by a silhouette of the chest-thrusting, Maori-esque statue that sits outside his winery. His estate vines are spread throughout the communes of Cellettes, Chitenay, and Fougères-sur-Bièvre.

Villemade's négociant cuvées are mostly easy-drinking weeknight wines. His range is completed by well-aged Cheverny and Cour-Cheverny estate wines, assembled from a mix of containers ranging from foudres to barrels to several types of amphora, including buried Georgian qvevri.

Since: 1995

Winemaker: Hervé Villemade

Vineyard surface: 22ha (54.4 acres)

Viticulture: Organic

Négociant work: Yes. The amount varies according to the vintage, but it can represent as much as two-thirds of production in some years. Grape purchases are bottled as separate cuvées from estate fruit.

Grapes: Sauvignon, chardonnay, gamay, and more

Appellations: Cheverny; occasionally other Loire appellations as négociant cuvées

Filtration/Fining: Entry-level bottlings of whites and rosés are usually filtered.

Sulfur: 10–20mg/L at bottling

LES VINS CONTÉS

Olivier Lemasson

CANDÉ-SUR-BEUVRON (LOIR-ET-CHER)

After Rennes wine retailer Éric Macé introduced him to Marcel Lapierre in the 1990s, Olivier Lemasson—then a sommelier—became a regular at Lapierre's estate at harvesttime, going on to work a full year at the famed Beaujolais domaine. It was a formative experience that would define his winemaking style. In 2002, Lemasson partnered with Loir-et-Cher vigneron Hervé Villemade to found Les Vins Contés, a natural wine négociant project. (The pair split amicably in 2006.)

"If I made wine in the Beaujolais, I'd do three cuvées of gamay, and I'd get bored. Here you have all the different grapes. Sometimes you don't succeed so well in the vinification, but at least you're not based on just one thing."

Lemasson favored short carbonic macerations and brief aging periods, releasing some wines as early as January following the harvest. Part of this was his own drinking preference, but one also sensed that for Lemasson, it was important that natural wines stay accessibly priced. This noble paysan instinct led his vast range of wines to define, for many natural wine drinkers, the *vin de soif* ("wine to quench thirst") style. Beer geeks might also know Lemasson as the source of the pineau d'aunis grapes that go into a certain ultra-rare cuvée of Cantillon.

Originally intended as a purely négociant project, Les Vins Contés grew to include vineyards in Touraine that Lemasson acquired from a former grape purveyor in 2016. Unbeknownst to even his closest friends, Lemasson struggled with depression. His suicide in the spring of 2021 extinguished a beloved leading light of the natural wine community.

THE WINE TO TRY

Vin de France–Poivre et Sel

Lemasson's signature aesthetic is best represented in Poivre et Sel, a crisp, textured blend of gamay and pineau d'aunis from sandy-flint soils. The winter-spice aromas of pineau d'aunis add a critical hint of complexity to an otherwise deliriously juicy wine.

Since: 2002. From 2002 to 2006, Lemasson made wine in partnership with Hervé Villemade.

Winemaker: Olivier Lemasson

Vineyard surface: 9ha (22.2 acres)

Viticulture: Organic

Négociant work: Yes; the majority of wines are produced with organic grape purchases.

Grapes: Sauvignon, romorantin, aligoté, menu pineau, and more

Appellations: All wines have been released as Vin de France since 2012.

Filtration/Fining: None

Sulfites: Sometimes a dose of about 20mg/L at bottling

LES CAPRIADES

Pascal Potaire and Moses Gadouche

FAVEROLLES-SUR-CHER (LOIR-ET-CHER)

Les Capriades is the winemaking project of veteran Cher valley winemaker Pascal Potaire and his business partner Moses Gadouche, two men who are dead serious about what most people consider a frivolous subject: pétillants naturels (*pét-nat*s for short; see page 113). Together they've helped make off-dry sparkling rosés a cult item.

Potaire credits the late Christian Chaussard, whom he met in 1995, for introducing him to the pét-nats that would become his passion. Potaire then worked for Nicolas Renard and later Clos Baudoin in Vouvray, before coming to the Cher valley in 2002 to work for Junko Arai of Domaine des Bois Lucas. In 2003, he took over vines formerly farmed by biodynamic natural wine pioneers Beatrice and Michel Augé and began producing his first wines. He continued working on a small scale until Gadouche joined as a partner in 2011, when the pair began supplementing their vineyard holdings with organic grape purchases from around the Cher valley. In 2013, the duo founded the wine-tasting salon Bulles au Centre, themed entirely around low-sulfite sparkling wines and ciders.

The pétillant naturel wine style has become far less rare since Potaire began producing it in the early 2000s—it's now in fashion from Santa Cruz to the Mosel. But Les Capriades was founded on the insight that *great* pét-nats are tricky enough to produce that they'd always remain uncommon. Potaire and Gadouche refine their approach tirelessly, right down to the bottle closure. (Since 2014 they've used nifty resealable ones called Zorks.) In recent years, they've expanded into producing cider from ancient varieties of apples and pears purchased in Normandy.

Pascal Potaire *(left)* and Moses Gadouche in their cellar

THE WINE TO TRY

See The Wines to Know: Pétillants Naturels, *page 113.*

Since: 2003

Winemaker: Pascal Potaire

Vineyard surface: 3ha (7.4 acres)

Viticulture: Organic

Négociant work: Yes. They purchase the equivalent of another 7ha (17.2 acres) from organic sources.

Grapes: Cabernet franc, chenin, sauvignon, and more

Appellations: Vin de France, Touraine

Filtration/Fining: None

Sulfites: None added

Christian Chaussard

A giant of Loire natural winemaking who died in 2012, Christian Chaussard made wine as Domaine La Saboterie in Vouvray from 1992 to 1998, and worked as a viticulture professor in Amboise from 1987 to 1999. He's credited with coining the term "*pét-nat*" to describe the *méthode ancestrale* sparkling wine method that he adapted to natural wine principles.

Chaussard's radical winemaking in the mid-1990s was ahead of its time. A series of difficult vintages forced him to declare bankruptcy in 1998. In an extraordinary show of support, Chaussard's peers—including Thierry Puzelat and Pascal Potaire—purchased his tanks of wine, only to return much of it to him as gifts. From 2002 until his death, he produced wine as Domaine le Briseau with his partner, Nathalie Gaubicher.

DOMAINE LE BRISEAU

Nathalie Gaubicher

MARÇON (THE SARTHE)

At the northernmost limit of French winemaking in the Sarthe, the Swiss actress turned vigneron Nathalie Gaubicher maintains Domaine le Briseau. A reference for natural Jasnières wine (no longer in AOC in recent vintages) and the cult pineau d'aunis grape, Domaine le Briseau is also a monument to the legendary career of Gaubicher's late ex-companion and estate cofounder, Christian Chaussard.

Le Briseau's reds from pineau d'aunis, côt, and gamay see long, whole-cluster maceration, with frequent foot-stomping, rendering them more extractive than has become the fashion in the region. The estate's whites see long presses; occasionally they're bottled with some residual sugar.

Since 2017, Gaubicher has divided her time between the Sarthe and the Languedoc, where she helps at the estate of her present companion, Emile Hérédia, a vigneron who formerly worked in the Coteaux du Loir before founding Domaine des Dimanches in the Hérault.

Since: 2002

Winemaker: Nathalie Gaubicher (since 2012)

Vineyard surface: 7ha (17.3 acres)

Viticulture: Biodynamic

Négociant work: Yes, under the label "Nana Vins et Cie"

Grapes: Chenin, pineau d'aunis, gamay, côt

Appellations: Vin de France; formerly Jasnières and Coteaux du Loir

Filtration/Fining: Kieselguhr filtration before bottling on all whites, and sometimes reds

Sulfites: Whites often receive 20mg/L at bottling.

L'ANGE VIN

Jean-Pierre Robinot

CHAHAIGNES (THE SARTHE)

With his Technicolor shirts and wiry hair, Jean-Pierre Robinot cuts a madcap figure. An energetic plumber turned wine journalist turned bistro owner turned vigneron, he has arguably done more than anyone to shape how we think about natural wine.

In 1983, at the dawn of what would become the natural wine movement, Robinot cofounded the influential French wine journal *Le Rouge & le Blanc* with Michel Bettane, a graduate of Steven Spurrier's famed Paris wine school Académie du Vin. A schism soon occurred between Bettane, who rewarded the more technical winemaking helping France's wine regions to exploit new export markets, and Robinot and his friend François Morel (see page 33), who defended a purer notion of agriculture and vinification. The conflict prefigured the natural wine movement.

"I prefer to drink a natural wine full of faults than a chemical wine. The chemical wine I won't drink; the faulty wine, I'll drink."

In 2001, when Robinot left his Paris bistro L'Ange Vin to become a vigneron in his native Chahaignes, he precipitated another schism, this time within the natural wine movement he'd helped create. He began promoting a radical vision of natural wine defined by zero sulfite addition, and organized an association of like-minded vignerons called Les Périphériques. To the generation of natural vignerons who began in the 1980s, Robinot and his friends were committing sacrilege by accepting wine flaws as normal features of a living wine.

Robinot's wines are as eccentric as he is: slender, savory reds from pineau d'aunis, often with a lash of volatile acidity, and long-aged, oxidative chenins that range from lumpen to divine. Never one to hurry a vinification, he follows the rhythms of his tanks and barrels, which are arranged in five ancient tunnels carved into the limestone. He is also adept at pétillant naturel vinfication, producing four cuvées that possess a chiming acidity to rival the greatest Champagne. Now over seventy, Robinot has lost none of the passion that has marked his charmed life in natural wine.

Robinot is in the habit of wearing a flashlight around his neck while conducting cellar visits.

THE WINE TO TRY

See Wines to Know: The Loire, *page 147.*

Since: 2001

Winemaker: Jean-Pierre Robinot

Vineyard surface: 7ha (17.3 acres)

Viticulture: Organic

Négociant work: Yes, under the label "Opéra du Vin"

Grapes: Chenin, pineau d'aunis, gamay

Appellations: Formerly Jasnières. Nowadays, all wine is bottled as Vin de France.

Filtration/Fining: None

Sulfites: None added

DOMAINE COUSIN-LEDUC

Olivier Cousin

MARTIGNÉ-BRIAND (MAINE-ET-LOIRE)

To understand the rift between the Loire's biodynamicists and its natural winemakers, consider the relationship between renowned biodynamic winemaker Nicolas Joly and his paysan counterpart, Olivier Cousin. In the 1980s and early 1990s, the two traveled together often, promoting biodynamic agriculture alongside Mark Angeli. But Cousin, a former wind-sailor who learned winemaking from his grandfather, grew frustrated with the hypocrisies of the bourgeois wine market Joly courted. Cousin renounced biodynamic certification in 1998, even as he continued to practice biodynamic farming. He cites an early 1990s visit to Sologne winemaker Claude Courtois as his first experience of a natural wine he truly adored.

"I got horses in 1988. That changed things. And in 1989, Mark Angeli came to see me because he thought I did biodynamics. I didn't know what it was. He said, 'Why do you plow your vines?' I said it was because I've always plowed my vines."

Today Cousin is a pillar of natural winemaking in Anjou, and has helped foster the careers of dozens of natural winemakers over the years. He's also renowned throughout France for his mastery of horse plowing. No fan of permaculture, he plows often, insisting on zero grass cover in April and November in particular.

In recent years, Cousin has ceded much of his original estate to his son Baptiste and his former cellarhand Sylvain Martinez, retaining only a small vineyard of cabernet franc on deep clay soils surrounding his house in Martigné-Briand. From this Cousin produces two rustic, bold red wines, Pur Breton and Vieilles Vignes, both aged in old oak barrels.

THE WINE TO TRY

Vin de France–Pur Breton

Olivier Cousin's younger-vine cabernet franc Pur Breton, often aged for over two years before release, is a frank, thoroughbred Anjou red, with deep flavors of cinder and cassis, at once workmanlike and divine.

Since: 1985

Winemaker: Olivier Cousin

Vineyard surface: 3ha (7.4 acres)

Viticulture: Organic and practicing biodynamic

Négociant work: None

Grapes: Cabernet franc

Appellations: Vin de France

Filtration/Fining: None

Sulfites: None added

LA FERME DE LA SANSONNIÈRE

Mark Angeli, Martial Angeli, and Bruno Ciofi

BELLEVIGNE-EN-LAYON (MAINE-ET-LOIRE)

Mark Angeli is a mason from Provence who became a pioneer in simultaneous revolutions in the Anjou wine world: biodynamic agriculture, dry white wine, and natural winemaking. In his telling, it all happened by accident. He drank wine for the first time in 1987. By 1990, he'd attended viticultural school in Sauternes and bought La Ferme de la Sansonnière, a historic Anjou estate dating to the fifteenth century. He converted it to biodynamics that year, influenced by Nicolas Joly. In 2001, Joly and Angeli cofounded the biodynamic wine-tasting series La Renaissance des Appellations.

"What's crazy is that among two thousand vignerons in Anjou back when I started here, there were none that made dry whites or unadulterated rosés."

—Mark Angeli

Mark Angeli with his plow horse

Angeli's role in the Anjou region's embracing dry white winemaking is ironic, for it was his passion for sweet wine that led him to study in Sauternes and come to Anjou in the first place. Angeli produced his first dry chenin in 1991, when a storm forced him to harvest before the grapes reached high enough sugar levels to produce sweet wine. The rest is history.

Today Angeli works alongside his son Martial, who joined the estate in 2008, and Bruno Ciofi, who formerly worked for Alsace's Jean-Pierre Frick and Domaine de la Pinte in the Jura.

THE WINE TO TRY

Vin de France–La Lune

A delicate yet voluptuous chenin, often with a touch of residual sugar despite its long fermentation and barrel aging (often over two years), La Lune is a fine introduction to the menagerie of pure Layon chenins Angeli helped foster.

Since: 1990

Winemakers: Mark Angeli, Martial Angeli, and Bruno Ciofi

Vineyard surface: 8ha (19.8 acres)

Viticulture: Biodynamic

Négociant work: None

Grapes: Chenin, grolleau gris, and more

Appellations: Vin de France

Filtration/Fining: Whites and rosés are filtered; reds are generally not.

Sulfites: Whites and rosés usually see a small dosage at bottling; reds usually do not. Total sulfur levels for all wines are displayed on the labels.

PRECURSORS

Nicolas Joly

Anjou native Nicolas Joly worked as an investment banker in New York and London before returning, at age thirty-two, to his family's wine estate, Château de la Roche-aux-Moines in Savennières, in 1977. He converted the estate to biodynamic agriculture in 1984, and would become the wine world's leading proponent of biodynamics, authoring numerous books on the subject.

Joly's contribution to the world of natural wine is more difficult to quantify. He has always worked adjacent to natural vignerons, progressing toward similar agricultural ends. Yet Joly advocates sulfite use in vinification and has avoided natural wine retail networks for his own wines. His colossal, unfiltered dry chenins were stylistically groundbreaking in the 1980s, 1990s, and 2000s. Yet these wines were rejected as oversulfited within the French natural wine market of the era. Joly's pedagogical approach to biodynamic farming, meanwhile, did not go down well among certain natural vignerons working in the peasant tradition. That he priced his wines far higher than his neighbors in the region added further insult.

Today, Joly is lauded on the export market for his influence on natural wine, but not within France. His Renaissance des Appellations association (see page 167) has nonetheless aided the cause of natural wine by promoting responsible agriculture and encouraging a rapprochement between conservative French appellation authorities and the vignerons who challenge them, both biodynamic and natural.

NATURAL HISTORY

ANJOU, SUGAR, AND NATURAL WINEMAKING

The sought-after dry chenins of Mark Angeli, Richard Leroy, and others are a relatively recent phenomenon in Anjou. A look at twentieth-century Anjou winemaking shows why this historical sweet-wine region was ripe for change—and why natural winemaking became the answer.

Faced with increasing demand in the postwar period, the region's winemakers embraced chaptalization to replicate high sugar levels without the risk of delaying harvest long enough to naturally obtain ripeness and botrytis. Soon enough, truly botrytized sweet wine from Anjou became rare. Chenin was farmed with herbicides and synthetic pesticides, harvested underripe, chaptalized, and yeasted, then sulfited and filtered to stop fermentation and ensure stability. By the mid-1990s, cheap sweet wine production had damaged the region's reputation, as well as the health of its vineyards. Land became available and affordable, as older winegrowers gave up in the face of cratering négociant prices.

For a new generation of organic and biodynamic farmers, like Olivier Cousin, Nicolas Joly, and Mark Angeli, the region's recent history was a vivid demonstration of the link between volume-driven synthetic chemical farming and a need for chaptalization and high sulfite dosage. Organics and biodynamics went hand in hand with a search for greater purity in winemaking. This initiated a focus on dry wines rather than sweet wines, for the latter practically require filtration and high sulfite dosage to prevent refermentation. Vignerons seeking to avoid those interventions turned to dry white wine—and in doing so, rediscovered one of France's great white wine terroirs.

DOMAINE MOSSE

Agnès, René, Joseph, and Sylvestre Mosse

SAINT-LAMBERT-DU-LATTAY (MAINE-ET-LOIRE)

René Mosse in his harvest kitchen

Known for their caustic wit and excellent cuisine, Agnès and René Mosse are a power couple of Anjou natural winemaking. They worked for Agnès's father at a wine shop in Tours from 1986 to 1992, which led them to take a course in winemaking in Amboise in 1993. There they met early Loire natural vignerons Christian Chaussard and Thierry Puzelat, who inspired the Mosses to work naturally when, in 1999, the couple purchased a 9.5-hectare (23.4-acre) estate in Saint-Lambert-du-Lattay. In the meantime, René had worked for Puzelat and done an internship in Burgundy at Domaine de Montille, where he took a shine to parcel vinification.

So it was that in the early 2000s, when Anjou was still unknown as a quality wine region, the Mosses turned out the parcel cuvées of organically farmed chenin that would become their calling cards: Le Rouchefer, Bonnes Blanches (later known as "Initials BB"), and Marie Besnard (which in recent years has been blended in with Le Rouchefer). The family's range has greatly expanded since sons Sylvestre and Joseph joined the estate in the late 2010s. It's the cabernet-based red wines that most animate René, now nearing retirement. He prefers destemming, longer macerations, and aging in old oak. Lately, he cedes most vinification to his sons, who favor a brighter, more fluid style, and who have laudably eliminated filtration on all wines.

"I'm a drinker of cabernet. I know it's not fashionable. Nowadays people like wines that are all finesse, delicacy, without tannins, without grip. I'm from the old school."

–René Mosse

THE WINE TO TRY

Anjou Rouge

The Mosses' Anjou Rouge is a handsome blend of cabernet sauvignon and cabernet franc that sees roughly three weeks' maceration in wooden vats and a year's aging in old barrels. With age, it can attain a surprising depth and minerality.

Since: 1999

Winemakers: René, Joseph, and Sylvestre Mosse

Vineyard surface: 15ha (37 acres)

Viticulture: Organic

Négociant work: Yes

Grapes: Cabernet franc, cabernet sauvignon, grolleau gris, and more

Appellations: Vin de France since 2011

Filtration/Fining: None since 2017; previously, whites and rosés were often filtered.

Sulfites: Most wines receive doses of 20–30mg/L at bottling.

JÉRÔME SAURIGNY

ROCHEFORT-SUR-LOIRE (MAINE-ET-LOIRE)

A native of the Deux-Sèvres department, south of Anjou, Jérôme Saurigny had completed enology studies in Bordeaux and was working as a cellar master in Pomerol when his brother introduced him to the visionaries behind Les Griottes, Patrick Desplats and Sébastien Dervieux. The meeting opened his eyes to the wonders of additive-free, sulfite-free vinification. Saurigny returned to Anjou and in 2007 he successfully crowdfunded the purchase of the estate of a retiring vigneron in Rochefort-sur-Loire. Since then his star has only risen as the most radical disciple of the Desplats school of fearless Anjou abstraction.

Early vintages displayed a deft hand with whole-cluster, ungassed carbonic maceration of reds, along with an impressive devotion to the long, careful aging required for producing unsulfited sweet chenin. Today Saurigny's wines more often than not blur the boundaries between red and white, as a series of frost catastrophes have inspired him to co-ferment his various grape varieties, as in his "Txindoki," a mesmerizing blend of chenin, sauvignon, grolleau, cabernet Franc, cabernet Sauvignon. Long vattings and extremely low yields lend the wines an intensity that rarely conflicts with an abiding purity and digestibility.

Since: 2005

Winemaker: Jérôme Saurigny

Vineyard surface: 6.5ha (16 acres)

Viticulture: Organic

Négociant work: None

Grapes: Grolleau, cabernet cauvignon, cabernet franc, and more

Appellations: All wines released as Vin de France

Filtration/Fining: None

Sulfites: None added since 2007

DIDIER CHAFFARDON

SAINT-JEAN-DES-MAUVRETS (MAINE-ET-LOIRE)

Modest, with a piercing gaze above his *Lord of the Rings* beard, Chaffardon cuts a singular figure in the Anjou natural wine scene, equally adept at producing grainy, structured reds as he is with rosé pét-nats and divine sweet chenin.

A circuitous career in viticulture took Chaffardon from his native Savoie to the Île de Porquerolles in Provence to Burgundy, where in 1994 the future Saint-Etienne wine retailer Jean-Jacques Maleysson introduced him to the natural wines of Pierre Overnoy and Domaine Gramenon. His former classmate Mark Angeli later drew him to Anjou for a job converting the Domaine des Charbottières to organics. Chaffardon worked for that estate until it was sold, prompting him to take on his own vines in 2006.

Unlike many in the Anjou natural wine scene, Chaffardon prefers to avoid carbonic maceration, instead practicing pigéage on whole-cluster fruit. Reds like "L'Incredule" stand out for their tannic sweep and salinity. He has progressively abandoned barrel aging on most wines over the years, preferring fiberglass tanks, save for certain sweet wines.

Since: 2006

Winemaker: Didier Chaffardon

Vineyard surface: 3.5ha (8.6 acres)

Viticulture: Organic

Négociant work: None

Grapes: Cabernet franc, grolleau, chenin

Appellations: All wines released as Vin de France

Filtration/Fining: None

Sulfites: None added

LES VIGNES DE BABASS

Sébastien Dervieux

SAINT-LAMBERT-DU-LATTAY (MAINE-ET-LOIRE)

Sébastien Dervieux—known by his nickname, "Babass"—is a guitar-playing, chain-smoking dropout who in the 2000s, alongside fellow Les Griottes vigneron Patrick Desplats, became a trailblazer of zero-sulfite vinification without filtration. Dervieux was the more pragmatic half of Les Griottes, believing in traditional organic farming and the necessity of plowing. When he and Desplats amicably split in 2010, Dervieux retained their clay-schist parcels atop a low plateau between Beaulieu and Rochefort-sur-Loire, bordering the vines of the Hacquet siblings, the pioneering sulfur-free winemakers for whom Dervieux and his companion Agnès Mallet had become unofficial caretakers.

Since 2013, Dervieux has been a chief organizer of the Angers natural wine salon Les Vins Anonymes (see page 167). From his cellar in Saint-Lambert-du-Lattay, Dervieux today produces an old-vine cabernet franc (Roc Cab'), a quenching, fruitful grolleau (Groll n' Roll), and manifold variations of magnificent chenin; sparkling, dry, and sweet, depending on the vintage.

"We were friends with Jo Pithon at the time, and René and Agnès Mosse, who had just arrived. And at the beginning we met Thierry Puzelat. And progressively we went a bit further than all of them."

THE WINE TO TRY

Vin de France–Roc Cab'

Babass's signature red, Roc Cab', is a semi-carbonic cabernet franc from seventy-year-old vines. It generally sees two weeks' fermentation on whole clusters before pressing; élevage occurs in fiberglass tank and usually lasts until mid-spring. The wine's blackberry fruit is at once lustrous and rough, the texture of raw silk.

Since: 2011. From 2000 to 2010, Dervieux made wine in partnership with Patrick Desplats under the domaine name Les Griottes.

Winemaker: Sébastien Dervieux

Vineyard surface: 4.2ha (10.4 acres)

Viticulture: Organic

Négociant work: None

Grapes: Cabernet franc, chenin, grolleau, gamay

Appellations: Vin de France

Filtration/Fining: None

Sulfites: None added

PATRICK DESPLATS

SAINT-LAMBERT-DU-LATTAY (MAINE-ET-LOIRE)

As one half of Les Griottes, Patrick Desplats was the partnership's clairvoyant visionary.

The son of a gardener in Sologne, Desplats was working for other estates in the 1990s when Loir-et-Cher vigneron Thierry Puzelat introduced him to natural wines. Throughout the 2000s, Les Griottes's wines saw success in the most avant-garde natural wine bars of Paris. Desplats's output since he and Sébastien Dervieux parted ways is like that of André 3000 since leaving OutKast: slim, indulgent, and wildly inconsistent, but there's nothing else like it on earth. The wines are lean, low-alcohol, mixed varietal, and brimming with an otherworldly luminosity.

Desplats spent the 2010s converting his vines in Saint-Lambert-du-Lattay to radical polyculture.

"Babass continues plowing. I totally abandoned plowing. That's why we took separate paths: we no longer agreed about farming. Today I can no longer conceive that the earth should ever be naked."

He began living in a shack without electricity amid his vines. He stopped plowing after his horse Caroline died, preferring to spare himself—and horses—the strain of turning the soil. He planted a plethora of new grape varieties in his vineyards. He began training his chenin vines to grow up trees, and soon Thierry Puzelat informed him that the practice was also used in Georgia, sparking what would become a passion for the post-Soviet

winemaking region. Desplats has embraced qvevri winemaking from the 2019 vintage on.

THE WINE TO TRY

Vin de France–Épona

Until 2019, Épona was a featherlight, spearmint-toned blend of direct-press young-vine chenin and pineau d'aunis (along with quite a few undeclared varieties), possessing a breadth and depth that belied its very low alcohol. As of the 2019 vintage, Épona has become a vivid, burnt-sienna-toned orange wine vinified and aged in qvevri.

Since: 2011. From 2000 to 2010, Desplats made wine in partnership with Sébastien Dervieux under the domaine name Les Griottes.

Winemaker: Patrick Desplats

Vineyard surface: 4.5ha (11.1 acres)

Viticulture: Organic

Négociant work: None

Grapes: Cabernet franc, grolleau, gamay, and more

Appellations: Vin de France

Filtration/Fining: None

Sulfites: None added

RICHARD LEROY

RABLAY-SUR-LAYON (MAINE-ET-LOIRE)

Richard Leroy is a loquacious and thoughtful vigneron whose tiny production of cult dry chenins has prompted a full-scale reevaluation of Coteaux du Layon terroir. He's also the rare vigneron whose wines have grown more radical in the wake of wine world stardom.

"Ten or twenty years ago, we weren't the same tasters. I tasted famous labels. I was in a certain world, a certain wine culture . . . But frankly, it has helped me to learn to take risks."

Originally from the Vosges, Leroy and his wife, Sophie, pursued a classical wine education in the 1980s, working for the Nicolas wine shop chain. When they left wine retail to work for banks in Paris, the couple maintained their passion for wine. In 1996, encouraged by organic Layon vigneron Joel Ménard, Leroy purchased a south-facing 2-hectare (4.9-acre) volcanic schist-soil chenin parcel known as Les Noëls de Montbenault, working on weekends and holidays to produce sweet wines. In 2000, after relocating to the region full-time, Leroy first produced dry chenin, inspired by the wines of Mark Angeli. By 2005, Leroy would stop producing sweet wine, preferring the purity—and lower sulfite addition—he could achieve with dry winemaking. He soon debuted a second dry wine, Les Rouliers, from a 0.7-hectare (1.7-acre) schist-gravel chenin parcel closer to the Layon river.

Leroy's wines changed signficantly in 2011, when he moved production to a new cellar in Rablay-sur-Layon and stopped filtering and adding sulfites to his wines.

Thanks to his strong connections in Burgundy, Leroy became an expert in the nuances of barrel aging. In recent years, he's ceased using sulfur wicks to clean his barrels, finding that even one application can mark wines for years. He's become a mentor to younger natural winemakers of the region, even as he credits them with broadening his palate and inspiring his own ever-more-natural practices.

THE WINE TO TRY

See Wines to Know: The Loire, *page 147.*

Since: 1996

Winemaker: Richard Leroy

Vineyard surface: 2.7ha (6.7 acres)

Viticulture: Organic and practicing biodynamic

Négociant work: None

Grapes: Chenin

Appellations: All wines have been released as Vin de France since 2008.

Filtration/Fining: None since 2011

Sulfites: None added since 2011

DOMAINE LA PAONNERIE

Agnès and Jacques Carroget

VAIR-SUR-LOIRE (LOIRE-ATLANTIQUE)

Agnès (*right*) and Jacques Carroget in front of their winery

The Coteaux d'Ancenis, situated north of the Loire river at the midpoint between Anjou and Muscadet, has never been known for ambitious wine production—at least not until the arrival of the Carrogets. The couple obtained organic certification for their largely schist-soil vineyards in 2000 and Demeter biodynamic certification in 2009; since then, they've gradually ceased sulfite addition and filtration in almost all wines.

Jacques Carroget has distinguished himself with his mentorship of a younger generation of natural winemakers, drawn to the area by low vineyard prices as well as by Carroget's own formidable example. Since effectively retiring in the 2010s, Carroget has ceded many vineyards to his daughter Marie, along with several other young winemakers of the region. (The estate counted 22 hectares/54.3 acres in the late 2000s.)

"We insist on the nonuse of products of synthetic chemicals. Because to accept the residues of synthetic chemistry in wine is horrible—especially in natural wine."

–Jacques Carroget

In recent years, Carroget founded and has presided over the consumer-producer association Syndicat de Défense des Vins Naturels (see page 117).

THE WINE TO TRY

Vin de France–Voilà du Gros Lot

The various types of the obscure Loire grape variety grolleau obtained their name, it is said, thanks to their tendency to yield *gros lots*, meaning "big lots." The Carrogets' slender, savory rosé of grolleau gris, Voilà du Gros Lot, provides a glimpse of refinement in this rustic and rather rare country grape.

Since: 1980

Winemakers: Agnès and Jacques Carroget

Vineyard surface: 6ha (14.8 acres)

Viticulture: Biodynamic

Négociant work: None

Grapes: Melon de Bourgogne, chenin, gamay, and more

Appellations: Coteaux d'Ancenis, formerly Muscadet

Filtration/Fining: None

Sulfites: None added

DOMAINE DE LA SÉNÉCHALIÈRE

Marc Pesnot

SAINT-JULIEN-DE-CONCELLES (LOIRE-ATLANTIQUE)

Soft-spoken and reflective, Marc Pesnot is an unlikely vigneron to revolutionize his region's winemaking. Yet that is what Pesnot's vivid, full-flavored melon de Bourgogne wines have done.

"I became conscious of natural winemaking in the mid-nineties, from meeting people like Philippe Laurent of Domaine Gramenon, and the Lapierres. I said to myself, 'I want to change to a different region. I want to do wines like them, where they are.' And they said to me, 'Try where you are. Maybe that'll work.'"

Pesnot is the fourth generation of winemakers in his family. He farmed conventionally upon taking over from his father in 1980. It wasn't until discovering the wines of Domaine Gramenon and Marcel Lapierre in the mid-nineties that he became aware of the natural wine movement in France. Meeting these winemakers—often through his brother, the artist and sommelier Denis Pesnot—led Pesnot far from the conventions of Muscadet. By 2001, he was converting to organics and had begun producing all his wines without additives other than sulfites. That same year, an encounter with Paris natural wine retailer Olivier Camus led to the creation of Pesnot's first wholly unsulfited, unfiltered cuvée, Chapeau Melon.

In 2004, Pesnot's wines were rejected from the Muscadet appellation for being low-cropped (Muscadet has a bizarre *minimum* yield requirement) and for going through malolactic fermentation, which almost all other producers block artificially with high sulfite doses and filtration. So Pesnot began bottling all wines as Vin de France. Today he continues to refine his approach to vinification. He conducts extremely slow presses of his harvest—from six to eighteen (!) hours, a technique that allows the grapes to macerate within the press and extracts key aromatic compounds from their skins.

THE WINE TO TRY

See Wines to Know: The Loire, *page 147.*

Since: 1980

Winemaker: Marc Pesnot

Vineyard surface: 20ha (49.4 acres)

Viticulture: Organic

Négociant work: None

Grapes: Melon de Bourgogne, folle blanche

Appellations: Wines have been released as Vin de France since 2004.

Filtration/Fining: None

Sulfites: Most wines receive doses of about 50mg/L at bottling.

TASTING DESTINATIONS
THE LOIRE SALONS

At the end of January, the world's natural wine importers, business owners, wine writers, and sommeliers converge on the Loire towns of Angers and Saumur for four large-scale wine tastings, along with a smorgasbord of smaller private tastings, often held at estates.

It's a big social occasion, for better and for worse. The US natural wine importer Zev Rovine put it best when he described the event as "wine prom." The region's handful of good restaurants are overrun. It often rains; sometimes it even snows. But the ritual of the Loire salons persists as a reunion for the most committed natural wine aficionados.

La Dive Bouteille

When: The Sunday and Monday of the last weekend of January and/or the first weekend of February

Who: A Who's Who of established natural winemakers from France and beyond

For: Open to the public

Where: Caves Ackerman, Saumur

La Dive Bouteille was founded on a small scale in 1999 at a Bourgeuil wine shop of the same name by vigneron Catherine Breton (see page 172), who in 2002 bequeathed the project to her friend the journalist (and later vigneron) Sylvie Augereau (see page 168). The tasting moved to the Cave de la Grande Vignolle in Montsoreau, then to the Normandy town of Deauville in 2005 to coincide with France's Omnivore Food Festival. La Dive Bouteille returned to Saumur in 2010, first to the freezing underground cellars of the Château de Brézé, then to its present location, Saumur's more spacious Caves Ackerman, in 2014.

The many site changes reflect the salon's growing popularity: the 2020 edition welcomed more than five

The Château de Brézé during La Dive Bouteille in 2011

thousand visitors. Augereau's evolving selection of vignerons at the labyrinthine tasting event has done more than trace the growth of natural wine—it has helped define it. Incidentally, the salon's name derives from a line from a 1546 work by the French author Rabelais, in which a character abbreviates the word "divine" in reference to a wine bottle. Here's to the Divine Bottle!

La Renaissance des Appellations/Les Greniers Saint-Jean

When: The Saturday and Sunday of the last weekend of January and/or the first weekend of February

Who: Established biodynamic winemakers from France and beyond, including a sizable contingent of natural winemakers

For: Open to the public, but entry is half price with an invitation

Where: Les Greniers Saint-Jean, Angers

La Renaissance des Appellations is a biodynamic wine association founded in 2001 by Savennières vigneron Nicolas Joly. Its January tasting in Angers is referred to by the name of its site, Les Greniers Saint-Jean, a majestic stone structure built as a hospital in the twelfth century. Recent editions have been organized by Joly's daughter Virginie, along with Anjou vigneron Mark Angeli.

Les Pénitentes

When: The Friday afternoon, Saturday, and Sunday of the last weekend of January and/or the first weekend of February

Who: Veteran natural winemakers selected by Thierry Puzelat, René Mosse, Hervé Villemade, and Pierre-Olivier Bonhomme

For: Wine professionals

Where: Hôtel de la Godeline, Angers

Les Pénitentes is named after its former site, the evocative Hôtel des Pénitentes, a seventeenth-century structure built as a reformatory for sex workers. Begun in 2012 as a satellite tasting to the Greniers Saint-Jean tasting, Les Pénitentes became an Angers showroom for natural winemakers whose practices and/or beliefs were not aligned with the Renaissance des Appellations set (for example, those who declined to obtain organic certification). In spirit, it resembles La Dive Bouteille on a smaller, more cliquey scale.

Les Vins Anonymes

When: The Sunday and Monday of the last weekend of January and/or the first weekend of February

Who: Younger, scrappier natural winemakers, about half from the surrounding Loire region

For: Wine professionals

Where: L'Abbaye du Ronceray, Angers

Les Vins Anonymes represents the unsulfited avant-garde of French natural winemaking. Founded in 2013 by Anjou vignerons Sébastien Dervieux and Jean-Christophe Garnier, Les Vins Anonymes attracts a younger generation of importers, wine retailers, and sommeliers, all keen to taste new vintages from a younger, more radical generation of natural winemakers. As of 2021, Dervieux and Garnier ceded organization of Les Vins Anonymes to a trio of younger natural winemakers: Damien Bureau, Cédric Garreau, and Adrien de Mello.

Inside La Dive Bouteille in Caves Ackermann, Saumur

Sylvie Augereau on La Dive Bouteille

For two decades, the wine journalist turned vigneron Sylvie Augereau has organized the natural wine tasting La Dive Bouteille, the world's most renowned natural wine tasting. A Saumur native with a comedian's easy grin and a twinkling blue gaze, Augereau began her career in wine as a peripatetic vineyard worker. Little by little, she has become a natural wine icon in her own right, publishing numerous books on the subject. In 2014, Augereau became a winemaker herself in the Saumurois village of Le Thoureil.

"I had my midlife crisis at thirty years old. I said to myself, 'What do I really like in life?' Wine. 'And what do I know how to do?' Write. So I put my computer and some clothes in my car and I went on the road."

"To write about wine, I felt I had to go work in vines. It was my peasant logic. I did a year as a harvest caravan with two friends. We started in Bandol, then we went to work with Marcel Richaud. Then the Beaujolais."

"I called Marcel Lapierre, and he said, 'I have fifteen minutes available.' So I stayed three and a half hours. And at the end he said, 'So you'll come back tomorrow? Because then we'll actually have a chance to speak.'"

"I met a parallel world of passionate, generous people. That's why I did La Dive Bouteille. I sought a way to give back everything that had been given to me. Catherine and Pierre Breton had done two editions. And Catherine says, 'If you want, you can take over the thing, because I'm struggling with it.' So I did something a little militant and funny with it. There was one year where I told all the vignerons to come in disguise."

"I've been asked to organize La Dive Bouteille in New York. But it's exhausting. Once per year is enough. Otherwise it becomes a business, something to make money."

PRECURSORS

Clos Rougeard

In the pantheon of natural wine, 11-hectare (27-acre) Saumur-Champigny estate Clos Rougeard sits atop Mount Olympus. The eight generations of vignerons who oversaw it before its 2017 sale to French billionaire Martin Bouygues never employed herbicides or pesticides, always farmed organically, and never used winemaking additives besides sulfites at bottling. Brothers Charly and Nady Foucault defended organic agriculture as far back as the 1970s, lending the movement their prestige at a time when it was derided in the wine press. Later they would mentor a raft of younger natural vignerons, from Sylvie Augereau to Sébastien Bobinet to Cyril Fhal (see page 296).

When the walrus-moustachioed Foucault brothers spoke, their words carried the authority of their wines, masterpiece cabernet francs (and one chenin) born of that kismet-like concordance between grand terroir (the deep limestone bed beneath Saumur's tuffeau plateau), sensitive farming, and discriminating cellar practices (only free-run juice was retained, with press juice sold to négociants). These symphonic Saumur-Champigny wines attained global stardom in the 2000s.

Charly Foucault passed away in 2015, and a family feud ensued between Nady Foucault and Charly's son Antoine Foucault, of nearby Domaine du Collier. Nady Foucault oversaw the sale of the estate and retired in 2018. Since 2017 Clos Rougeard's wines have been made by the enologist Jacques Antoine Toublanc.

DOMAINE BOBINET

Sébastien Bobinet and Émeline Calvez

SAUMUR (MAINE-ET-LOIRE)

Sébastien Bobinet has been producing natural wine in Saumur since 2002, yet he still has the air of a wunderkind about him. His range of wines—produced with the help of his partner, Émeline Calvez, who joined him at the estate in 2011—unites his diverse influences. Some wines are deep and intense, in the Saumur lineage of Clos Rougeard; others show the carbonic-maceration influence of Bobinet's friend and neighbor, the early natural wine bistro owner and Beaujolais fan Bernard Pontonnier (see page 35). In recent years, Bobinet has also become known for throwing one of the best after-parties of La Dive Bouteille at his tasting room, a few minutes' drive down the road.

Since: 2002

Winemaker: Émeline Calvez has taken the lead in vinification in recent years.

Vineyard Surface: 3.8ha (9.4 acres)

Viticulture: Organic

Négociant work: Yes

Grapes: Cabernet franc, chenin. Négociant work includes pineau d'aunis, gamay, and côt.

Appellations: Saumur Filtration

Fining: None

Sulfites: Occasional small doses of 20–30mg/L at assemblage or before bottling

ALEXANDRE BAIN

TRACY-SUR-LOIRE (NIÈVRE)

In the fifteen years since he founded the estate, Alexandre Bain's vivid, unfiltered, additive-free sauvignons have become a reference for the potential of natural winemaking in the Pouilly-Fumé appellation, the right-bank brother of Sancerre.

"Ninety-nine percent of the time, people harvest sauvignon unripe, to keep those fresh green herbaceous aromas and to keep acidity. I'm against that. I find the wines are more appetizing and interesting when they're much riper."

—Alexandre Bain

A native of nearby Pouilly-sur-Loire, Bain wasn't born with vines in his family. He studied winemaking in Beaune before working as a vineyard manager in Menetou-Salon. As he prepared to start his estate in 2007, his cousin introduced him to Sancerre natural vigneron Sébastien Riffault. Upon learning that Bain sought to make natural wine, Riffault invited him to join a workshop in horse plowing conducted by Anjou natural vigneron Olivier Cousin. The experience proved foundational: both Bain and Riffault have practiced horse plowing ever since.

Bain's range comprises seven parcel-specific estate wines from clay-limestone, flint, and marl soils. All wines see long aging periods in containers ranging from foudres to barrels to steel and fiberglass vats. The core of his production—the cuvées Mademoiselle M, Pierre Précieuse, and L. d'Ange—are named for his three children. Stylistically, he seeks ripeness without botrytis, precision without additives or filtration, and the complexity of long aging without oxidation. He believes CO_2 in vinification is too industrial, but doesn't hesitate to use a machine that captures ambient nitrogen to help prevent oxygen contact during racking. It's this kind of ambition and openness to experiment in the name of quality that distinguish Bain's winemaking, which often brushes greatness.

THE WINE TO TRY

See Wines to Know: The Loire, *page 147.*

Since: 2007

Winemaker: Alexandre Bain

Vineyard surface: 11ha (27.2 acres)

Viticulture: Biodynamic

Négociant work: Bain does négociant cuvées in low-yield years, from organic or biodynamic fruit. The origins of the grapes are identified by the French regional department number on the label.

Grapes: Sauvignon blanc

Appellations: Vin de France and Pouilly-Fumé, depending on the vintage

Filtration/Fining: None

Sulfites: Usually 10mg/L at bottling, but occasionally none added

SÉBASTIEN RIFFAULT

SURY-EN-VAUX (CHER)

Sancerre's lone wolf of natural wine, Sébastien Riffault took over his father's estate after working for wine retailers in London and Paris. His initiation to natural wine came in the early 2000s through the wines of Domaine du Prieuré Roch, which he tasted while working at Lavinia in Paris.

When he returned to Sancerre in 2004, Riffault converted to organics and later to biodynamics. He began horse plowing after taking a course with Anjou vigneron Olivier Cousin. Riffault soon eliminated sulfite addition and filtration for all but his young-vine cuvée Les Quarterons, becoming one of the most radical natural winemakers of his generation. An early participant in L'Association des Vins Naturels (see page 116), Riffault became its president in 2013 and oversaw its transition to a more strict charter.

"Why natural wine? Because there's no pollutants for the planet or for me. That's important to respect. It's important to be transparent about that."

Sébastien Riffault's wines are atypical both due to the industrial nature of most Sancerre and to Riffault's unique winemaking style, which prizes botrytis and extreme ripeness along with very long aging (three years!) in tank and a variety of old oak containers (foudres and barrels of various sizes). Only Les Quarterons gestures to the organoleptic norms of the Sancerre appellation. His other cuvées are profound, oxidative, changeling wines. Riffault's sole red is a heavily botrytized dry pinot noir, wiry and strange.

THE WINE TO TRY

Sancerre–Saulétas

Saulétas is from old vines on Kimmeridgian limestone. In favorable vintages, its ginger-toned fruit can show impressive salinity and grace.

Since: 2004

Winemaker: Sébastien Riffault

Vineyard surface: 10ha (24.7 acres)

Viticulture: Biodynamic

Négociant work: None

Grapes: Sauvignon, pinot noir

Appellations: Sancerre

Filtration/Fining: Only the cuvée Les Quarterons is filtered.

Sulfites: Only the cuvée Les Quarterons sees 10mg/L at bottling; all other wines are unsulfited.

DOMAINE BRETON

Catherine, Pierre, Paul, and France Breton

BENAIS (INDRE-ET-LOIRE)

Catherine Breton outside the salon La Dive Bouteille

Catherine and Pierre Breton were a young winemaking couple from Touraine when they discovered early natural wine in the late 1980s at one of Marcel Lapierre's lamb roasts in the Beaujolais. After taking on Pierre's family vines in Bourgeuil in 1989, they embraced organic and later biodynamic viticulture. (Their wines have been certified organic and biodynamic since 1991 and 1999, respectively.) In 1994, they released their first—and only—unsulfited cuvée, a crunchy Bourgeuil called Nuits d'Ivresse (French for "nights of drunkenness").

Today the Bretons head a family enterprise spread between cellars in Pierre's native Bourgeuil and Catherine's native Vouvray. They also maintain a seasonal wine shop on the Atlantic coast island of Île d'Yeu. Their iconic "Épaulé Jeté" poster (see page 397) is posted on the walls of seemingly every natural wine bar from Tokyo to Tbilisi. Catherine Breton also helped found the world's most famous natural wine salon, La Dive Bouteille, in 1999.

The Breton wine range—a few venerated, age-worthy estate reds and a broad slate of easy-drinking, accessible wines from both estate and purchased fruit—reflects the couple's own aesthetic differences. Catherine prefers lighter, more Beaujolais-style wines, while Pierre's tastes hew closer to the historical Bourgeuil tradition of extractive, long-macerated cabernet franc. (In 2016, their son Pierre produced his first wine, a Vouvray from vines in Catherine's family, and their daughter France joined the estate in 2018.) The most commercially savvy acolytes of Marcel Lapierre, they've played a key role in promoting awareness of organic farming and purity in winemaking worldwide.

"Marcel [Lapierre] used to fight with Pierre, because Pierre used to say it wasn't possible to make wine without sulfur. Marcel told us we just needed to return to the vineyards to do more work. It's the opposite of conventional farming, which says it's man who does things to nature on his own schedule. In fact, one must reestablish oneself at the whim of nature."

—Catherine Breton

THE WINE TO TRY

See Wines to Know: The Loire, *page 147.*

Since: 1985

Winemakers: Catherine, Pierre, Paul, and France Breton

Vineyard surface: 17ha (42 acres)

Viticulture: Biodynamic

Négociant work: Yes

Grapes: Cabernet franc, chenin, sauvignon, grolleau

Appellations: Bourgeuil, Chinon, Vouvray, IGP Val de Loire

Filtration/Fining: Whites, rosés, and sparkling wines are filtered; reds are usually not.

Sulfites: All wines except Nuits d'Ivresse receive small doses at assemblage and/or before bottling.

THE LOIRE

Natural Wine Dining

Loire cuisine is no easier to summarize than its diverse winemaking culture. Riverside villages are known for their freshwater fish, often in simple butter preparations, while inland areas enjoy many of the same staple meats as the rest of rural France (chicken, veal, pork, etc.). Specialties to keep an eye out for, if you're traveling in the Loire, include sautéed eel, Paris mushrooms grown in the caves of Saumur, and, of course, the region's enormous diversity of raw goat's-milk cheeses like Selles-sur-Cher and Sainte-Maure-de-Touraine.

Here are some of the unmissable natural wine restaurants of the Loire. As elsewhere in France, reservations are a must—particularly during the Loire salons.

La Route du Sel

LE THOUREIL

A family business run by native Saumur chef Marie Monmousseau and her husband, Daniel Eastcott, La Route du Sel offers an idyllic riverside terrace alongside its tasteful Loire cuisine and extensive natural wine list.

Chez Rémi

ANGERS

Since 2006, chef Rémi Fournier's Chez Rémi has been the top dining destination for wine professionals and vignerons during the Loire salons, thanks to its sophisticated cuisine and Fournier's longtime support of natural wine.

Le Chat

COSNE-COURS-SUR-LOIRE

With its sprawling interior courtyard, sincere bistro cuisine, and impressive natural wine list, Laurent Chareau's Le Chat has been a destination for travelers to Sancerre and Pouilly-Fumé since its opening in 2009.

Comptoir Archimède

SAINT-AIGNAN-SUR-CHER

Opened in 2018 in the shadow of the Collègiale Saint-Aignan church by Belgian restaurateurs Maxime Herbert and Sinem Usta, Comptoir Archimède is a jewelbox-size bistro known for seasonal market cuisine and natural wines from the Cher valley.

Comptoir Archimède in Saint-Aignan-sur-Cher

Legends in the Making

Christophe Foucher in his cellar

Many corners of the Loire—particularly Anjou and the Cher valley—have seen droves of new natural vignerons establish themselves in recent years, drawn as much by the region's supportive and diverse natural wine community as by its still-low vineyard prices. The influx is such that the younger estates listed here are better considered veterans than upstarts. They represent some of the best of a fast-maturing scene, the most talented acolytes of figures like Olivier Cousin, Mark Angeli, and Richard Leroy.

Christophe Foucher, La Lunotte

COUFFY (LOIR-ET-CHER)

When you first taste them, Christophe Foucher's unsulfited, unfiltered sauvignon and menu pineau wines can seem to come out of nowhere. In a sense, they did: Foucher left a career teaching handicapped children to become a winemaker in an unheralded corner of the Touraine appellation, renting and progressively purchasing vines from his (now former) father-in-law. Working naturally was already in the air when he went to viticultural school in Amboise in 2001; among his fellow students were Bertrand Jousset and Pascal Simonutti. Foucher, in his understated way, went further than all of them, ceasing all sulfite addition in the 2010s.

Kenji and Mai Hodgson, Vins Hodgson

RABLAY-SUR-LAYON (MAINE-ET-LOIRE)

Kenji and Mai Hodgson are a low-key, keenly intelligent Canadian Japanese couple who interned at wineries throughout France, Canada, and Japan before establishing their own small winery in the Coteaux du Layon in 2010. Former acolytes of Mark Angeli, the Hodgsons work more radically than their onetime mentor, producing unsulfited, unfiltered chenins of mesmerizing purity that rival the work of their friend and neighbor Richard Leroy.

Lise and Bertrand Jousset

MONTLOUIS-SUR-LOIRE (INDRE-ET-LOIR)

Lise and Bertrand Jousset might be considered heirs to the Catherine and Pierre Breton school of commercially savvy Loire Valley natural winemaking: the Joussets' vast, varied production of Montlouis and Touraine wines isn't the most radical of the region, but it's among the most successful and well liked. The couple founded their estate in 2004, expanding it to 11 hectares (27.2 acres) despite a succession of tough vintages in the 2010s due to frost and hail. (They also produce négociant wines.) Keep an eye out for their cuvée Singulier, a single-vineyard Montlouis from more than one-hundred-year-old vines that packs a memorable intensity despite light filtration.

Laurent Saillard

POUILLÉ-SUR-CHER (TOURAINE)

A transatlantic natural wine veteran, Laurent Saillard was among the influential diaspora of the New York City restaurant Balthazar (see page 413). He ran the early Brooklyn natural wine restaurant Ici from 2004 to 2008, before decamping to the Loir-et-Cher. He worked with his then partner, Noëlla Morantin, before becoming a vigneron in his own right in 2012, taking on parcels including sauvignon rented from influential organic estate Clos Roche Blanche. Saillard produces a slim range of craftsmanlike, unfiltered wines, as well branded as they are well made.

Noëlla Morantin

THÉSÉE (LOIR-ET-CHER)

A linchpin of the Loir-et-Cher winemaking community, Noëlla Morantin put in time working with Agnès and René Mosse, Marc Pesnot, and Philippe Pacalet before replacing Pascal Potaire as winemaker at importer Junko Arai's Loir-et-Cher estate Domaine des Bois Lucas. Morantin founded her own estate in 2008, and has since made a name for herself with juicy, low-sulfite reds from côt, cabernet franc, and gamay, along with easy-drinking filtered whites from sauvignon.

François Saint-Lô

BERRIE (VIENNE)

Living in an isolated commune with his friends and cellarhands, François Saint-Lô is among the most ambitious and sophisticated young winemakers in France. Before producing his own wines, he gained experience working for Olivier Cousin and Eric Dubois of the historic Saumur estate Clos Cristal. He converted a former quarry in Berrie into his breathtaking, naturally refrigerated wine cellar, where he produces bright, quaffable reds from cabernet franc and grolleau and wiry, high-acid chenin. Everything is unsulfited and unfiltered.

Kenji (*left*) and Mai Hodgson outside their cellar

A tractor makes its way up Jura vigneron Stéphane Tissot's vineyards.

6. The Jura

The Jura is a verdant, hilly region at the far east of France, named for the mountain range that separates it from Switzerland. Just an hour's drive from the Burgundy wine town of Beaune, the Jura is a world of its own: sparsely populated, pastoral, and home to a wealth of unique culinary and winemaking traditions, from comté cheese to its savory, long-aged Vin Jaune.

At just under 2,000 hectares (4,942 acres) of vines, the Jura is among the smallest of all the French wine regions, and was long considered a backwoods cousin of Burgundy. Yet in recent years, it has become a touchstone for natural winemaking, winning international acclaim for intensely characterful natural wines that resemble nothing else on the market.

The Jura's reds—pinot noir; dark yet buoyant trousseau; and pale, shimmery poulsard—possess a wild, ethereal quality all their own. Its whites, primarily from chardonnay and the native savagnin, are distinguished by the saline tang of the local yeasts. In the traditional Jurassien white winemaking method, these yeasts are left to form veils upon the liquid surface of wine while it ages in containers that are not kept filled, or "topped up," as is the norm in other regions. Interestingly, the Jura's natural winemakers come down overwhelmingly on the side of innovation, practicing topped-up Burgundian-style aging on most white wines.

The region's natural wine icon is the Pupillin vigneron Pierre Overnoy. Today he is known as much for his lifelong perseverance in traditional organic farming as for his willingness, in 1984, to embark on a path of innovative natural winemaking—and topped-up aging of white wines—with the itinerant natural winemaker Jacques Néauport (see page 28). Along with such visionary vignerons as Stéphane Tissot, the late Pascal Clairet, and Jean-François Ganevat, Overnoy has helped nurture a community of forward-thinking natural vignerons who make the Jura the unlikely hot property it is today.

Total vineyard surface:
2,000ha (4,942 acres)[1]

Rate of organic agriculture:
37 percent[2]

Key grapes:
Savagnin, chardonnay, trousseau, poulsard/ploussard, pinot noir

1 France Ágrimer 2019
2 Agence Bio 2020

Wines to Know

The Jura is almost unique in that if you have expectations of its wines, they probably already refer to many of its esteemed natural estates. The region languished in relative obscurity until its close-knit natural wine community ignited an international infatuation. Nowadays connoisseurs the world over clamber for the region's slender, savory reds from poulsard, trousseau, and pinot noir. Even more prized, for their age-worthiness, are the region's inimitable whites from savagnin and chardonnay, imprinted with the unique pecorino-like notes of the region's native yeasts.

Domaine Ganevat | Côtes du Jura Savagnin–Les Chalasses Marnes Bleues

There is no picking one clear vintage-to-vintage winner from the Ganevats' dizzying range of wines, which numbers more than one hundred cuvées in a given year. The quality is stratospheric throughout. Les Chalasses Marnes Bleues is representative of Jean-François Ganevat's exacting, long-aged Burgundian style, applied to almost ninety-year-old vines of savagnin vert (an old, high-acid, low-yield variant of savagnin) on blue marl soil. The results are a savagnin of shining purity and ripe, citron-inflected fruit that yields to notes of smoke and soft white pepper.

Domaine de la Tournelle | Arbois–Trousseau des Corvées

Perhaps because ploussard is so impossibly light, and trousseau is comparatively dark, many wine drinkers believe that trousseau must be naturally rich. It's not. The greatest trousseaus hew to the example of the late Pascal Clairet's Trousseau des Corvées, showing a keen acidity to complement brambly, dark-cherry fruit, finishing on a characteristic and delicious bitter note. The wine is destemmed by hand and macerated for between ten and thirty days, with daily foot-stomping, before a year's aging in old oak foudres. The Les Corvées vineyard site is known for its thin, gray marl soils, as well as its evocative name, which historically referred to unpaid labor for a lord or king.

Maison Overnoy-Houillon | Arbois-Pupillin

For a prime example of the mind-bending potential of the ploussard grape (also known as poulsard), find a bottle of Maison Overnoy-Houillon's Arbois-Pupillin. Ploussard is a loose-bunched pale-skinned grape that yields wines that people often confuse for rosé. Its acidity, indeed, resembles that of a white wine, while its glimmering raspberry fruit is joyously red. Maison Overnoy-Houillon's is vinified in a variety of tanks, depending on the vintage; it is distinguished by its extended, monthslong vatting. Crisp in youth, the wine grows gossamer with age, turning a sublime flaming-orange color without oxidizing.

MAISON OVERNOY-HOUILLON

Pierre Overnoy and Emmanuel Houillon

PUPILLIN

Pierre Overnoy (*left*) and Emmanuel Houillon in Pierre's kitchen

Pierre Overnoy is a soft-spoken octogenarian who has become a global natural wine icon. In a collaboration with Jacques Néauport that began in 1984, Overnoy introduced sulfite-free winemaking to the Jura and proved that great Jura white wines could be produced with the Burgundian practice of topping up barrels.

Overnoy began mentoring Emmanuel Houillon, the cellarhand who would become his successor, in 1990, when Houillon was fourteen. Together, Overnoy, Houillon, and Néauport created a template for natural winemaking in the Jura, drawing from Néauport's technical expertise and his experiences alongside Marcel Lapierre in the Beaujolais. Yet Overnoy's winemaking is anything but Beaujolais-style. His ploussard is hand-destemmed and undergoes extremely long macerations, before aging in a variety of containers (foudres, steel tanks, barrels, cement eggs). Since 2018, it has contained a small percentage of young-vine trousseau. White wines from savagnin and chardonnay (sometimes blended) are aged extensively in various containers. Whites are distinguished by subtlety, savoriness, and long, graceful acidity. Often the same lot of wine will be released in several successive bottlings, years apart. Barrels are not sulfur wicked between use; other wines are racked directly into emptied barrels to keep them in constant use.

"Jacques Néauport and I thought, why not top up? We thought it could have a longer capacity to age. Our first topped-up wine was in 1985. It was a big controversy at the time."

—Pierre Overnoy

THE WINE TO TRY

See Wines to Know: The Jura, *page 179.*

Since: 1968 officially, although Overnoy has been involved in the family estate since the 1950s.

Winemakers: Pierre Overnoy and, since 2001, Emmanuel Houillon

Vineyard surface: 6.5ha (16 acres)

Viticulture: Biodynamic

Négociant work: None

Grapes: Ploussard, savagnin, chardonnay, and more

Appellations: Arbois-Pupillin, Vin Jaune

Filtration/Fining: None

Sulfites: None added, except in rare circumstances

DOMAINE BORNARD

Philippe and Tony Bornard

PUPILLIN

Philippe Bornard joining the band at the Vive les Vins Libres! tasting at Paris restaurant Quedubon

Pupillin vigneron Philippe Bornard experienced a late-career renaissance in 2005 when, at the age of fifty, he left his local cave cooperative after thirty years. Bornard was already steeped in the world of natural wine thanks to his friendship with Pierre Overnoy and his own experience producing wine for private consumption. The wines Bornard finally released received instant acclaim: they were bold, natural expressions of well-farmed Pupillin marl terroir.

Bornard's fascinating wines push stylistic boundaries, often through entrapped CO_2 or a stirring rush of volatility. Equally memorable is Bornard's outsize personality—he's the guitar-wielding life of any wine tasting. (In France, he's actually more famous outside the wine scene, thanks to his 2012 appearance on *L'amour est dans le pré*—"Love Is in the Field"—a long-running French dating show featuring farmers.)

Philippe's son Tony Bornard began making wines alongside him in 2013, and officially took the reins of the family estate in 2017. Somewhat surprisingly, it was Philippe who persuaded Tony to experiment with Georgian-style qvevri winemaking on savagnin in 2016, not the other way around. Loire vigneron Thierry Puzelat had gifted a qvevri to their mutual friend, the Beaujolais vigneron Jean Foillard. Foillard had no desire to use it and regifted it to Philippe, who was happy to give it a try.

"My father spent thirty years in the cave cooperative. So in fact he's an old guy and a young vigneron. He's enjoying himself now."

—Tony Bornard

THE WINE TO TRY

Arbois-Pupillin–La Chamade

From seventy-year-old ploussard vines planted on red marl by Philippe's father, La Chamade sees hand-destemming and long macerations of two to three months in foudre, before further aging in foudre. The result is a structured, granular ploussard, with raspy tannicity to its fig-inflected red fruit.

Since: 2005 (But the estate has existed for generations. Philippe Bornard has produced small quantities of wine since 1973, selling the bulk of his harvest to the cave cooperative until 2005.)

Winemaker: Tony Bornard since 2017. Formerly Philippe Bornard.

Vineyard surface: 11.5ha (28.2 acres)

Viticulture: Biodynamic

Négociant work: None

Grapes: Ploussard, savagnin, chardonnay, trousseau

Appellations: Arbois-Pupillin, Côtes du Jura. But most wines released as Vin de France since 2018.

Filtration/Fining: None

Sulfites: Usually a microdose of 10–20mg/L at bottling, but not always. Vin Jaune is usually sulfited.

DOMAINE DE LA TOURNELLE

Évelyne and Pascal Clairet

ARBOIS

Domaine de la Tournelle is the estate of Évelyne and the late Pascal Clairet, a couple of savvy agronomy advisors who, from the early 2000s until Pascal's suicide in 2021, came to exemplify contemporary Jura natural winemaking. The estate's unsulfited, unfiltered production has long been a model of rigor in both farming and winemaking.

The estate practices hand-destemming for all reds (save petit béclan), and age their white wines for no less than two years. Vinification begins in a separate cellar space in nearby Montmalin, before wines are transferred in tank to an impressive, churchlike vaulted cellar in Arbois for aging. Whites age primarily in barrel before assemblage in foudre; reds age primarily in foudre. White wines almost all see topping up; they are incisive, sophisticated chardonnays and savagnins that all bear the nuances of their terroirs.

> **"I was good friends with the Barrals, the Lapierres, and all the team from the Beaujolais in the 1990s. Theirs were the wines that we liked to drink, and we wanted to know how it was done. Well, we knew how it was done, but we had to be ready to do it ourselves."**
>
> —Pascal Clairet

The Clairets took part in the early years of L'Association des Vins Naturels, alongside friends like Alsace's Christian Binner. Among the alumnae of Domaine de la Tournelle are several younger vignerons who've become stars in their own right, including Etienne Thiébaud of Domaine des Cavarodes.

THE WINE TO TRY

See Wines to Know: The Jura, *page 179.*

Pascal Clairet in his cellar

Since: 1991

Winemaker: Évelyne Clairet. Pascal Clairet until 2021.

Vineyard surface: 8ha (19.8 acres)

Viticulture: Biodynamic

Négociant work: None

Grapes: Ploussard, savagnin, chardonnay, and more

Appellations: Arbois

Filtration/Fining: None since 2003

Sulfites: None added since 2003

DOMAINE ANDRÉ ET MIREILLE TISSOT

Bénédicte and Stéphane Tissot

MONTIGNY-LES-ARSURES

Stéphane Tissot in his cellar in Arbois

Since the 2000s, Domaine André et Mireille Tissot has become a beacon for natural farming and winemaking in the Jura. Behind the estate's transformation is Stéphane Tissot, who oversees all farming in addition to vinification of the estate's vast range of wines. He takes an experimental approach, conducting tests of rare grape varieties (like chardonnay rose) or ultra-high vine densities. In 2013, encouraged by his friend Thierry Puzelat, he began experimenting with amphora aging of poulsard.

Since 1999, Tissot has emphasized parcel cuvées. His range of precise whites, from limestone and lias and trias clay soils, reveals an aesthetic kinship with the more conservative biodynamic winemakers of Alsace: the wines are all plate filtered and lightly sulfited before bottling. Tissot's reds, from poulsard, trousseau, and pinot noir, are as natural as they come: hand-destemmed (except for pinot noir), fermented and aged mostly in foudre, bottled without sulfites or filtration. Also of interest is Tissot's unique range of parcel-specific Vins Jaunes, which he presents en masse at the annual Le Nez dans le Vert tasting (see page 185).

"When we talk about Vin Jaune, we often talk more about vinification than terroir expression. But we've been working organically for over twenty years. We want to demonstrate that the terroir shows through this oxidative vinification."

—Stéphane Tissot

THE WINE TO TRY

Arbois–DD

An homage to Tissot's father's historical winemaking style, DD is a blend of hand-destemmed poulsard and trousseau with whole-cluster pinot noir, half from limestone and half from trias clay soils. It's a deep, barritone Jurassien red, richer than most of its peers.

Since: 1962 (Stéphane Tissot took over in 1990)

Winemaker: Stéphane Tissot

Vineyard surface: 50ha (123.5 acres)

Viticulture: Organic since 1999. Biodynamic since 2004.

Négociant work: None

Grapes: Poulsard, savagnin, chardonnay, and more

Appellations: Arbois, Vin Jaune, Côtes du Jura, Crémant du Jura

Filtration/Fining: Whites see plate filtration before bottling. Reds have been unfiltered since 2003.

Sulfites: 30mg/L before bottling for white wines. None added for most reds.

DOMAINE DE LA PINTE

Pierre and Vincent Martin

ARBOIS

Domaine de la Pinte is a historic Arbois estate that has helped anchor the Jura natural wine scene throughout the last two decades. Founded in the 1950s by present owner Vincent Martin's grandfather, Roger Martin, Domaine de la Pinte was initially intended to produce exclusively Vin Jaune, but the estate diversified throughout the decades.

Roger Martin's son Pierre oversaw the estate's conversion to organic viticulture in 1999 and then to biodynamics in 2009—both courageous moves, given the estate's substantial surface. Along with their neighbors at Domaine André and Mireille Tissot, who converted to biodynamics around the same time, Domaine de la Pinte ushered in a sea change in Jura viticulture during the 2000s.

But producing truly natural wine requires risk-taking, and Domaine de la Pinte's winemaking remains conservative by contemporary standards. Most wines—even reds—see filtration, along with levels of sulfite addition more typical of the biodynamic-but-not-natural wine scene. Longtime winemaker Bruno Ciofi left in 2016 to join Mark Angeli at La Ferme de la Sansonnière. Today winemaking is run by Jura native Emmanuelle Goydadin.

Since: 1950

Winemaker: Emmanuelle Goydadin since 2019

Vineyard surface: 34ha (84 acres)

Viticulture: Organic since 1999. Biodynamic since 2009.

Négociant work: None

Grapes: Chardonnay, savagnin, melon à queue rouge, and more

Appellations: Arbois, Vin Jaune, Crémant du Jura, Arbois-Pupillin

Filtration/Fining: Most wines see diatomaceous earth filtration.

Sulfites: Most wines see sulfite addition of about 40mg/L.

DIDIER GRAPPE

SAINT-LOTHAIN

A former industrial mechanic and the son of hotelier school officials, Didier Grappe began his estate in his spare time, planting new parcels in his native Saint-Lothain throughout the early 2000s. He cites the ambient influence of local legend Pierre Overnoy in inspiring his embrace of natural vinification. His local clay terroirs, by his own estimation, are better suited to whites than reds, so whites predominate in his range of unsulfited wines. Grappe employs old barrels for aging uniquely on his rare un-topped-up (*non-ouillé*) white wines, with all other wines fermenting and aging in steel in his very cold cellar.

Since 2015, Grappe has become a vocal advocate of resistant grape varieties, aka hybrids, planting small surfaces of old ones each year (see page 69). Red and white hybrid varieties co-ferment in his rare Pif Purple, a reed-thin but characterful expression of these half-forgotten grapes.

Since: 2001

Winemaker: Didier Grappe

Vineyard surface: 4ha (9.9 acres)

Viticulture: Organic

Négociant work: None

Grapes: Chardonnay, savagnin, trousseau, and more

Appellations: All wines are released as Vin de France.

Filtration/Fining: None

Sulfites: None added since the mid-2000s

TASTING DESTINATIONS
LE NEZ DANS LE VERT

When: The Sunday and Monday of the last weekend of March

Who: Almost everyone who's anyone in organic farming in the Jura

For: Open to the public on Sunday. Monday is reserved for wine professionals.

Where: It rotates each year. In 2019, the festival was held in the Royal Saltworks of Arc-et-Senans.

Founded by former Domaine de la Pinte winemaker Bruno Ciofi in 2011, Le Nez dans le Vert is the Jura's annual organic and natural wine-tasting salon, a two-day festival that unites forty-eight vignerons, ranging from legends like Pierre Overnoy to under-the-radar newcomers like Thomas Popy. Its name, which translates to "The Nose in the Green," is a pun on the French for how one sniffs wine—with your nose in the *verre*, or glass.

Recent events have attracted more than a thousand visitors, including many renowned natural winemakers from the Beaujolais, who come to support their Jurassien peers. When Ciofi moved to the Loire in 2016, he ceded organization of subsequent salons to his vigneron friends Stéphane Tissot and Jean-Etienne Pignier.

Le Nez dans le Vert 2019 at Domaine de la Pinte

DOMAINE GANEVAT

Anne and Jean-François Ganevat

ROTALIER

The Ganevat family has farmed vines in the south Jurassien hamlet of La Combe since 1650. But only in the past two decades has the virtuosic winemaking talent of fourteenth-generation vigneron Jean-François Ganevat launched the estate into the global wine firmament.

The estate was for sale without buyers when Ganevat returned from a career as a cellar master in Chassagne-Montrachet to take it over in the late 1990s. Ganevat converted the estate to organics, partly influenced by his late friend Henri-Frédéric Roch. By 2006, Ganevat had ceased filtering and adding sulfites on all wines. He uses a range of small wooden tanks for red fermentations, and has adopted the use of refrigerated trucks during harvest to chill grapes before vatting. Ganevat's white wines generally do not see must settling after press; instead, they go right into barrels or amphorae, where most undergo very long, topped-up aging on lees.

> **"I have friends who've already bottled their vintage by December. I don't know how to do that and, personally, I don't want to do that. The wines, I let them make themselves and I let them live."**
>
> —Jean-François Ganevat

Jean-François Ganevat in his cellar

A series of low-yield vintages in the 2010s led Ganevat and his sister Anne to create a négociant business in 2014. Don't bother trying to follow the exact grape blends on the cross-regional Vin de France négociant wines—they change according to Ganevat's whims.

In 2021, Ganevat shocked the wine world when he sold the estate to Russian investor Alexander Pumpyanskiy. Ganevat will continue managing the estate under its new ownership.

THE WINE TO TRY

See Wines to Know: The Jura, *page 179.*

Since: 1998 in its current form

Winemaker: Jean-François Ganevat

Vineyard surface: 12ha (29.6 acres)

Viticulture: Biodynamic

Négociant work: Yes, since 2014

Grapes: Chardonnay, savagnin, poulsard, trousseau, pinot noir, and many more

Appellations: Estate wines are Côtes du Jura. Négociant work includes Arbois, Château-Chalon, and more, along with many Vin de France wines.

Filtration/Fining: None since 2005

Sulfites: None added since 2006

DOMAINE DES CAVARODES

Etienne Thiébaud

CRAMANS

Tall as Gandalf and spare with words, Cramans native Etienne Thiébaud is the visionary wunderkind of the Jura, his authoritative winemaking and abiding support of younger peers belying his own relative youth. Thiébaud followed viticultural school in Beaune (where he studied alongside Alsace's Catherine Riss, page 255) with several years working for the late Pascal Clairet at Domaine de la Tournelle before founding his own Domaine des Cavarodes in 2007 at the age of just twenty-three. He quickly made a name for himself within the region for the sensitivity of his farming, which today includes horse plowing and many biodynamic treatments. His vineyards, spread throughout sites in Arbois, Liesle, and Mouchard, include many obscure local varieties (like the high-acid enfariné), as well as an experimental planting of mondeuse. His cellar work shows the influence of his former employers at Domaine de la Tournelle: whites from chardonnay and savagnin age for two winters in a mix of foudres and *demi-muids,* while poulsards and trousseaus see hand destemming and macerations from ten days to a month before aging in foudre. These latter are unparalleled in the region for their lucent, smiling red fruit. In recent years, Thiébaud has also experimented with weeklong, whole-cluster maceration of savagnin.

In the decade and a half since his debut, Thiébaud has quickly become influential in his own right, helping mentor a newer generation of natural Jura vignerons including Arnaud Greiner, Morgane Turlier, and Thomas Popy. In everything from his locavore habits to his almost comical aversion to the wine media, he evinces a deep attachment to the rural community of the Jura. When not in the cellar or his vineyards, he can be found playing the drums in a local reggae combo.

THE WINE TO TRY

Trousseau–Les Lumachelles

Thiébaud credits the suppleness and violety sweep of his masterful trousseau Les Lumachelles to the infinitesimally subtle oxygenation of foudre aging. His modesty surely masks other factors, for few other trousseaus of the Jura match its beauty and opulence.

Since: 2007

Winemaker: Etienne Thiébaud

Vineyard surface: 8ha (19.8 acres)

Viticulture: Organic

Négociant work: None

Grapes: Poulsard, trousseau, enfariné, and more

Appellations: Côtes du Jura, Arbois, IGP de Franche Comté

Filtration/Fining: None

Sulfites: None added since 2014

DOMAINE LABET

Romain, Charline, and Julien Labet

ROTALIER

Along with his neighbor Jean-François Ganevat, Rotalier vigneron Julien Labet represents the height of the Burgundy-trained winemaking vanguard of the Sud Revermont, the Jura's rustic south. Julien trained at famed Meursault estate Domaine Ramonet before returning to the Jura in 1999. From 2003 on, he worked simultaneously at the family estate and at his own separate estate, established so he could practice his beliefs in organic farming and eliminating sulfites in winemaking. Upon his parents' retirement in 2012, Julien, along with his brother Romain and sister Charline, converted the entirety of the family estate to organic farming.

Today the Labets produce around thirty cuvées each year, including many parcel selections (for which their father, Alain, was a pioneer in the 1990s), with whites almost all aged in the topped-up Burgundian style, in barrel rather than foudre. Both whites and reds are known for masterful precision and piercing acidity. Julien Labet gives the impression of having been shell-shocked by the upsurge in global demand in recent years for Jura wine at large, and his wines in particular. (Only his brother Romain evinces the congeniality of a bygone era of Jura paysanry.) The family has nonetheless been instrumental in supporting a new generation of natural vignerons in the Jura, including Christian Boulanger and Mathieu Allante.

Since: 1974. Julien began a separate estate in 2003, uniting it with the family estate in 2013.

Winemaker: Julien Labet

Vineyard surface: 13ha (32.1 acres)

Viticulture: Organic from 2014

Négociant work: None

Grapes: Chardonnay, savagnin, poulsard, and more

Appellations: Côtes du Jura, Vin Jaune

Filtration/Fining: None

Sulfites: Sometimes a small dose at bottling

LE MOUTHEROT

Georges Comte

ÉMAGNY

A Besançon native who prospered in the prosthetics industry, Georges Comte lived for years in Dijon, where his passion for wine led to friendships with famous Burgundy winemakers of the day. When the chance arose to purchase a once renowned but neglected clay-limestone hill east of Besançon called Le Moutherot, Comte became a hobby winemaker in the almost nonviticultural Doubs region north of Arbois.

He hired a farmer named Henri Colin to farm vines according to organic principles and took a portion of the harvest as rent. This began in 1988, and eventually expanded to 8 hectares (19.8 acres) of vines, almost all chardonnay, with a small amount of pinot noir. Unconstrained by financial necessity, Comte produced his wines without sulfites or other additives from the get-go. He found support in the founder of Besançon natural wine shop Les Zinzins du Vin, Fabrice Monnin (see page 280). Comte's methods include two years' barrel aging and extensive bottle aging before release. He works out of an ancient farmhouse that looks like it was destroyed and reassembled by a succession of tornadoes.

Comte's wines became cult items at high-end natural wine restaurants, cherished for how they mimic the sinuous, mineral profile of classic Côte d'Or wine in a zero-sulfite alternate reality.

Since: 1988

Winemaker: Georges Comte

Vineyard surface: 8ha (19.8 acres)

Viticulture: Practicing organic

Négociant work: In some sense, it is all négociant work, since Comte uses rent grapes from a tenant vigneron.

Grapes: Chardonnay, pinot noir

Appellations: IGC du Doubs

Filtration/Fining: None

Sulfites: None added

Natural Wine Dining

Once upon a time, the Jura had trouble marketing its diverse and characterful wines. Not so its local cuisine, which is one of the most renowned in France. Jura cooking is based around dairy from the region's pastures, along with chicken from nearby Bresse. Famous Jura cheeses include comté and the obscenely rich Mont d'Or. Cream serves as the base of the region's iconic dish, *coq au vin jaune aux morilles*, or chicken with Vin Jaune and morel mushrooms.

With its famed cuisine, pastoral beauty, and nearby lakes, the Jura doesn't lack for tourism. But the region is thin on dining options. There are nonetheless a handful of crucial standbys to nourish a traveling natural wine aficionado.

Le Bistrot des Claquets

ARBOIS

The center of the Arbois social scene, Le Bistrot des Claquets incarnates the French village bistros of yore, serving just one plat du jour at lunch only, along with a simple, hearty appetizer buffet. The endearing owner Rachel Gariglio's natural wine selection highlights rare cuvées from the region's young up-and-comers.

Maison Jeunet

PORT LESNEY

A historic two-Michelin-star restaurant founded by Jean-Paul Jeunet in the heart of Arbois, Maison Jeunet has been run by the Belgian chef Steven Naessens since 2016, and moved location to the Château de Germigney in Port Lesney in late 2021. It remains a destination for its encyclopedic Jura wine list and its unforgettable chicken with Vin Jaune and morel mushrooms.

Le Bistrot de la Tournelle

ARBOIS

For two months each summer, Domaine de la Tournelle hosts a terraced wine bar on the idyllic riverside interior courtyard of the winery. It offers salads at lunchtime and little more than terrines, rillettes, and sausage in the evening, yet still comprises a valuable addition to the Jura dining landscape.

Left: Bistrot des Claquets proprietor Rachel Gariglio

Below: Le Bistrot des Claquets in Arbois

Legends in the Making

In the Jura, it's striking how many young estates are already producing wines that rival those of their more famous forebears in quality—and, less fortunately, in scarcity. A succession of frost vintages in the 2010s, combined with the unabating Jura mania in natural wine circles, has made everyone's wines hard to come by.

Here are the leading lights of a succeeding generation of Jura vignerons.

Renaud Bruyère and Adeline Houillon

PUPILLIN

Renaud Bruyère and Adeline Houillon began producing small quantities of wine in 2011, after work experience with Stéphane Tissot and Pierre Overnoy, respectively. (Adeline is the sister of Overnoy's successor Emmanuel Houillon.) Today the couple farm their 4 hectares (9.9 acres) organically, and both whites and reds are produced without sulfites or filtration. The wines often show a teeth-stripping acidity, but when encountered at the right moment, they justify their prices, among the highest in the region.

Domaine de Saint Pierre

Fabrice Dodane

MATHENAY

An intense mountain of a man, Fabrice Dodane began working as cellar master of Domaine de Saint Pierre in 1989 and purchased the estate upon the death of its owner in 2011. Having already converted to organic viticulture in 2008, he embraced natural vinification, with unsulfited pinot noirs earning applause from peers like Philippe Bornard and Jean-François Ganevat. Severe frost damage in recent years (2017, 2019, and 2021) unfortunately meant Dodane's wines became very scarce right when they attained widespread acclaim.

Domaine des Miroirs

Kenjiro Kagami

GRUSSE

A quiet Japanese émigré armed with a showstopping résumé, Kenjiro Kagami established his Domaine des Miroirs in 2011 with vines sourced by a former employer, Jean-François Ganevat. There Kagami put into practice his years of experience working for other renowned masters: Bruno Schueller in Alsace (see page 252), Thierry Allemand in the northern Rhône (see page 235), and Domaine Comte Georges de Vogüé in Chambolle-Musigny. Kagami practices no-till organic farming, risking extremely low yields in search of greater balance in vines and in the wines. Early vintages of these low-yield, long-aged, unsulfited, unfiltered wines were forbiddingly reductive. It's nonetheless worth trying these profound, experiential wines, wherever you might find them.

Renaud Bruyère in his cellar

Domaine Les Bottes Rouges

Jean-Baptiste Menigoz and Florien Klein Snuverink

ABERGEMENT-LE-PETIT

A former teacher, Jean-Baptiste Menigoz made wine alongside his friend Raphael Monnier before establishing his own estate in 2011. In 2014, he partnered with Florien Klein Snuverink, former proprietor of Amsterdam restaurant Café Schiller, who had come to the Jura to learn winemaking. Today the duo farm 7 hectares (17.3 acres) of vines. Acolytes of Stéphane Tissot, Menigoz and Klein Snuverink filter and lightly sulfite white wines, while reds generally see neither filtration nor sulfite addition. Their work with pinot noir marries a crystalline lightness with rare juiciness.

Ratapoil

Raphael Monnier

ARC-ET-SENANS

A gregarious, fast-talking history teacher by trade, Raphael Monnier made wine as a hobby winemaker (a *ratapoil*, in local slang) with his friend Jean-Baptiste Menigoz from 1999 until 2009, when Monnier and his wife set up a real winemaking business with 2 hectares (4.9 acres) of vines in Arbois. The estate's small surface, combined with frost troubles, has resulted in a free-form range of wines, with cuvées evolving according to the vintage. In low-yield vintages, Monnier blends a memorable, soulful, multivarietal red from all his parcels, aptly named Partout ("everywhere").

François Rousset-Martin

NEVY-SUR-SEILLE

François Rousset-Martin is another Burgundy-influenced up-and-comer making his mark with low-sulfite, topped-up whites on Jura terroir. His work is distinguished by the terroir of Château-Chalon, the Jura's historical grand cru. Rousset-Martin inherited the vines from his father, a microbiologist in Burgundy. After establishing his estate in 2007, Rousset-Martin vinified just 3 hectares (7.4 acres) of an available 10 hectares (25 acres), selling the rest to the cave cooperative as he perfected his craft. As of 2018, he vinifies the entire 10 hectares. His topped-up Côtes du Jura Savagnin–Cuvée du Professeur is already a reference for boldly ripe savagnin from gray-clay Château-Chalon soils.

Florien Klein Snuverink (*left*) and a harvester cleaning up at the close of the last day of harvest

Thomas Popy

MESNAY

Wry, press-shy, and intelligent, Thomas Popy worked for several natural wine maestros of his native Beaujolais (Julie Balagny, Jean-Claude Lapalu, Michel Guignier) before decamping to greener pastures in the Jura. After studying at the viticultural school in Montmorot and working at Maison Overnoy-Houillon and Les Dolomies, he founded his estate in 2015, working out of a temporary cellar near the vineyards of Emmanuel Lançon of Domaine des Murmurs. Since 2017, Popy has employed a small cellar beneath his home in Mesnay, on the outskirts of Arbois. From barely more than 2 hectares (4.9 acres) of vines, his tiny production of unsulfited, unfiltered chardonnay, savagnin, and trousseau disappears quickly into the cellars of Jura natural wine aficionados. (His resonant and angular savagnin Sous la Roche Maldru is a highlight.)

Looking south over the Chassagne quarry from Domaine Derain's En Remilly vineyard in Saint-Aubin

7. Burgundy

Burgundy is hallowed ground in the wine world. Thanks to the ecclesiastical winemaking that occurred after the arrival of Christianity, the region benefits from an unparalleled legacy of scholarship about its vaunted clay-limestone terroir. Many of the most expensive, sought-after wines on earth are born here.

On the one hand, success has had a conservatizing effect on the region's wine culture. The region embraced synthetic fertilizers, herbicides, and pesticides with gusto in the 1970s and '80s, prompting the agronomist Claude Bourguignon's famous observation in 1992 that the soils of the Sahara had more life in them than those of Burgundy's vineyards. Today many famous Burgundy vineyards are owned by wealthy families with little day-to-day involvement in wine production. And the amount of money at stake in both farming and vinification discourages the risk-taking inherent to farming and making wine naturally.

Yet Burgundy's combination of wealth and wine scholarship has led many renowned estates to maintain links to natural methods of farming and vinification. The most famous example is Domaine de la Romanée-Conti, which adopted biodynamic farming in the 1990s under the direction of Aubert de Villaine and Henri-Frédéric Roch.

The trailblazers of natural vinification in Burgundy were Saint-Aubin's Dominique Derain, who began vinifying naturally in 1987 with the aid of famed enologist Max Léglise, and Marcel Lapierre's nephew, Philippe Pacalet, who in the 1990s would forge an influential template for great natural Burgundy at Domaine Prieuré Roch.

It was not until the 2000s that a more populous natural winemaking scene emerged in Burgundy, spearheaded by vignerons outside the high-money circuit of the Côte d'Or: in the Mâconnais, Philippe Valette, Julien Guillot, and Gilles and Catherine Vergé; in Northern Burgundy, Alice and Olivier de Moor, the Montanet family, and the former wine retailer Nicolas Vauthier. Meanwhile, Philippe Pacalet, working as a solo négociant since 2001, and the more recently installed vigneron-négociant Frédéric Cossard have worked to demonstrate what natural vinification can reveal on the region's most celebrated—and highly valued—terroirs.

Total vineyard surface:
33,700ha (83,275 acres)[1]

Rate of organic agriculture:
37 percent[2]

Key grapes:
Chardonnay, aligoté, pinot noir

1 France Agrimer 2019
2 Agence Bio 2020

BURGUNDY

Wines to Know

To familiarize yourself with the natural wine of Burgundy is to expand your purview beyond the regional stereotypes of lean chardonnay and prim pinot noir. Natural Chablis is fleshier and fuller-fruited than its conventional counterparts, while retaining the zone's signature stoniness. In the Côte de Nuits and the Côte d'Or, natural iterations of the famous reds will be conspicuously lighter and more bristly, largely unmarked by extractive innovations like cold soaking (a prefermentary maceration at low temperature).

Alice and Olivier de Moor | Chablis–Côteau de Rosette

Bordered by a forest, Rosette is a 1-hectare (2.5-acre) marne soil parcel that Olivier de Moor planted in 1989. Its wine shows the concentration and glittering minerality brought by low-yielding vines on the very thin soils at the top of the slope, rounded out by fleshier fruit from the rich soils at the bottom. Barrel fermented and aged for twelve to sixteen months, it's the most austere and powerful wine of the de Moors' range—yet by the artificial standards of filtered, yeasted, underripe, overcropped Chablis, it's almost opulent.

Clos des Vignes du Maynes | Mâcon Rouge–Cuvée 910

Embodying the experimentalist nature that natural winemaking has brought to Burgundy, Julien Guillot's Cuvée 910 is a short-macerated blend of old-vine chardonnay, pinot fin (an ancient tight-bunched variant of pinot noir), and gamay à petits grains (an ancient small-berried variant of gamay) from his best parcels. The wine is an homage to the historical winemaking methods of the Clunisien monks: a razor-fine elixir of sappy red fruit and candied citrus. The wine's translucent, brickish color belies an admirable intensity.

Maison Valette | Mâcon-Chaintré–Vieilles Vignes

Philippe Valette's signature wine is his bold, ginger-toned Mâcon-Chaintré, from the family's clay-limestone parcels, organically farmed since the early 1990s. Valette's divisive style involves courageous ripeness and long, unsulfited élevage (two to three years for this cuvée; much longer for certain others). Critics call the style oxidative. Natural wine aficionados recognize it as ambitious, patient, pure winemaking, of a class rarely encountered in the Mâconnais.

Philippe Pacalet | Pommard Premier Cru–Les Arvelets

Philippe Pacalet's Pommard Premier Cru–Les Arvelets, from a steep, red-clay atop limestone site, is a fine-boned, saline pinot fin with firm cherry fruit and flecks of winter spice. Like all Pacalet's reds, it's made with two- to three-week whole-cluster fermentation, with daily pigéage, before aging for just under a year and a half on lees in neutral oak barrels.

ALICE AND OLIVIER DE MOOR

COURGIS

Alice (*right*) and Olivier de Moor in their Rosette vineyard

Alice and Olivier de Moor are the thoughtful, unpretentious enologists behind the marquee natural wine estate of Chablis. The de Moors make pure, full-fruited Chablis like none other on the market, along with sterling aligoté.

The pair met working at a conventional Chablis estate, and planted three sites in 1989 that would go on to form the basis of their top cuvées: Rosette, Bel Air, and Clardy (the latter two are blended into one cuvée). Their first true harvest was in 1995, after which they made quick strides toward additive-free winemaking. Encouraged initially by American importer Peter Vezan and the Paris wine shop Caves Legrand, the de Moors were introduced to the wider natural wine scene around 2001 through Troyes wine shop Aux Crieurs de Vin.

"There's a strong demand for Chablis, but it's often a compromise between a famous name and a moderate price. There's a resistance to change in the market, and this gets echoed in the minds of the vignerons."

—Olivier de Moor

Today Olivier handles vineyard duties and label design, while Alice takes the lead in vinification. All estate wines save the Bourgogne Aligoté (and some Bourgogne Chitry) see fermentation and aging in used oak barrels. In 2009, the couple began a négociant project titled Le Vendangeur Masqué ("The Masked Harvester"). In 2017, the de Moors acquired vines in two premier crus, Mont de Milieu and Vau de Vey.

THE WINE TO TRY

See Wines to Know: Burgundy, *page 195.*

Since: 1995

Winemaker: Alice de Moor

Vineyard surface: 10ha (25 acres)

Viticulture: Organic since 2005

Négociant work: Yes, since 2009, under the label "Le Vendangeur Masqué"

Grapes: Chardonnay, aligoté, sauvignon, and more

Appellations: Chablis, Chablis Premier Cru Mont de Milieu, Chablis Premier Cru Vau de Vey, and more

Filtration/Fining: None, except in rare cases on négociant work

Sulfites: 20–30mg/L at bottling

VINI VITI VINCI

Nicolas Vauthier

AVALLON

A cofounder of influential Troyes cave à manger Aux Crieurs de Vin (see pages 342 and 349), Nicolas Vauthier played a key role in the development of the natural wine scene in the 2000s, encouraging nearby vignerons like Emmanuel Lassaigne and Alice and Olivier de Moor early in their careers. Having studied winemaking in the Beaujolais in his youth, in 2008 Vauthier decided to sell his stake in Aux Crieurs de Vin to establish a négociant winemaking business. He apprenticed himself to Burgundy natural winemaker Philippe Pacalet for a year before setting up shop in the Yonne, where the lone famous appellation, Chablis, tends to hog the limelight from undervalued terroirs around Auxerre and Tonnerre.

"The young public have a lot less prejudices about wine. It's more the old folks who say, 'Oh, it sparkles. Your wine is going bad!'"

Thanks to longtime friendships with other prominent natural wine retailers—including the founders of Le Verre Volé in Paris and La Part des Anges in Nice—Vauthier had no trouble finding clients for his bold, unfiltered, and often unsulfited wines. Early vintages saw as many misses as hits, but as Vauthier (or "Kikro," as everyone knows him) has rediscovered his craft, his wines have become increasingly elegant. They are "little wines," produced with the scrupulous care typically devoted to *grands vins*. In his aligoté one can discern the purity of a bygone era of winemaking, before winemakers began rushing "little wines" to market. Almost as refreshing as Vauthier's wines are their absurd labels, which feature a circus of men, women, and intersex figures, nude save for the odd cosplay item.

THE WINE TO TRY

Bourgogne Coulanges-la-Vineuse–Chanvan

Chanvan, from mid-slope in a steep, stony clay-limestone parcel south of the town of Auxerre, exemplifies Vauthier's light touch in vinification, combining a pleasant stalkiness with crisp, quenching raspberry fruit.

Since: 2009. Vauthier had an influential career in wine retail from 1998 to 2008.

Winemaker: Nicolas Vauthier

Vineyard surface: He purchases from about 12ha (29.6 acres) of vines.

Viticulture: Mostly practicing organic, some certified organic

Négociant work: Yes—it's all négociant work.

Grapes: Pinot noir, gamay, sauvignon, and more

Appellations: Irancy, Bourgogne Epineuil, Bourgogne Coulanges-la-Vineuse, often Vin de France

Filtration/Fining: None

Sulfites: Sometimes up to 10mg/L if necessary, but more often none added

FRANÇOIS ECOT

MAILLY-LE-CHÂTEAU

A spry, motormouthed jack-of-all-trades, François Ecot fell into Parisian natural wine circles in the early 1990s, embarking on an eccentric and influential career, first as a wine importer, then as a vigneron. In 1998, he enrolled in winemaking studies in Beaune, going on to work with Thierry Allemand (see page 235) and Hervé Souhaut (see page 233) in the northern Rhône. With his then partner, Jenny Lefcourt, he founded Jenny & François Selections in 2000, playing a significant role in introducing the United States to natural wine. In 2002, Ecot planted his own vines on land he cleared himself in his native Mailly-le-Château,

"I got into wine in 1989. I began to learn there were certain wines that didn't give me a headache. But it wasn't just the absence of a headache. In a bodily sense, when I drank them, I didn't feel the same."

producing his first vintage of estate wine in 2005. Ever the innovator, he planted a fascinating blend of rare local varieties (abourriou and césar) and traditional varieties (gamay and pinot noir), along with a few from farther afield (pineau d'aunis and sangiovese).

Ecot supplements the tiny yields he derives from his own parcels with grape and wine purchases from other estates in the Yonne, the Mâconnais, the Gard, and beyond. His wines are recognizable for their unsulfited purity and for his sensitive touch in vinification; like his friend and neighbor Nicolas Vauthier, he applies painstaking attention to the production of "little wines" from modest corners of France.

THE WINE TO TRY

Vin de France–L'Insolent

Ecot harvests the unusual grape blend that makes up his signature L'Insolent in successive passes. The singular wine has the opacity of the césar grape, the rusticity of cool-vintage sangiovese, the fruit of young pinot noir, the aromatic spice of pineau d'aunis, the often fierce reduction of gamay. Grainy and dense, the wine requires careful decanting.

Since: 2005

Winemaker: François Ecot

Vineyard surface: 2ha (5 acres)

Viticulture: Organic

Négociant work: Yes

Grapes: Pinot noir, pinot beurot, gamay, and more

Appellations: All wines are released as Vin de France.

Filtration/Fining: None

Sulfites: None added

LA SOEUR CADETTE

Famille Montanet

SAINT-PÈRE

From their winery beside the basilica town of Vézelay, the Montanet family makes pristine northern Burgundies under three labels. Domaine de la Cadette is the estate founded by Jean and Catherine Montanet in 1987, which they converted to organic agriculture in 1999. Domaine Montanet-Thoden is the estate founded by Catherine Montanet after the couple split. La Soeur Cadette, meanwhile, is the label used for négociant wines from northern Burgundy and the Beaujolais. Since joining the estate in 2010, the Montanets' son Valentin has taken over winemaking for all three labels.

The soils around Vézelay are variants of Chablisien clay-limestone. Throughout the 2000s, the Montanets' production became a standard-bearer for this neglected corner of the Yonne, where most wines are produced for bulk sale. All but one of their pristine, high-toned chardonnays are aged in stainless steel and bottled with a light filtration, while red wines, a mix of whole-cluster and destemmed fruit, are fermented in steel tanks before aging in neutral oak barrels. The wines are modest beauties, well priced for Burgundy, and produced to standards that exceed the norms of many more famous appellations.

Since: 1987

Winemaker: Valentin Montanet since the mid 2010s. Formerly Jean Montanet.

Vineyard surface: 21ha (51.9 acres)

Viticulture: Organic since 1999

Négociant work: Yes, from purchased grapes from Mâcon-Villages, Chénas, and Juliénas

Grapes: Chardonnay, melon de Bourgogne, pinot noir

Appellations: Bourgogne Vézelay

Filtration/Fining: Whites see filtration.

Sulfites: Light sulfite doses before bottling

JEAN-CLAUDE RATEAU

BEAUNE

Jean-Claude Rateau was among the first biodynamic vignerons in France, converting his father's 1.5 hectares (3.7 acres) of vines in 1979. Today Rateau farms 10 hectares (25 acres), producing a modest range of Burgundies whose reputation for value and traditionalism belies a mild experimental bent in recent years. Rateau divides his time between Burgundy and Alsace, where his girlfriend, Sylvie Spielmann, maintains her family estate. An Alsatian influence shows in micro-cuvées of late-harvest and skin-macerated chardonnay, and in Rateau's practice of including pinot beurot and pinot blanc in his white vine plantings.

Rateau's wines ferment in enameled steel before aging in used oak barrels. Reds ferment destemmed on a small layer of whole-cluster fruit, throughout fifteen- to twenty-day macerations. His vines comprise a patchwork of Beaune terroirs, from iron-rich soils in Les Beaux et Bons to black, manganese-heavy soils in Les Prévoles.

Despite being active in biodynamic circles since 1979, Rateau has never sought certification, preferring autonomy in his work as a paysan practitioner of Rudolf Steiner's philosophy.

Since: 1979

Winemaker: Jean-Claude Rateau

Vineyard surface: 10ha (25 acres)

Viticulture: Organic, practicing biodynamic

Négociant work: None

Grapes: Chardonnay, pinot noir, pinot blanc, and more

Appellations: Hautes Côtes de Beaune, Beaune, Gevrey-Chambertin

Filtration/Fining: Most whites and some reds see filtration before bottling. Rarely the premier crus.

Sulfites: 10mg/L after malolactic fermentation and 10mg/L before bottling

PHILIPPE PACALET

BEAUNE

Burgundy négociant Philippe Pacalet's influence on natural wine extends almost as far as that of his uncle, the Beaujolais vigneron Marcel Lapierre. Following Lapierre, Pacalet frequented the Chauvet residence from 1986 to 1989, working in exchange for meals and winemaking know-how. Pacalet then studied chemistry and microbiology in Lyon and enology in Dijon, while working alongside his family in the Beaujolais. In 1991, Jacques Néauport introduced him to Burgundy scion Henri-Frédéric Roch, who sought a cellar master for his Vosne-Romanée estate. Over the next decade at Domaine du Prieuré Roch, Pacalet would establish his legend, proving natural winemaking could make *grands vins.*

Pacalet avoids sulfite addition during fermentation and aging, and decries the tannin addition widely practiced in Burgundy. His red winemaking is characterized by whole-cluster fermentation with daily pigéage. He emphasizes extended lees contact, going so far as to roll barrels during aging. He avoids racking and pumping. During the 1990s, as Pacalet's fame grew, he advised many other key natural wine estates, including Bordeaux's Château Meylet (see page 307), Thierry Allemand in Cornas, Hervé Souhaut in Saint-Joseph, and Domaine Gramenon (see page 230) in the Drôme.

In 2001, Pacalet left Prieuré Roch and founded a business that combines négociant winemaking with vineyard management services. Pacalet thus oversees the vines from which he purchases fruit. He says most of his work is practicing organic, but he prioritizes the wishes of his vineyard-owner clients, applying chemical treatments when necessary—to fight mildew, for example. For him, such is the price of admission to purchase and vinify grapes from famed appellations. He produces close to a hundred wines per year, ranging from Saint-Aubin to Chambolle-Musigny to Échézeaux. The zenith of his production is a Ruchottes-Chambertin of staggering, symphonic complexity. Pacalet's range is not limited to *grands vins.* He also produces crémant and no less than three cuvées of Beaujolais primeur for the Asian market, produced with his cousin, the Beaujolais négociant Christophe Pacalet.

"If you're going to learn guitar, it's better to learn with Jimmy Page than with some random guy. Jules Chauvet, and even my uncle Marcel Lapierre, they were my Jimmy Page. I was good friends with Philippe Laurent of Domaine Gramenon, too. And Jacques Néauport. I grew up with them. I'm a bit of a hybrid, their child."

Today Pacalet occupies a curious place in the pantheon of natural wine. His agnostic approach to organic farming, chaptalization, acidification, and yeast nutrients place him, like Jules Chauvet, in a lineage of winemaker's winemakers, rather than one of advocates of natural wine as such. Yet his influence and authority within the world of natural wine remains unquestionable.

THE WINE TO TRY

See Wines to Know: Burgundy, *page 195.*

Since: 2001

Winemaker: Philippe Pacalet

Vineyard surface: Varies, but generally just over 10ha (24.7 acres)

Viticulture: 90–95% practicing organic, according to Pacalet

Négociant work: Yes—it's all négociant work.

Grapes: Chardonnay, pinot noir, gamay, and more

Appellations: A plethora of Côte d'Or and Côte de Nuits appellations, plus Chablis, Moulin-à-Vent, Chénas, and more

Filtration/Fining: None

Sulfites: Sometimes 15mg/L at the end of élevage

Philippe Pacalet on Bringing Natural Wine to Burgundy

Burgundy négociant Philippe Pacalet's career has taken him from the Beaujolais to the northern Rhône to the heights of Burgundy, crossing the paths of many of natural wine's most significant figures along the way.

"I was raised in the Beaujolais, with my grandparents in Villié-Morgon. I had the luck to grow up with Marcel Lapierre. In 1978, when I was fourteen, he started to make the first barrels of wine without sulfites. It was five barrels. Until 1985, it was experimental.'"

"Later I worked for Nature & Progres. I did consulting in the northern Rhône valley, to develop organic agriculture and encourage vinification that was, let's say, a bit more natural. Two things happened to me then: I met Hervé Souhaut and Thierry Allemand, and I fell in love with syrah."

"Jacques Néauport and Marcel were like a couple. They made a good team, because Néauport had an intellectual side, focused on wine tasting. Meanwhile, I frequented Jules Chauvet's all the time between 1986 and 1989. He'd pay me in lunch at a little restaurant. On Saturdays, we were allowed to dine with the Chauvets—because it was the old France, you know."

"Roch had ideas. He wanted to make wine that way, and he gave me my chance. I arrived and we threw out all the sacks of tannins—which he had because at Romanée-Conti, they were adding tannins at the time. To make red Burgundy without sulfur is good, but to make Clos de Vougeot or the grand crus of Vosne-Romanée that way, you really had to have balls."

"I went to see Henri-Frédéric Roch in 1991. I brought a bottle of wine, a 1989 Morgon. That was my CV. I said, 'I know how to do this.' And they liked it. They hired me. They were surely more intelligent than me, but they'd never made wine in their lives. In 1991, I was twenty-five or twenty-six years old, but I'd made wine since I was fifteen."

DOMAINE DE CHASSORNEY

Frédéric Cossard

SAINT-ROMAIN

Hedonistic and highly opinionated, Frédéric Cossard is a larger-than-life figure in Burgundy, half paysan vigneron, half wine impresario. Cossard studied in the dairy industry before working for a decade as a courtier, arranging wine sales between winegrowers and négociants. Upon founding his estate in 1996, he began producing wine without additives, guided by his years of tasting in Burgundy. Reds ferment whole-cluster in wooden vats without punching down or pumpovers. Cossard formerly fermented both whites and reds in barrel, but in recent years he has switched to concrete eggs and Georgian qvevri. Keep an eye out for his qvevri-aged reds, which display the majesty of some of Burgundy's grandest appellations (Morey-Saint-Denis, Gevrey-Chambertin, etc.) in a wholly new light.

"Working with milk and making cheese with raw milk justified making wines without additives for me. Because cheese is a milieu that is much more fragile than wine."

Cossard's Burgundies possess an exquisite immediacy and finesse. His négociant work in the Beaujolais, done in conjunction with the sons of his friend Georges Descombes, is straightforward, crushable gamay. Following frost losses in 2016, Cossard expanded négociant work to the Languedoc and the Jura, as well as experimenting with skin-macerated whites.

THE WINE TO TRY

Auxey-Duresses–Les Crais

Les Crais derives not from chardonnay, but from old-vine, massal-selection pinot blanc. It shows a maritime profile, with a bright, seashell salinity to its vertical white fruit.

Since: 1996

Winemaker: Frédéric Cossard

Vineyard surface: 10.5ha (26 acres)

Viticulture: Practicing organic

Négociant work: Yes, since 2006, under the label "Frédéric Cossard"

Grapes: Pinot noir, chardonnay, pinot blanc; négociant work includes savagnin, poulsard, gamay, and more

Appellations: Saint-Romain, Auxey-Duresses, Savigny-les-Beaune, and more

Filtration/Fining: None

Sulfites: Depends on the year and the cuvée; small doses of 10–30mg/L are added when it's necessary to save a tank.

DOMAINE DERAIN

Dominique Derain, Julien Altaber, and Carole Schwab

SAINT-AUBIN

A native of the Côte Chalonnaise, Dominique Derain founded his estate with his then wife, Catherine, in 1987, after working as a cellar master for conventional estates in Meursault, Alsace, and Chablis. An audience with biodynamic researcher Maria Thun led Derain to embrace biodynamics (although he gave up Demeter certification in 1997). He was also influenced by the famed enologist Max Léglise, who, like Jules Chauvet, cautioned against overusing sulfites. Léglise accompanied Derain for the estate's first five vintages.

"After ten years in the wine industry, I said to myself, 'It's cheating, to put products in wine.' We sell terroir, but it's just sugar and wood and filtration."

—Dominique Derain

Dominique Derain (*left*) and Julien Altaber during the 2019 harvest in their En Remilly parcel

Vineyard acquisitions and rentals in Saint-Aubin premier cru parcels and in prestigious appellations like Pommard and Gevrey-Chambertin made the Derains some of the most high-profile natural vignerons in Burgundy for years. Derain avoided carbonic maceration, preferring to destem certain layers of the harvest, and practiced daily pigéage on his wooden vats. Both whites and reds age in used oak.

Dominique Derain officially retired in 2016, ceding management of the estate to his longtime cellarhand Julien Altaber. Altaber vinifies in the spirit of his mentor, and produces his own range of excellent négociant Burgundies under the "Sextant" label.

THE WINE TO TRY

Saint-Aubin–En Vesvau

Unmistakably structured, evoking pear, flint, and lanolin, Domaine Derain's En Vesvau derives from his first Saint-Aubin vineyard, a cool, southwest-facing slope behind the village.

Since: 1988

Winemaker: Julien Altaber since 2016

Vineyard surface: 8.5ha (21 acres)

Viticulture: Organic since 1989. Practicing biodynamic.

Négociant work: Derain produces wine in Chile and the Roussillon. Altaber produces négociant wine under the "Sextant" label.

Grapes: Pinot noir, chardonnay, aligoté, pinot beurot

Appellations: Saint-Aubin, Bourgogne, Gevrey-Chambertin, and more

Filtration/Fining: None in recent years

Sulfites: Sometimes none added, but more often 15mg/L at assemblage

DOMAINE DU PRIEURÉ ROCH

Famille Roch

PRÉMEAUX

Domaine du Prieuré Roch was the passion project of late Burgundy visionary Henri-Frédéric Roch, a scion of the Leroy négociant family, who co-own Domaine de la Romanée-Conti. Roch founded Prieuré Roch in 1988 with parcels of Vosne-Romanée vines purchased from Domaine de la Romanée-Conti. To perform the viticultural and winemaking work at his new estate, he hired Philippe Pacalet. Thus began ten years of experimentation in additive-free winemaking on the most hallowed terroirs of the Côte de Nuits. In 1992, Roch became comanager of Domaine de la Romanée-Conti, a role he occupied in parallel with his work at Prieuré Roch until his death in 2018.

> **"We prefer to say that at Prieuré Roch, the terroir is the winemaker. We are just keepers."**
>
> –Antonio Quari

Throughout the 1990s and 2000s, the Prieuré Roch wines were highly pedigreed outcasts in the world of fine Burgundy. Yet Roch's unassailable social status helped give his wines—and, by extension, natural wine itself—a strange authority. Clients farsighted enough to have invested in the wines in the early years have been rewarded by the estate's ascension, throughout the late 2000s and 2010s, into the cult stratosphere.

As Prieuré Roch's fame has grown, so has its vineyard surface, extending to 16 hectares (39.5 acres). But the Prieuré Roch legend remains with its premier cru and grand cru wines from Vosne-Romanée and Nuits-Saint-Georges.

Much has evolved since Philippe Pacalet left in 2001. Estate comanager and former cellar master Yannick Champ, who began as an intern in 2002, has since named his own cellar master, Antonio Quari, a Lake Como native who began as a harvester in 2010. The vinification period at Prieuré Roch is still overseen by Pacalet's erstwhile cellarhand Krzysztof Andrzejewski. The estate has made a practice of refusing cellar visits since 2017. Fermentation and vinification today take place in a cellar in Prémeaux, a five-minute drive from the cellar in which the wines are aged. Almost all red wines see whole-cluster fermentation with daily pigéage in wooden tanks, with maceration lasting generally around three weeks. Aging is conducted on lees without racking in oak barrels.

THE WINE TO TRY

Vosne-Romanée–Les Clous

The most immediate of Prieuré Roch's rather intellectual Vosne-Romanée wines, Les Clous stands out for its vivid red fruit and delicate aromas of graphite and porcini mushroom.

Since: 1988

Winemaker: Antonio Quari, in collaboration with Krzysztof Andrzejewski and Yannick Champ. Founder Henri-Frédéric Roch died in 2018. From 1991 to 2001, Philippe Pacalet was winemaker.

Vineyard surface: 16ha (39.5 acres)

Viticulture: Practicing biodynamic

Négociant work: None

Grapes: Pinot noir, chardonnay, gamay, aligoté

Appellations: Vosne-Romanée, Ladoix, Nuits-Saint-Georges, and more

Filtration/Fining: Direct-press whites have seen diatomaceous earth filtration in recent years.

Sulfites: None added during vinification. Wines generally receive doses of 20–30mg/L before bottling.

FANNY SABRE

POMMARD

Fanny Sabre began her wine career in 2000 as a protégé of Beaune négociant Philippe Pacalet; her father had died suddenly, leaving her with a coveted estate in Pommard at age sixteen. (She recalls neighbors showing up two days after the funeral to place offers to buy the vineyards.) Pacalet purchased the estate's fruit for his négociant winemaking business for the next five years, while Sabre worked for him and learned from his methods, converting her estate to organic viticulture all the while. She struck out on her own on 2006, and soon earned renown for her refined and supple Pommards.

Like her erstwhile mentor, Sabre practices whole-cluster fermentation in wood tanks for her red wines. She uses a system of frequent chilled pumpovers to slow fermentation and keep temperatures down. White wines age without racking or lees stirring in large 500-liter barrels, a surface-to-wine ratio which she says allows her to use 50 percent new oak without overly marking the wines. Red wines age in 30 percent new oak. Sabre is not averse to filtering white wines before bottling, and adds sulfites in microdoses in successive stages, rather than adding any at bottling. Sabre takes pains to keep her prices accessible, even for renowned appellations. Nor are her humbler wines anything to sniff at: both her Bourgogne rouge and her Bourgogne aligoté derive from vines within Pommard.

Since: 2006

Winemaker: Fanny Sabre

Vineyard surface: 6.5ha (16 acres)

Viticulture: Organic

Négociant work: None

Grapes: Chardonnay, pinot noir, aligoté

Appellations: Pommard, Volnay, Bourgogne, and more

Filtration/Fining: Early release whites sometimes see filtration.

Sulfites: 5mg/L after malolatic fermentation, and again after two to three months of aging

SARNIN-BERRUX

Jean-Pascal Sarnin and Jean-Marie Berrux

MONTHÉLIE

Sarnin-Berrux is a négociant house specializing in naturally vinified Burgundy, along with wines from the Rhône and the Beaujolais. Parisian financial officer Jean-Pascal Sarnin and Savoy-raised PR worker Jean-Marie Berrux met in the village of Saint-Romain, where Sarnin had moved to join his wife at the time, and where Berrux was working for a conventional estate. They bonded over a shared love of natural wine, which Sarnin had discovered at Paris caves à manger like Le Verre Volé and La Muse Vin.

All Sarnin-Berrux négociant purchases are in grapes, rather than juice or must. Whites see a slow press before barrel aging on fine lees. (Of particular interest are Berrux's two solo wines from his own vines in the Bourgogne appellation, Le Petit Tetu and Le Tetu, which show a structure and complexity to rival Puligny-Montrachet.) Reds see partial destemming and fermentation at ambient temperatures in wooden vats, with a modest rhythm of pigéage, before aging in used oak barrels.

The Sarnin-Berrux wines have been fixtures in Paris's natural wine scene for over a decade, offering many budding natural wine aficionados their first experience of natural wines from Burgundy's more vaunted appellations.

Since: 2007

Winemakers: Jean-Pascal Sarnin and Jean-Marie Berrux

Vineyard surface: Jean-Marie Berrux has 1ha (2.5 acres).

Viticulture: Berrux is practicing biodynamic. Grape purchases are mostly practicing organic.

Négociant work: All save Berrux's Le Tetu and Le Petit Tetu are négociant wines.

Grapes: Chardonnay, pinot noir, aligoté, gamay, syrah

Appellations: Saint-Romain, Beaune, Bourgogne Aligoté, and more

Filtration/Fining: None

Sulfites: Light sulfite dosage before bottling

CLOS DES VIGNES DU MAYNES

Julien Guillot

SAGY-LE-HAUT

Gifted with a lineage of organic advocacy, historic Clunisien vineyard sites, and a superhuman constitution, Mâconnais vigneron Julien Guillot is among the most dynamic natural winemakers of his generation. After careers in acting and construction management, Guillot took over his family estate in the Mâconnais in 1998. The Clos des Vignes du Maynes was already established in early organic wine circles, thanks to the farsighted refusal of his grandfather (and estate founder) Pierre Guillot to use synthetic chemicals, and to the instrumental role played by Julien's father, Alain Guillot, then president of the French national federation of organic agriculture, in the passage of France's organic labeling legislation. Julien Guillot initiated a conversion to biodynamic agriculture, reducing sulfite doses to their current low levels.

Guillot has expanded the estate beyond its initial 7 hectares (17.3 acres), leasing and acquiring parcels in Pouilly-Fuissé, the Côte Chalonnaise, and farther afield. He also began making négociant wines from the Beaujolais, including a stellar biodynamic Morgon from the Charmes climat. Guillot's red wines are mostly fermented in wood tanks, with layers of destemmed and whole-cluster fruit. He does pumpovers but no pigéage; aging occurs in foudre, barrel, and tank. White wines are aged in a mixture of barrel and tank, depending on the cuvée. The top wines of Guillot's range are Aragonite, an austere, old-vine chardonnay; Manganite, an intense gamay from a small-berried variant of the variety; and Auguste, a pinot noir from the rare pinot fin variant. Grapes from all three of these parcel cuvées go into Guillot's 910, a red-white blend made in the ancient style of the local Clunisien monks.

In addition to his significant wine production, Guillot produces tiny quantities of sought-after natural spirits, including long-aged Fine de Bourgogne and Marc de Bourgogne, along with a few experimental gins.

THE WINE TO TRY

See Wines to Know: Burgundy, *page 195.*

Since: 1952. Julien Guillot took responsibility for the estate in 2001.

Winemaker: Julien Guillot

Vineyard surface: 18ha (44.4 acres)

Viticulture: Biodynamic certified since 2005

Négociant work: Yes

Grapes: Chardonnay, aligoté, gamay, and more

Appellations: Mâcon-Villages, Saint-Véran, Pouilly-Fuissé, and more

Filtration/Fining: Occasional Kieselguhr filtration on early bottlings of white wines

Sulfites: 10–20mg/L at racking and/or bottling for whites; usually none added for reds

DOMAINE ROBERT-DENOGENT

Nicolas and Antoine Robert

FUISSÉ

Jean-Jacques Robert was among the first of Marcel Lapierre's friends to begin applying natural wine principles outside the Beaujolais—in this case, *right* outside the Beaujolais, in the Mâconnais village of Fuissé. There Robert bucked regional norms of overcropping, early harvesting, chaptalization, and commercial yeast use, instead producing expensively oaked parcel cuvées of chardonnay from old vines. The wines caught the attention of US importer Kermit Lynch, who became the estate's biggest client.

The estate didn't expand its following in natural wine circles beyond the Mâconnais and the Beaujolais. The wines are perhaps appreciated for being ripe, oaky chardonnays first, and natural ones second. Today Robert's charismatic sons, Antoine and Nicolas, continue the family's proto-natural approach of organic farming with abiding efforts to minimize sulfur use. White wines see débourbage before alcoholic and malolactic fermentation in barrels, followed by long aging—from eighteen to thirty months—in mostly new oak barrels. In 2012, the Roberts debuted their first gamay, the Cuvée Jules Chauvet, from the master's former vines.

Since: 1988

Winemakers: Nicolas and Antoine Robert. Formerly Jean-Jacques Robert.

Vineyard surface: 12.5ha (30.9 acres)

Viticulture: Organic; will be certified biodynamic by 2023

Négociant work: Yes, from organic growers in the Mâconnais

Grapes: Chardonnay, gamay

Appellations: Pouilly-Vinzelles, Mâcon-Fuissé, Pouilly-Fuissé, and more

Filtration/Fining: None

Sulfites: Lots destined for larger production wines see 10mg/L at press. Most wines also receive 10mg/L after malolactic fermentation and 5mg/L at bottling.

CATHERINE AND GILLES VERGÉ

VIRÉ

Catherine and Gilles Vergé are a couple of garrulous Mâconnais vignerons who have become unlikely leaders within the radical fringe of natural winemaking. They left the local cave cooperative (which Gilles's grandfather founded) in 1998, and found a following in Paris with their unclassifiable chardonnays. From their very old vines (70 to 120 years) within the Viré-Clessé appellation, the Vergés produce a menagerie of long-fermented, long-aged parcel cuvées, in which the customary flavors and aromas of Mâconnais chardonnay are twisted and refracted by what you might call creative oxidation. Grapes are harvested very ripe, before direct pressing, débourbage, and fermenting and aging in enameled steel tanks. During aging, in accordance with certain lunar phases, the Vergés open the rubber-corked tops of their tanks for specific spans of time to encourage oxidation.

The couple insist that their wines be sold only when ready to drink—so wines up to two decades old remain on the market. In 2012, the Vergés founded the natural wine association Les Vins S.A.I.N.S., which unites some of the most militant zero-zero vignerons of Europe (see page 118).

Since: 1998. Catherine and Gilles were winegrowers within the cave cooperative for fifteen years before then.

Winemakers: Catherine and Gilles Vergé

Vineyard surface: 4ha (9.9 acres)

Viticulture: Practicing organic

Négociant work: None

Grapes: Chardonnay

Appellations: Vines are within the Mâcon-Villages or Viré-Clessé appellations but are nowadays bottled as Vin de France.

Filtration/Fining: None

Sulfites: None added

MAISON VALETTE

Cécile and Philippe Valette

CHAINTRÉ

Maison Valette is a touchstone for natural white winemaking, renowned for its regal, long-aged chardonnays. A second-generation sharecropper, Gérard Valette converted to organic agriculture (uncertified) with the aid of his sons Baptiste and Philippe in the early 1990s. Philippe Valette, who took over part of the estate in 1992, would prove a visionary vigneron in his own right, installing a regime of biodynamic treatments, gradually eliminating sulfite addition during élevage, and forswearing filtration starting in 2006. For a long time, the estate's production was a confusing mix of wines farmed and vinified by Gérard and Baptiste, wines farmed and vinified by Philippe, and wines farmed by Gérard and Baptiste but vinified by Philippe, all with slightly different labels. As of 2016, the range has been united under the name Maison Valette. Wines are produced and aged throughout three cellar sites in Chaintré.

"For me, human intervention represents zilch for the quality of a wine. It's the work of the vines, the life of the soil, that will make the difference of terroir."

—Philippe Valette

Maison Valette's grapes are harvested at maximum ripeness. Neutral oak barrels are used to ferment and age all wines save the Mâcon-Villages, which ferments and ages in glass-lined cement tanks. Aging periods range from two years (for the Mâcon-Villages) to up to twelve years (!) for the oxidative Clos du Monsieur Noly. Once fixtures throughout the natural wine scene, the Valette wines have become rarer in recent years, as their increasing fame has coincided with a series of extremely low-yield vintages throughout the 2010s due to frost, mildew, and hail.

Philippe Valette outside his home

THE WINE TO TRY

See Wines to Know: Burgundy, *page 195.*

Since: 1977

Winemaker: Philippe Valette

Vineyard surface: 15ha (37 acres)

Viticulture: Practicing organic since 1990

Négociant work: None

Grapes: Chardonnay

Appellations: Mâcon-Chaintré, Mâcon-Villages, Pouilly-Fuissé, formerly Pouilly-Vinzelles, Viré-Clessé

Filtration/Fining: No filtration since 2006

Sulfites: A dose of around 20mg/L at bottling. Since 2015, certain cuvées have seen zero added sulfites.

BURGUNDY

Natural Wine Dining

Burgundian cuisine is as famed as its wines, embodied in dishes so iconic that they've become clichés of French Cooking 101. Unsurprisingly, quite a few involve wine, like the famous boeuf Bourguignon. (In the parlance of French cuisine, the suffix *à la Bourguignonne* or *Bourguignon* basically just means red wine is involved.) Just as famous are *escargots de Bourgogne*, a specific, environmentally protected race of snail oven-roasted with parsley, garlic, and heaps of butter.

The natural wine restaurants in Burgundy—more youthful than the tourist traps that dot the region—eschew these classics in favor of the flavors of Japan, Spain, and Italy. The best employ farm-raised meat and vegetables from within the region to create a rural vanguard of savvy natural wine dining. Here are the highlights.

La Dilettante

BEAUNE

Laurent Brelin (or "Lolo," as everyone calls him) runs this idiosyncratic, reservations-recommended wine bistro with his wife, Rika. Diners build meals around croque-monsieurs and *karaage* (Japanese fried chicken) while enjoying Brelin's mighty natural wine selection.

Le Soleil

SAVIGNY-LES-BEAUNE

Documentary filmmaker (and Dilettante alum) Lola Taboury-Bize founded Le Soleil in early 2020 in Domaine Simon Bize's former bed-and-breakfast, which offers an expansive interior courtyard. Taboury-Bize's scintillating wine list highlights cult natural winemakers alongside back vintages from her family's estate. Chef Laila Aouba trained at Le Saint Eutrope in Auvergne (see page 224) and brings a moving finesse to her vegetable-driven menus.

La Ferme de la Ruchotte

BLIGNY-SUR-OUCHE

Chef, farmer, and heavy metal fanatic Frédéric Ménager's lunch-only table d'hôte in Bligny-sur-Ouche is a gastronomic destination, one of very few restaurants where you can dine on meat from animals raised by the chef himself.

La Dilettante in Beaune

Legends in the Making

Yann Durieux of Recrue des Sens

Until as recently as the late 2000s, the natural wine scene in Burgundy comprised no more than a handful of estates. In the years since, a host of youthful, small-scale natural wine estates have taken root, and a surprising number of expatriate winemakers from the US, Australia, and England have joined Burgundy's blossoming natural wine community. Here are a few of the most exciting.

Renaud Boyer

MEURSAULT

The most radical, sulfite-forswearing organic vigneron working in the Côte d'Or, Renaud Boyer farms 3 hectares (7.4 acres) inherited from his uncle, the early organic vigneron Thierry Guyot, in the appellations of Bourgogne, Beaune, Saint-Romain, and Puligny-Montrachet. Short and resolute, Boyer has the air of a principled outcast, and indeed, he is among the only Burgundy winemakers who'll profess to drawing inspiration from an Alsatian—Bruno Schueller (see page 252), to be precise. Boyer's reds are piercing, crystalline pinots; his whites, at their best, are as vertical and luminous as watchtowers.

La Maison Romane

Oronce de Beler

NUITS

Oronce de Beler worked in Paris and London for the French wine institution the Revue de Vin de France before establishing a boutique négociant natural winemaking project in Vosne-Romanée in 2004. Since then, he's diversified into offering horse-plowing services and, more recently, raising black pigs and making craft beer. Save for two cuvées from the Hautes Côtes de Nuits, de Beler's wines mostly derive from renowned appellations like Chablis, Gevrey-Chambertin, and Vosne-Romanée. Reds see partial or full destemming. All wines are bottled without filtration or sulfite addition.

Recrue des Sens

Yann Durieux

VILLERS-LA-FAYE

Burgundy wunderkind Yann Durieux put in a decade of work with the likes of Domaine du Prieuré Roch and Julien Guillot before he began producing wine from family vines in the Hautes Côtes de Nuits in 2010. Since his first wine, an expressive aligoté called Love and Pif, Durieux has gone on to found a sprawling, Philippe Pacalet–like business, working as a négociant and a vineyard contractor on renowned terroirs throughout the Côte de Nuits. His expensive wines are all unsulfited, unfiltered, and bottled as Vin de France; their names give only the faintest reference to their terroirs of origin.

Le Grappin

Andrew and Emma Nielsen

BEAUNE

Andrew and Emma Nielsen changed careers from advertising and banking, respectively, to embrace the lifestyle of micro-négociant winemakers in Beaune. In 2011, they set up shop in the former cellar of Philippe Pacalet, after an internship with Domaine Simon Bize in Savigny (among other conventional estates in California, Australia, and New Zealand). Savvy and charismatic, the Nielsens are linchpins of the expat natural winemaking scene, impressing all comers with their work ethic and soft, detail-driven touch with the wines of Savigny-les-Beaune, Beaune, and the Beaujolais.

Vin Noé
Jon Purcell

GAMAY

A bearded Southern Californian with the bearing of a Norse god, Jon Purcell interned for Philippe Pacalet before adopting the life of a tractor driver in Burgundy, working for conventional estates. He began making négociant wine in borrowed cellar space in Auxey-Duresses in 2014, and distinguished himself with a ripe, extractive, yet impeccably natural (unsulfited and unfiltered) winemaking style. As of the 2020 vintage, he's produced Saint-Aubin and Puligny-Montrachet from vines rented from retired natural vigneron Jean-Jacques Morel.

Domaine de la Cras
Marc Soyard

PLOMBIÈRES-LÈS-DIJON

Jura native Marc Soyard worked for six years for Jean-Marc Bizot in Vosne-Romanée before the opportunity arose in 2014 to manage this 8-hectare (19.8-acre) estate just outside Dijon. Here, a ten-minute drive from the city center, Soyard has made his name among the most talented vignerons of his generation, producing masterful low-sulfite, unfiltered pinot noir and chardonnay from terroirs that were neglected for a generation. Keep an eye out for his unsulfited wines, made under the "Equilibriste" label.

Domaine Dandelion
Morgane Seuillot and Christian Knott

MAVILLY-MANDELOT

A native of the Hautes Côtes, Morgane Seuillot worked for several years in Australia before returning to Burgundy and purchasing a small surface of vines. Her husband, Christian Knott, maintains a day job as the cellar master of Domaine Chandon de Briailles (where he also produces a range of immaculate natural wines). Together as Domaine Dandelion, they make stellar pinot noir and aligoté, with an artisanal approach that extends to handmade paper labels and beeswax seals.

Jon Purcell devatting a small tank of Côtes de Nuits

The view from Vincent and Marie Tricot's hilltop vineyards

8.
Auvergne

An overlooked, half-industrialized region in central France, Auvergne has in the last decade become the unlikely ground zero for a radical natural wine new wave seeking to explore its varied volcanic terroir.

Despite its history of winemaking, Auvergne represents sort of a blank canvas. Its wine production peaked before the turn of the twentieth century, when its winegrowers benefited from increased demand due to the earlier onset of phylloxera farther south. The vine pest soon made its way to Auvergne, however, and the region's viticulture, never very renowned or quality-oriented, was abandoned in favor of industrial grain agriculture. Many Auvergnats quit agriculture to work in factories, or departed for Paris, where they retain a strong hold on the city's bistro culture.

Put it this way: If you make natural wine in Burgundy or the Loire, you're working in the shadows of the great wine estates of yesteryear. If you make natural wine in Auvergne, you're working in the shadows of grain silos and tire factories.

That's what the IT guy turned pioneering vigneron Jean Maupertuis did when he began making natural wine in Auvergne in 1993. A few years later, he partnered with Stéphane Majeune and Eric Garnier to form Domaine de Peyra, a groundbreaking collaboration that continued until 2006. Its tiny, unsulfited wine production was like the Velvet Underground of Auvergne natural winemaking, attaining massive influence via the more successful estates it inspired. Today early Peyra acolyte Patrick Bouju has become a wine-media sensation, while fellow traveler Pierre Beauger's intense, exotic elixirs are cult items from Singapore to San Francisco.

The region's natural wine renaissance is, for now, a storm in a wineglass: a lot of media hubbub and international fandom for what is ultimately a minuscule amount of wine. What's stopping Auvergne's natural winemakers from making more wine? Mostly it's an entrenched reluctance among the locals to sell land. There aren't a lot of vines, and even virgin plots are fiercely guarded by local families in hopes the land will be approved for new construction. Urban sprawl and industrial grain agriculture often outcompete the region's shining potential for quality winemaking.

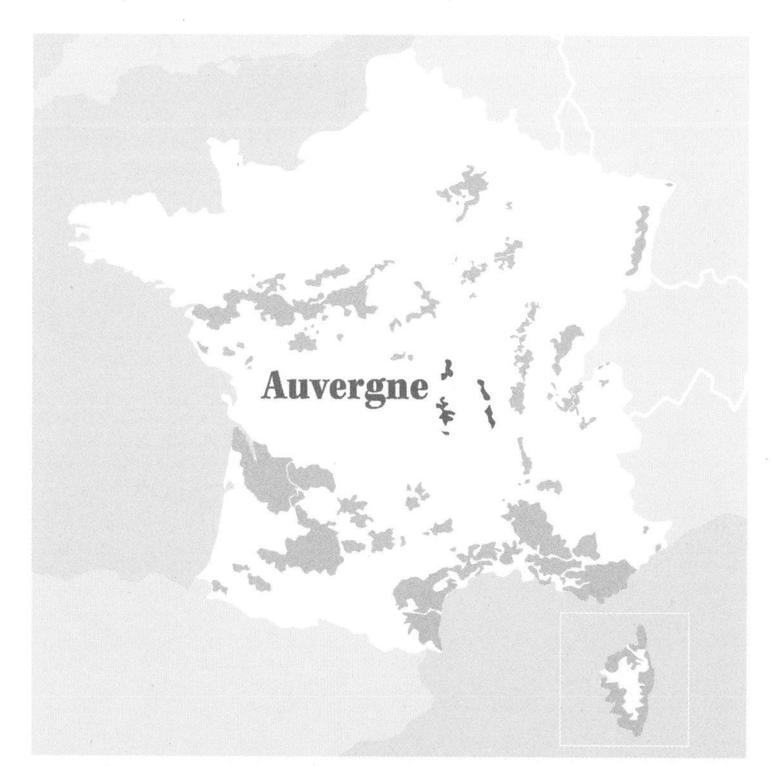

Total vineyard surface:
800ha (1,977 acres)[1]

Rate of organic agriculture:
19 percent[2]

Key grapes:
Gamay d'Auvergne, gamay, pinot noir, chardonnay

1 France 3
2 Agence Bio 2020. Puy-de-Dome department figures from Vitisphere.

AUVERGNE

Wines to Know

Reds predominate in Auvergne: gamay in all styles, from pale and fruity to inky and gnarly, along with a fair amount of pinot noir. Connoisseurs prize the native gamay d'Auvergne for its ashy, reductive notes, a reflection of the local volcanic terroir. Whites from the region are rare, limited to chardonnay and more recent plantings of sauvignon and pinot gris. In Auvergne, however, expect the unexpected: freethinking vignerons like Pierre Beauger and Aurélien Lefort are forever pushing stylistic boundaries.

Jean Maupertuis | Vin de France–Les Pierres Noires

Pathbreaking Auvergne winemaker Jean Maupertuis's Les Pierres Noires shows what old-vine gamay d'Auvergne can give when vinified cool carbonic, in the manner of the great natural winemakers of the Beaujolais. Low extraction and intracellular fermentation lend more dark-cherry fruit to gamay d'Auverge, without ever masking its native sappiness or firm structure.

Patrick Bouju | Vin de France–La Bohème

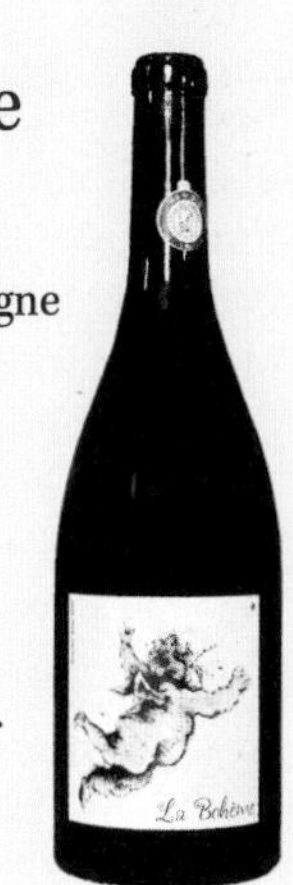

The flagship estate wine from Auvergne natural winemaking icon Patrick Bouju, La Bohème is a destemmed, old-vine gamay d'Auvergne that tends to macerate for upwards of two months before barrel aging for over a year. In youth, it can show a deep, forbidding reduction. Look for old vintages, though—they uncoil to reveal a savory majesty.

Pierre Beauger | Vin de France–Rotten Highway

The uninitiated tend to mistake Pierre Beauger's white wines first for pétillants naturels (thanks to their rustic bottle-cap closures and clear glass), and then for orange wines, thanks to their vivid, squash-skin hues. But they belong to no discernible category. Rotten Highway, from Beauger's young-vine sauvignon, seems to derive its opulent, concentrated tropicality from extreme ripeness and some drying out of the grapes before press. It's nominally a dry wine. But it possesses an intensity reminiscent of a Recioto di Soave.

JEAN MAUPERTUIS

SAINT-GEORGES-SUR-ALLIER

Middle-aged, modest, and soft-spoken to the point of inaudibility, Jean Maupertuis was the earliest of Auvergne's natural winemakers, the critical link between Marcel Lapierre's founding generation of natural winemakers in Villié-Morgon and their successors in Auvergne.

In 1991, Maupertuis left a job in IT to pursue a winemaking degree at Mâcon-Davayé, just north of the Beaujolais. There he studied alongside Rennes wine retailer Éric Macé, who introduced him to Marcel Lapierre's circle of vignerons. Maupertuis credits Lapierre's cellarhand Didier Ferret with showing him the finer points of natural vinification. Maupertuis's own career began in 1993—six years before his later, more influential work in collaboration with Eric Garnier and Stéphane Majeune at Domaine de Peyra.

Today Maupertuis still practices a variant of the whole-cluster, sulfite-free vinification he learned in the Beaujolais. But unlike his Beaujolais forebears, he doesn't add sulfites at bottling or refrigerate the harvest to cool it down before vatting. Reds like La Plage and Les Pierres Noires can be a little gnarly and reduced in their youth, revealing the full breadth of their spiced, dark-cherry fruit only after several years. Maupertuis also produces one of France's most exquisite pétillants naturels in his gamay d'Auvergne Pink Bulles.

THE WINE TO TRY

See Wines to Know: Auvergne, *page 215.*

Jean Maupertuis (*left*) chatting with his neighbor Patrick Bouju after a near-miss from a hailstorm

Since: 1993

Winemaker: Jean Maupertuis

Vineyard surface: 5.5ha (13.6 acres)

Viticulture: Organic

Négociant work: None

Grapes: Pinot noir, gamay d'Auvergne, gamay, chardonnay

Appellations: All wines released as Vin de France.

Filtration/Fining: None

Sulfites: None added

Domaine de Peyra

Stéphane Majeune pouring wine at a tasting at Caves Augé in Paris

Founded in 1999 by early Auvergne vignerons Stéphane Majeune, Eric Garnier, and Jean Maupertuis, Domaine de Peyra led only a brief, roller-coaster existence of just six years. But it made a lasting impression on natural wine fans in France, confirming the potential of Auvergne's forgotten terroirs.

At the estate's start, Maupertuis (who still produces wine today) and Garnier were both recently installed vignerons. Maupertuis had already been initiated into natural wine circles thanks to an internship with Morgon vigneron Marcel Lapierre. Majeune, a young Auvergnat friend of Garnier's, followed in Maupertuis's footsteps, attending winemaking school in the Mâconnais and interning for Lapierre acolytes Jean-Paul Thévenet and Guy Breton. Maupertuis, Garnier, and Majeune joined forces in 1999, separately farming a combined 12 hectares (29.7 acres) of almost exclusively gamay vines but sharing equipment and winemaking decisions. Maupertuis handled administration, while Garnier took the lead in the cellar, and Majeune took up commercial responsibilities.

Influenced by natural Beaujolais, the Peyra gamays were unfiltered and unfined, and saw a light semi-carbonic maceration. But the Peyra wines went further. The trio ceased adding sulfites after their first vintage, and bottled parcel cuvées, seeking out the nuances of their newfound terroirs. The wines were ambitious in every sense except pricing. Peyra found fans in influential buyers like Nicolas Carmarans of Café de la Nouvelle Mairie (see page 37), Olivier Camus at Le Baratin (see page 34), and Pierre Jancou (see page 43).

The financial risk inherent in natural winemaking is never easy for vignerons working alone. For a partnership like Peyra, it proved impossible. After severe hail losses in 2003, it was decided that Majeune would purchase the shares of the other two partners. He continued for two more years, but ceased production in 2006. Majeune went on to found the Auvergne wine salon Les 10 Vins Cochons (see page 222), and today runs a bistro called Mauvaises Herbes in the Yonne town of Sens. Garnier, embittered by the failure of Peyra, also stopped producing wine. Jean Maupertuis continues as a vigneron in Saint-Georges-sur-Allier. Today the wines of Peyra are cult commodities among natural wine geeks—in no small part because Paris restaurateur Ewan Lemoigne purchased the entirety of the remaining stock when the estate folded.

PIERRE BEAUGER

MONTAIGUT-LE-BLANC

Mercurial, innovative, and possessed by his convictions, Auvergnat vigneron Pierre Beauger has become an icon for radical natural wine believers the world over. His unique, highly artisanal wines represent a through-the-looking-glass moment for many drinkers: to taste them is to confirm the possibility of abstraction in winemaking.

Beauger led a peripatetic career in vineyard and cellar work during the 1990s, crossing paths with Thierry Puzelat during winemaking school and later at Domaine de la Tour du Bon in Bandol before going on to work in California and Chile. In 2002, he began producing wine from rented vines not far from Clermont-Ferrand. His two key wines from this era, Champignon Magique and V.I.T.R.I.O.L., from chardonnay and gamay, respectively, showed an unmistakable style: low yields, diabolical ripeness, and a willingness to bottle juice that was perceptibly *alive*. Beauger's quest for purity met a perfect collaborator in Paris restaurateur Pierre Jancou, who introduced those early wines to an international audience.

Having never gotten along with the owner of his vineyards, Beauger gave them up after 2010 to focus solely upon his own plantings of pinot noir, pinot gris, sauvignon, and syrah near the village of Montaigut-le-Blanc. In this second stage of his career, his wines are marked by an eye-popping sumptuousness, seemingly the result of a regimen of air-drying late-harvested grapes.

"I do just a little skin maceration on whites. Everyone does it now. When everyone does something, I change. I bicycle, for example. When everyone bicycles, I'll drive a car."

THE WINE TO TRY

See Wines to Know: Auvergne, *page 215*.

Pierre Beauger (*left*) in his garden with his neighbor and acolyte Mito Inoue

Since: 2002

Winemaker: Pierre Beauger

Vineyard surface: 1.5ha (3.7 acres)

Viticulture: Organic

Négociant work: Yes, after hail in 2013, with grapes purchased from Bruno Schueller, Stéphane Bannwarth, and Bruno Allion

Grapes: Formerly chardonnay and gamay d'Auvergne. Since 2010, sauvignon, pinot gris, syrah, and pinot noir.

Appellations: All wines are released as Vin de France.

Filtration/Fining: None

Sulfites: None added

DOMAINE DE LA BOHÈME

Patrick Bouju

SAINT-GEORGES-SUR-ALLIER

With his sunny disposition and indefatigable curiosity, Patrick Bouju is the de facto ringleader of Auvergne's natural wine circus, the most enterprising of the region's vignerons.

Originally from Tours and allergic to sulfites, Bouju discovered the natural wines of Clos du Tue-Boeuf and Marcel Lapierre as a chemistry student in Rennes in the mid-1990s. Following military service in southern Burgundy, he studied winemaking in Beaune, before moving to his then wife's native Auvergne. There he set about acquiring vines and making wine on an amateur scale, becoming an early contemporary of Jean Maupertuis and Domaine de Peyra. He began producing wine on a commercial scale in 2002, but until 2008 retained a part-time job working for ÉDF, the French electricity giant.

Compared to most of his peers, Bouju is largely self-taught when it comes to winemaking. Beyond a total rejection of sulfite addition, his approach is undogmatic and instinctual. For his estate reds, from vineyards scattered throughout Corent, Billom, Lempdes, and beyond, he prefers machine destemming and long macerations, often up to two months, before aging in old oak. Wines like La Bohème, Lulu, and Violette are tightly coiled creations that benefit from significant bottle aging. Low yields, hail, and sheer demand have made these intense gamay d'Auvergne and pinot noir wines rare in recent years.

Today Bouju is better known for his vast and varied négociant winemaking, which includes fruit purchased throughout France, along with collaborations with Greek vigneron Jason Ligas and the US rapper Action Bronson.

These négociant wines are often cross-regional assemblages employing a panoply of vinification and aging vessels, to inconsistent effect. Yet Bouju's own enthusiasm has led a whole new generation of natural wine drinkers to embrace his freestyle, exploratory natural winemaking.

THE WINE TO TRY

See Wines to Know: Auvergne, *page 215*.

Since: 2002

Winemaker: Patrick Bouju

Vineyard surface: 7ha (17.3 acres)

Viticulture: Organic

Négociant work: Yes, lots, from throughout France and even elsewhere in Europe

Grapes: Estate cuvées are from gamay d'Auverge, chardonnay, and pinot noir, along with several forgotten local varieties. Négociant work runs the gamut of the grapes of France and beyond.

Appellations: All wines released as Vin de France.

Filtration/Fining: None

Sulfites: None added

DOMAINE DE L'ARBRE BLANC

Frédéric and Caroline Gounan

SAINT-SANDOUX

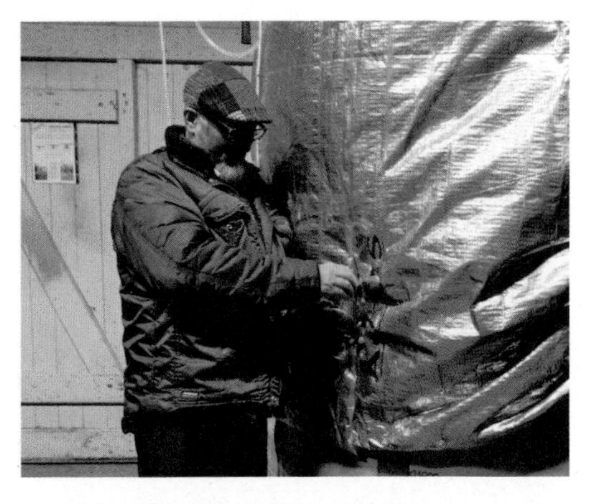

Fréderic Gounan pulling sample of Les Fesses

Half hard-nosed realist, half madcap uncle, Frédéric Gounan left a career in motorcycle production in 1999 to devote himself to winemaking in his native Saint-Sandoux. He had a hunch it was good terroir based on childhood memories of the taste of the potatoes that grew there.

While dividing his time between a winemaking course in Beaune and planting the pinot noir that would form the heart of his small estate, Gounan fell under the influence of the natural and biodynamic winemakers of turn-of-the-century Burgundy, including Dominique Derain, Jean-Claude Rateau, and Emmanuel Giboulot, for whom he interned. From 2000 to 2009, Gounan produced old-vine gamay d'Auvergne from rented vines. Finding the goblet-trained vines difficult to work, he gave them up in 2010, by which time his pinot noir had come into production. A half hectare (1.24 acres) of sauvignon and pinot gris came into production in 2013; it goes into his evocatively named orange wine Les Fesses (translation: "the butts"—it's the actual cadastral name of the vineyard). His partner, Caroline, a horticulturalist, joined the estate in 2008, taking the lead in vineyard work.

Gounan crops his lyre-trained pinot noir vines somewhat high. He practices about two weeks' partially destemmed maceration in a wood tank before aging his wines for two years, first in tank, then in old oak barrels. His favorite barrels become Les Grandes Orgues, while the rest become Les Petites Orgues. Gounan does no négociant work. He spends his free time on other pursuits, such as constructing the cute handcrafted drag racer in which he and Caroline travel.

"You have grapes that fall in a tank, that get a bit crushed, that start to transform, that turn into wine. If you don't intervene, it'll go to the end of the process, which means vinegar. We try to find an equilibrium before then."

—Frédéric Gounan

THE WINE TO TRY

Vin de France–Les Petites Orgues

Les Petites Orgues is the Gounans' calling card; a nervy, unctuous pinot with character to spare. Its eminent drinkability testifies to Gounan's light touch in vinification.

Since: 2000

Winemaker: Frédéric Gounan

Vineyard surface: 1.6ha (4 acres)

Viticulture: Organic

Négociant work: None

Grapes: Pinot noir, sauvignon, pinot gris. Formerly gamay until 2009.

Appellations: All wines released as Vin de France

Filtration/Fining: None

Sulfites: None added

VINCENT AND MARIE TRICOT

ORCET

In the Clermont-Ferrand suburb of Orcet, Vincent and Marie Tricot enjoy an admirable working situation. Their 4.5 hectares (11.1 acres) of vines have long been farmed organically, and are all situated on one volcanic-debris-influenced clay-limestone hill.

The Tricots' initiation to natural wine came from Brouilly's Patrick "Jo" Cotton, a longtime friend of both their families who began experimenting with zero-sulfite vinification in 1996. Vincent worked for Patrick's brother Guy Cotton (father of Côte de Brouilly wunderkind Pierre Cotton) before working in conventional farming and vinification at estates in the Rhône, Costières de Nîmes, and Chile. In 2003, the Tricots purchased their estate from a retiring organic vigneron. They began, from their first vintage, producing zero-sulfite cuvées, and were encouraged by their peers of the era, Pierre Beauger and the trio behind Domaine de Peyra.

> **"I don't aim to make wines that age for ten years. I'm proud of wines that hold up after five or six years. I like to make good, modest, bistro wines. That way I don't have pretentious guys coming to visit."**
>
> —Vincent Tricot

Today Vincent estimates that 95 percent of their wines see zero sulfites added. For reds, they practice carbonic maceration on gamay, and sometimes partial destemming on pinot noir. Whites, sometimes cold fermented, are produced without skin maceration.

THE WINE TO TRY

Vin de France–Les Milans

Les Milans is a carbonic-macerated gamay d'Auvergne blended, after pressing, with semi-carbonic-macerated pinot noir. Aged in old barrels, it's a soulful Auvergne rouge that retains its makers' Beaujolais sensibilities.

Vincent Tricot leading a tasting

Since: 2003

Winemaker: Vincent Tricot

Vineyard surface: 4.5ha (11.1 acres)

Viticulture: Organic

Négociant work: Yes, in certain vintages

Grapes: Pinot noir, gamay d'Auvergne, chardonnay, and more

Appellations: All wines are released as Vin de France.

Filtration/Fining: None

Sulfites: 95% of wines see zero sulfites added, according to Vincent Tricot.

TASTING DESTINATIONS
AUVERGNE NATURAL WINE SALONS

Auvergne's natural wine salons are for the diehards. What these tastings lack in accessibility, however, they gain in a sense of community.

Les 10 Vins Cochons

When: The first week of December

Who: Stéphane Majeune and knife-maker and charcutier Emmanuel Chavassieux founded the salon in 2003, but left in 2016; the salon is now organized by an association of volunteers.

For: Open to the public.

Where: The location shifts.

The first and most influential natural wine salon of Auvergne, Les 10 Vins Cochons translates to either "The Ten Pig Wines" or "The Divine Pigs." It attracts natural wine fanatics willing to trek to isolated rural locations in December.

La Fête du Vin

When: Late July

Who: Founded by chef Harry Lester, the salon is now run by an association of local natural vignerons.

For: Open to the public.

Where: The location shifts.

La Fête du Vin unites the natural winemakers of Auvergne for a festival and dinner executed with typical Harry Lester panache.

Les Vins au Vert

When: Late January

Who: The salon brings together a handful of radical natural winemakers who prefer to remain apart from the simultaneous professional salons in the Loire.

For: Open to the public.

Where: Glaine-Montaigut

Les Vins au Vert is held in Pierre Beauger's village, and is the rare salon at which he still presents wines.

PRECURSORS

Harry Lester

Auvergne wouldn't be the natural wine destination it is today without British chef Harry Lester. Leaving a lauded career at his London gastropub Anchor & Hope in 2007, Lester moved with his family to the Auvergne village of Chassignolles to take over a bed-and-breakfast. Thus was born the Auberge de Chassignolles (see page 224). Lester's research into quality local ingredients brought him into contact with the region's nascent winemaking community, which embraced his rigorous and inventive market cuisine.

Lester moved to Clermont-Ferrand and opened his bistro Le Saint Eutrope (see page 224) in 2013. A sister wine bar, Le Quillosque, followed on the same street in 2015. The restaurants share a kitchen, and they remain the only dining destinations in Clermont-Ferrand.

Harry Lester, tasting in the Jura

LE PICATIER

Christophe and Géraldine Pialoux

SAINT-HAON-LE-VIEUX

Le Picatier vignerons Christophe and Géraldine Pialoux are not quite in Auvergne. They're in the Côte Roannaise, a small appellation outside the town of Roanne, stranded, culturally, between Auvergne and the Beaujolais. The thin-soiled, sandy-granite, sparsely planted terroir more resembles the nonvolcanic stretches of Auvergne than the Beaujolais, and the Pialouxs' highly characterful wines bear out the comparison.

"I'm more peasant than farmer. The farmers are what killed peasant culture. I don't defend farmers."

—Christophe Pialoux

Christophe Pialoux was raised in the Auvergne town of Brioude and met Géraldine, a Côte Roannaise native, during landscaping school. They moved to southern Burgundy afterward, where Christophe found work for the Burgundy négociant Maison Joseph Drouhin. He grew disgusted with the chemical agriculture and vinification practiced there, preferring the approach of neighbors like Dominique Derain and Julien Altaber. The Pialouxs practiced low-sulfite vinification, organic farming, and nonfiltration since establishing their own estate in 2007. A series of catastrophic vintages from a mixture of hail, frost, and dryness, however, led them to abandon all sulfite addition—they figured there was so little wine, why even bother? This second stage of their wine production has been more rewarding: insanely low yields resulted in concentrated, radically pure wines. Today their chardonnay is deep, oxidative, gassy, often with a lick of volatility. It's all the more exciting for it.

Géraldine (*left*) and Christophe Pialoux at Le Picatier

Since 2011, the Pialouxs have also abandoned plowing and mowing in favor of rolling grass cover to preserve humidity in increasingly dry seasons.

THE WINE TO TRY

Vin de France–Auver-Nat-Noir

A divinely sappy country pinot with the targeted force of a cruise missile, Auver-Nat-Noir possesses a depth and purity that exceed those of much of the Côte de Nuits.

Since: 2007

Winemaker: Christophe Pialoux

Vineyard surface: 8ha (19.8 acres)

Viticulture: Organic, without plowing

Négociant work: Yes

Grapes: Chardonnay, pinot noir, gamay

Appellations: Wines are all released as Vin de France.

Filtration/Fining: None

Sulfites: None added since 2012

AUVERGNE

Natural Wine Dining

Auvergne's limited dining scene is a reality check for the region's latter-day renaissance. There are just three excellent restaurants in the region, all founded by British chef Harry Lester, and they all keep peculiar opening hours. Don't hesitate to plan your trip around them.

Le Saint Eutrope

CLERMONT-FERRAND

At the supremely tasteful, understated bistro he founded in 2013, Harry Lester revisits French country cooking with devoted mastery and flashes of quiet irreverence, like the house-fermented bean curd paste that accompanies a duck pot-au-feu. His seven-hour-roasted lamb shoulder is a succulent, textured masterpiece, and the radical wine list unites the crème de la crème of the Auvergne and beyond.

Le Quillosque

CLERMONT-FERRAND

Situated beside Le Saint Eutrope, and sharing its kitchen, is Le Quillosque, a wine bar that offers essentially the same cornucopia of stellar cuisine, perfect service, and zero-zero wines in a more informal setting. On Saturday afternoons—the last service of the week for Lester's restaurants—you usually encounter a handful of Auvergne vignerons propping up the bar.

Auberge de Chassignolles

CHASSIGNOLLES

When Lester left Chassignolles in 2013, he sold this seasonal bed-and-breakfast and restaurant to his friend Peter Taylor, former owner of Bristol restaurant Riverstation. Taylor has ably upheld the unique Chassignolles style with a series of seasonal service and kitchen teams assembled from natural wine fans throughout the world. Auberge's rooms are simple, sunlit, elegant; its restaurant is filled with a sophisticated crowd of industry types, for whom it functions as a peaceful, insidery getaway. It's open from May until early October each year.

The Auberge de Chassignolles

AUVERGNE

Legends in the Making

It is imposingly difficult to acquire vines in Auvergne, but this has not discouraged the influx of young vignerons drawn to the region in recent years by its underexplored terroir and the charisma of its tight-knit natural wine community. Many of these new winemakers produce wine destined to travel no farther than a handful of Parisian restaurants. Yet several younger vignerons have already established reputations as future greats.

Aurélien Lefort examining a maceration of négociant roussanne inside his cellar in Madriat

François Dhumes

ORCET

Low-key Auvergne native François Dhumes began producing wine in 2006 while doing vineyard work for friends like Vincent and Marie Tricot and Patrick Bouju. Today he farms 5 hectares (12.3 acres). From his home and winery in an industrial hangar on the outskirts of Orcet, Dhumes produces about eight cuvées, including a range of nuanced, unsulfited reds that comprise a master class in the differences between Beaujolais gamay and old-vine gamay d'Auvergne. Keep an eye out for Minette, a revelatory Burgundian-style maceration of gamay d'Auvergne aged in old oak.

Aurélien Lefort

MADRIAT

Parisian-born, with a degree in fine arts, Aurélien Lefort cuts an original figure among the paysan farmers of the natural wine world. After internships with Loir-et-Cher biodynamic natural vigneron Michel Augé and Patrick Bouju, Lefort took on a few scrappy parcels of remote vines and released his first wines in 2012. Today he farms 3 hectares (7.4 acres) and indulges in the occasional négociant wine with grapes from as far afield as Savoie and Abruzzo. Lefort's wines are unique, thanks to late harvests, radically long maceration times, and high proportions of direct-press juice in his red wines.

No Control

Vincent Marie

VOLVIC

A towering, tattooed Bad Religion fan from Normandy, Vincent Marie worked in sales and marketing and nurtured a longtime love of natural wine before diving headlong into a winemaking degree in Alsace in 2012. There he cultivated internships and work experiences with the likes of Bruno Schueller (see page 252) and Patrick Meyer (see page 250), cementing his beliefs about biodynamic agriculture, sulfite-free vinification, and oxidative aging. Attracted by the natural wines he knew from Auvergne, Marie acquired vines north of Clermont-Ferrand and produced a tiny first vintage in 2013. He presently farms 4.5 hectares (11.1 acres), producing a range of artisanal spirits and a chardonnay alongside intense, bristly reds from gamay, syrah, and pinot noir.

A view of the Cornas vineyards
of Thierry Allemand

9. The Rhône

The Rhône wine region is a bugle-shaped zone extending southward from Vienne to Avignon. From the steep, craggy granite zones in the north, one passes through the epic, rocky greenery of Ardèche into the baking, sandy flatlands surrounding Avignon.

Each subregion of the Rhône has its own history within natural wine culture.

René-Jean Dard and François Ribo were classmates of Jacques Néauport in Beaune, and their stubborn devotion to additive-free winemaking set them apart from the rest of their generation in the northern Rhône.

Their influence in Rhône natural wine is rivaled only by that of Michèle Aubéry and her late husband, Philippe Laurent, who began producing iconic grenaches at Domaine Gramenon in the Drôme in the late 1980s. A decade later, their neighbor Marcel Richaud in the Vaucluse would follow them in embracing natural vinification, aided by the enologist Yann Rohel (page 323).

Around the same time, in southern Ardèche, the peripatetic natural winemaker Jacques Néauport began advising Gérald and Jocelyne Oustric of Domaine le Mazel. The Oustric siblings soon converted their 30-hectare (74-acre) estate to organic viticulture and natural vinification. Then, as now, they sold a portion of their grapes to upstart natural winemakers in the region, laying the groundwork for a natural wine renaissance in Ardèche. Their brother-in-arms throughout these years was vigneron Gilles Azzoni, who had begun producing natural wines on the other side of Ardèche.

Meanwhile, at the cooperative Cave des Vignerons d'Estézargues in the Gard—the hinterlands between Nîmes and Avignon, divided, administratively, between the Languedoc and the Rhône—cellar master Jean-François Nicq encouraged the cave's winegrowers to convert to organic viticulture. During his tenure from 1989 to 2001 at Estézargues, Nicq, a friend of Thierry Puzelat, introduced natural winemaking to a winegrower named Eric Pfifferling. In 2002, Pfifferling founded his Tavel estate Domaine l'Anglore, which would soon redefine natural rosé vinification and inspire a generation of acolytes.

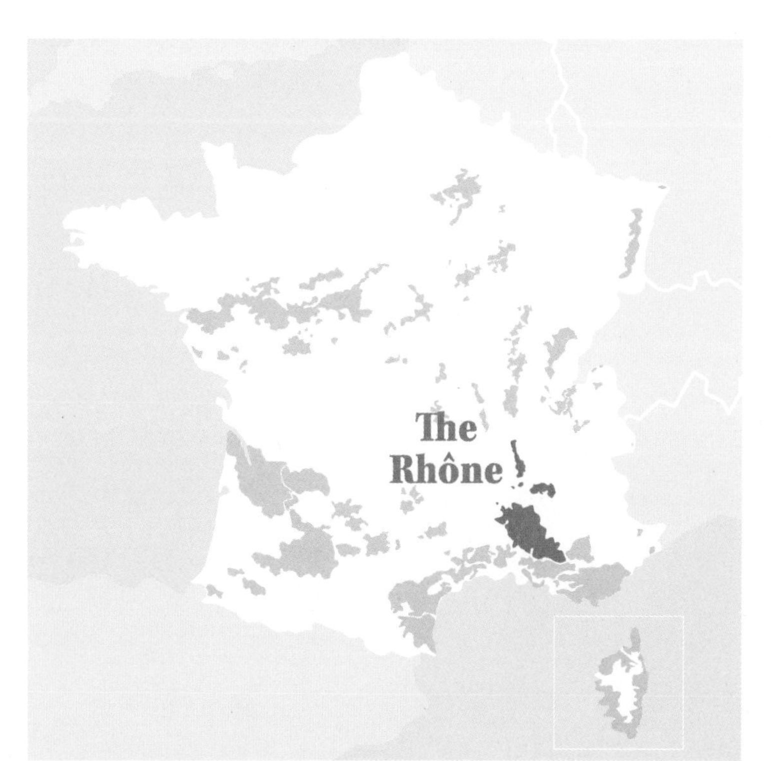

Total vineyard surface:
67,628ha (167,112 acres)[1]

Rate of organic agriculture:
13 percent[2]

Key grapes:
Viognier, roussanne, marsanne, syrah, grenache, cinsault, mourvèdre, counoise, gamay

1 Syndicat Côtes du Rhône 2019
2 Vins Rhône 2020

Wines to Know

From the natural wine heavyweights of the northern Rhône, expect a unique marriage of power and buoyancy, a more airborne style of syrah than you might be used to. In Ardèche, you'll encounter fresh, rugged, value-driven reds from just about every French grape you can think of, from carignan to pinot noir, and brisk, brightly acid whites from chardonnay, sauvignon, and viognier. More massive reds are found in the Drôme and Vaucluse, from Marcel Richaud's suave Cairanne to Domaine Gramenon's immortal, wide-screen grenaches. Finally, Domaine l'Anglore and its many followers in the northern Gard serve up luminescent, Lapierre-inspired reds and rosés.

Dard et Ribo | Crozes-Hermitages–C'est le Printemps

Dard et Ribo's early-release C'est le Printemps derives from their flatter, less grand terroirs. This fluid, fruit-driven syrah is made in precisely the way they make their more cult cuvées: a roughly two-week whole-cluster maceration, with gentle pigéage throughout. C'est le Printemps sees shorter aging than the duo's other wines, but showcases the signature buoyancy and lift they bring to this intense grape.

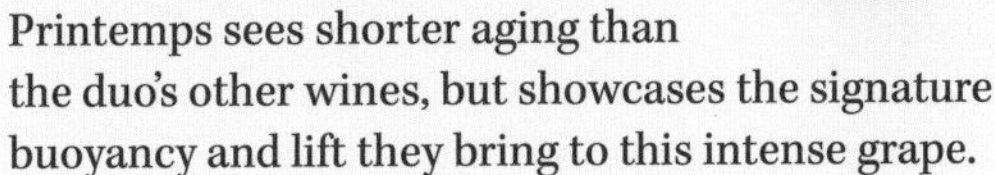

Domaine l'Anglore | Tavel

A rosé in name only, the Pfifferling family's renowned Tavel is, in fact, a short, cool-carbonic maceration of mostly grenache, supplemented with cinsault, mourvèdre, carignan, and clairette. The grenache is anchored by a prized parcel of more than one-hundred-year-old vines on sandy soils, inherited from Eric Pfifferling's grandmother. Dark-colored yet luminous, intensely fruited and almost free of tannins, the Anglore Tavel is a benchmark rosé that hearkens back to an era before widespread malo blockage and sterile filtration.

Domaine Gramenon | Côtes du Rhône–La Sagesse

From a blend of three parcels, two predominantly limestone and one with more sand influence, La Sagesse is among the iconic grenaches of France, showcasing the grape's more muscular, concentrated side, with thick aromas of black currant, wild thyme, and strawberry. It's often bottled with bold levels of CO_2 that integrate sublimely with age.

Domaine le Mazel | Vin de France–Cuvée Charbonnières

The Oustrics' steel-tank-fermented and tank-aged chardonnay will be, depending on the vintage, off-dry, slightly sparkling, volatile, oxidative, or some combination of the four. Yet the vivid, almost brittle, yet succulent fruit of well-farmed Ardèchois chardonnay always shines through.

DOMAINE GRAMENON

Michèle Aubéry and Maxime Laurent

MONTBRISON-LE-LEZ

Philippe Laurent (*left*) and Michèle Aubéry

Domaine Gramenon's structured, pure Côtes du Rhône wines possess the mysterious authority of ancient monoliths.

Organic from the get-go upon purchasing their then 12-hectare (29.6-acre) estate in 1978, Philippe Laurent and Michèle Aubéry intuitively avoided the era's incipient enological sophistication. When they began presenting their wines at tasting salons in France in the early 1990s, they met Marcel Lapierre and his collaborator Jacques Néauport. Laurent and Aubéry and their profound Côtes du Rhône wines grew widely admired among the era's small network of natural winemakers. Tragedy struck not a decade later, however, when Philippe Laurent died in a tractor accident. Aubéry, a former nurse, took up the herculean task of continuing alone—and did so with panache, converting the estate to biodynamic practices in the wake of the heat-wave 2003 vintage. Her son Maxime joined the estate in 2006.

Domaine Gramenon's powerful wines have always been worlds apart from the silkier, cool-carbonic style of their peers in the Beaujolais. The estate's reds are lightly refrigerated before vatting in concrete, mostly whole-cluster. Without foot-treading, they extract an impressive fruit intensity, which persists through ten months' aging in a mixture of demi-muid and used barrels. (In recent years, they have begun to experiment with concrete eggs.) Wines are barreled by gravity into an aging cellar carved into the rock beneath the winery. According to Aubéry, every vinification and élevage choice aims to preserve the innate complexity of the estate's unique terroir, a pre-Alpine plateau of sand and clay interwoven with limestone gravel of ancient sea deposits.

"We made wines without additives instinctively. We felt we had a terroir that was quite exceptional. And we wanted to keep its purity."

—Michèle Aubéry

THE WINE TO TRY

See Wines to Know: The Rhône, *page 229*.

Since: Estate founded in 1978. The first bottled vintage was 1990.

Winemakers: Michèle Aubéry and Maxime Laurent. Formerly Philippe Laurent.

Vineyard surface: 22.5ha (55.6 acres)

Viticulture: Organic since founding. Biodynamic since 2006.

Négociant work: Maxime Laurent purchases fruit for wines released under his name.

Grapes: Grenache, syrah, and more

Appellations: Côtes du Rhône, Vinsobres

Filtration/Fining: None

Sulfites: Sometimes small doses after racking and/or at bottling

DOMAINE RICHAUD

Marcel, Claire, and Thomas Richaud

CAIRANNE

Marcel Richaud felt the change in the air in the 1990s, when natural wine began to make itself felt in the Paris bistros of the early "*bistronomie*" chefs (see page 37). Already the head of a large, practicing-organic estate, Richaud wanted to produce wine more naturally, and found a guide in the enologist Yann Rohel. Rohel worked with Richaud for a decade beginning in 1999, helping to eliminate filtration and sulfite addition in the estate's red wine vinification. Richaud now practices full destemming and vinifies at cool temperatures (never over 25°C/62°F), with minimal pumpovers thanks to the use of robo-pigéage.

Among the chief advocates of the cru recognition for his native Cairanne, Richaud insisted on including, in the appellation charter, a minimization of herbicides and an outright ban on machine harvesting (this counts as great progress in the Vaucluse). Today he works alongside his son Thomas and his daughter Claire, who joined the estate in 2015.

The Richauds notably adhere to boldly ripe harvest timing, despite the heat-wave vintages of recent years.

Marcel Richaud in his tasting room

"It's easy to show you one wine that'll make you fall on your knees and cry because it's sublime. To make the production of the whole estate at the same level of excellence is hard."

—Marcel Richaud

THE WINE TO TRY

Cairanne

The grapes that form Richaud's Cairanne see destemmed maceration in cement tanks of just over two weeks, before aging in a mix of barrel and demi-muid. Evoking licorice and black currant, flecked with the pepperiness of drought-season herbs, it's a muscular red for drinkers who expect nothing less from the southern Rhône.

Since: 1974. Natural vinification began around 1999.

Winemaker: Marcel Richaud. Claire and Thomas Richaud joined the estate in 2015.

Vineyard surface: 60ha (148.2 acres)

Viticulture: Organic

Négociant work: Yes, since 2018, under the name Famille Richaud

Grapes: Grenache, syrah, mourvèdre, and more

Appellations: Cairanne, Rasteau, Côtes du Rhône

Filtration/Fining: Reds are not filtered, save for one primeur. Whites and rosés see tangential filtration.

Sulfites: Wines see a dosage of 30mg/L before bottling.

DARD ET RIBO

René-Jean Dard and François Ribo

MERCUROL-VEAUNES

The fact that as young vignerons in 1983, René-Jean Dard and François Ribo decided to partner up is a testament to the huge challenge of producing wine naturally in the northern Rhône. Their unusual arrangement allowed the two visionaries to sustain their ideals throughout the 1980s and '90s, a time when winemaking in the northern Rhône and beyond was adapting itself to the tastes of foreign critics and a hungry export market. Dard and Ribo stayed the course, producing thoroughbred syrah the way Dard's late father had: without additives, without filtration, and usually without sulfites.

René-Jean Dard in the duo's cellar

"There were two of us at the estate, which is already a lot. So we didn't need other people. We were always independent of everything."

—René-Jean Dard

Dard and Ribo met in 1981 at winemaking school in Beaune, where they were classmates with Jacques Néauport. Practically alone among their generation of winemakers, however, Dard and Ribo were not influenced by Néauport, nor by his master, Jules Chauvet. Instead, they represent an older school: those who willfully continued the great winemaking practices of decades past. Such beliefs put them, almost by accident, in the company of natural winemakers.

Today the pair farm 8.5 hard-won hectares (21 acres) of vines, of which about 35 percent are white grapes, the rest syrah. Soft-spoken Ribo takes the lead in vineyard work. The irascible Dard takes the lead in vinification and what little commercial work the pair still do. Reds are vinified whole-cluster, and see macerations of roughly two weeks with gentle pigéage. Whites are pressed direct and generally ferment in barrel. Never overoaked, never overextracted, always evincing a signature balance of frankness and finesse, Dard and Ribo's wines are touchstones for natural wine in the northern Rhône.

THE WINE TO TRY

See Wines to Know: The Rhône, *page 229.*

Since: 1983

Winemakers: René-Jean Dard and François Ribo

Vineyard surface: 8.5ha (21 acres)

Viticulture: Practicing organic (uncertified)

Négociant work: Dard releases négociant wines vinified with other vignerons as Les Champs Libres.

Grapes: Syrah, roussanne, marsanne

Appellations: Saint-Joseph, Crozes-Hermitage, Hermitage

Filtration/Fining: None

Sulfites: Case by case: For whites, they will sometimes add sulfites on the harvest. Some reds are bottled entirely without sulfites.

DOMAINE ROMANEAUX-DESTEZET

Hervé and Béatrice Souhaut

ARLEBOSC

> **"Often when winemakers destem wines, they extract far too much. The whole of the bunch is necessary to arrive at a certain finesse."**
>
> —Hervé Souhaut

A trained biologist hailing from Paris's tony western suburbs, Hervé Souhaut never planned to enter winemaking. His career is the combined result of two fortuitous meetings. The first was with his wife, Béatrice, whose family possessed vineyards in the hills above Saint-Joseph, and a small plot within Saint-Joseph itself. The second was with a young Philippe Pacalet, who encouraged Souhaut to explore the potential of his in-laws' terroir. Pacalet pointed Souhaut in the direction of Dard et Ribo, then the area's lone natural wine estate, where Souhaut would work for two years before founding his own estate.

Souhaut began producing gamay from his in-laws' vines in 1993, along with Saint-Joseph from purchased fruit. (The same year, he planted the white parcel that yields the lush Romaneaux-Destezet blanc.) His ascent to the cult stratosphere truly started in 1999, when he produced his first vintage of the diaphanous Saint-Joseph–Sainte-Epine. Fourteen years later, the Souhauts were able to begin farming Les Cessieux, another, larger parcel from Béatrice's family, which yields a richer, more muscular Saint-Joseph. Souhaut's winemaking shows the influence of Pacalet and Dard: reds see cool three-week whole-cluster vatting in small wood tanks, and wines undergo pumping only once, at assemblage. The Souhauts' entire range, from its humblest IGP l'Ardèche gamay to the divine Sainte-Epine, reflects the keen intelligence and fine sensibilities of its authors.

THE WINE TO TRY

Saint-Joseph–Sainte-Epine

The rare Sainte-Epine is the ethereal zenith of Saint-Joseph syrah, with finely etched tannins that carry its delicate, spiced–black cherry fruit.

Hervé Souhaut in his Sainte-Epine parcel of Saint-Joseph

Since: 1993

Winemaker: Hervé Souhaut

Vineyard surface: 14.5ha (35.8 acres)

Viticulture: Organic

Négociant work: Yes

Grapes: Syrah, gamay, roussanne, and more

Appellations: Saint-Joseph, IGP l'Ardèche

Filtration/Fining: None

Sulfites: A small dose before bottling

MAISON STÉPHAN

Jean-Michel, Romain, and Dorian Stéphan

TUPIN-ET-SEMONS

Jean-Michel Stéphan in his Côte Rôtie syrah vines in Tupin

Presenting wines at a tasting in Brussels in 1996, Côte-Rôtie vigneron Jean-Michel Stéphan encountered Beaujolais vignerons Marcel Lapierre and Yvon Métras, who led him to discover the community that had formed around the additive-free style Stéphan had adopted instinctively upon founding his estate in 1991. Already farming organically since 1995, Stéphan would study the writings of Beaujolais wine scientist Jules Chauvet and come to adopt unsulfited cool-carbonic vinification on his syrah starting from the 1997 vintage.

Today Stéphan farms 10 hectares (25 acres), five in Côte-Rôtie and five spread among the nearby areas of Chasse-sur-Rhône and Isère that his sons planted in anticipation of their joining the family business. (As of 2020, there is also a tiny, newly obtained parcel of Condrieu.) Yields are very low, generally just over 20 hectoliters per hectare.

Stéphan practices whole-cluster cool-carbonic maceration of two to three weeks in cement tanks and steel tanks, with the cap submerged in the juice. Côte-Rôtie wines, except the steel-aged Les Binardes, age two years in used barrels.

A new phase at the estate began in 2021, when the family moved production into a brand-new, vastly expanded cellar space, which will also house an épicerie and a gastronomic restaurant.

"It was no problem farming Côte-Rôtie as long as there were old folks, strong people, who worked with hoes. When the younger farmers arrived, herbicides were easier for them."

—Jean-Michel Stéphan

THE WINE TO TRY

Côte-Rôtie–Coteaux de Bassenon

From a steep parcel of very old serine—a local variant of syrah—coplanted with syrah and 20 percent of viognie, Coteaux de Bassenon shows the fluidity and grace Côte-Rôtie can achieve with a masterful cool-carbonic maceration.

Since: 1991

Winemaker: Jean-Michel Stéphan

Vineyard surface: 10ha

Viticulture: Organic since 1995

Négociant work: Some forthcoming in future vintages

Grapes: Syrah, serine, viognier, and more

Appellations: Côte-Rôtie, Condrieu. Other wines released as Vin de France.

Filtration/Fining: None

Sulfites: None added, except in 2014

THIERRY ALLEMAND

CORNAS

Born without family vines, Thierry Allemand assembled his vineyard holdings in Cornas over four decades, clearing scrubland, restoring crumbling stone terraces, and doing his own plantings. For the first two decades, he supported himself by working as an electrician. Today his Cornas is credited with cementing the legend of the small appellation, placing it, alongside Côte-Rôtie and Hermitage, directly in the crosshairs of the international luxury wine market.

Allemand's path toward natural vinification was a familiar one. His career in winemaking began working for Cornas legend Robert Michel. It was while working for Michel that Allemand befriended Paris restaurateurs Bernard Pontonnier and Jean-Pierre Robinot, who hipped him to the low-sulfite wines of Marcel Lapierre. Allemand soon stopped adding tannins and tartaric acid (as was the norm in Cornas at the time). He began experimenting with low- and zero-sulfite vinification in 1992, encouraged by his friend Philippe Pacalet. Today Allemand vinifies whole-cluster with moderate pigéage in open-topped wood and steel vats. After press wines descend with lees via gravity for a two-year passage in barrel and demi-muid. Everything is vinified parcel by parcel and blended before bottling. Chaillot derives from younger vines, while Reynard contains older-vine fruit from sandier, more broken-down granite.

"We always say syrah is a reductive grape, that you need to aerate it. But my wine in barrel isn't reduced. Even two years on lees, and it's not reduced. Because there's no sulfites. But if you put a half gram on the harvest, it really reduces."

THE WINE TO TRY

Cornas–Chaillot

Allemand's Cornas wines are ornate broadswords of syrah, long and sturdy, revealing striking floral and saline detail as they age. Chaillot derives from younger vines on granite with shades of limestone.

Since: 1982. Estate bottling began in 1991. Natural vinification since the mid-1990s.

Winemaker: Thierry Allemand

Vineyard surface: 6ha (42.8 acres)

Viticulture: Practicing organic

Négociant work: None

Grapes: Syrah

Appellations: Cornas

Filtration/Fining: None

Sulfites: 10g/L at assemblage before bottling

DOMAINE LE MAZEL

Gérald and Jocelyne Oustric

VALVIGNIÈRES

Brother-and-sister duo Gérald and Jocelyne Oustric took responsibility for their father's large southern Ardèche estate in 1983 at a time when the estate's harvest went to the local cave cooperative. In 1985, long before he began bottling wine, Gérald visited the Beaujolais at the suggestion of his wife, who had been harvesting there with Jean-Paul Thévenet. In the Beaujolais, Oustric met Lapierre and Jacques Néauport. For years, Oustric would visit the Beaujolais after harvest with Néauport and participate in vinification. Lapierre and Néauport even attempted to help Oustric convince his fellow Ardéchois cooperateurs that native-yeast fermentation and low-sulfite vinification could be practiced on the scale of a cave cooperative. When that effort failed in 1997, the Oustrics began vinifying and estate bottling their own wine with the aid of Néauport. (Tasting that first vintage would convince their neighbor Gilles Azzoni to embrace natural vinification.) By 2000, they had obtained organic certification for the estate, and had begun vinifying its entire then 30 hectares (74.1 acres) without sulfites or filtration. This put them in a position to help install like-minded vignerons beside them, which they did, notably ceding vines and loaning cellar space to Andréa Calek, Sylvain Bock, and Anders Frederik Steen.

"If you produce small volumes of natural wine, it's really easy to sell. If you do twenty thousand bottles, you're permanently in demand. Us, we make fifty to sixty thousand bottles, depending on the year."

—Gérald Oustric

Gérald Oustric preparing a tasting outside his winery

In early years, Gérald hewed closely to Néauport-style Beaujolais, employing cold-carbonic maceration followed by aging in used oak. Nowadays he vats reds just below ambient temperatures, practicing a long, often ninety-day carbonic maceration in steel or concrete before extended aging in steel—however long it takes for sugars to finish, often up to three years or more. Don't look to Le Mazel wines for vinification finesse. They are, rather, vital country wines par excellence, inexpensive and surprisingly age-worthy.

THE WINE TO TRY

See Wines to Know: The Rhône, *page 229*.

Since: 1983. Estate bottling began in 1997.

Winemaker: Gérald Oustric

Vineyard surface: 20ha (49.4 acres)

Viticulture: Certified organic since 2000

Négociant work: None

Grapes: Syrah, cabernet sauvignon, and more

Appellations: All wines released as Vin de France since 2009

Filtration/Fining: None

Sulfites: None added

LE RAISIN ET L'ANGE

Gilles and Antonin Azzoni

SAINT-MAURICE-D'IBIE

Gilles Azzoni measuring the sugar density of a must

> **"In 2001, I made wines that were very volatile. In 2002, there was bitter rot. So I blended the two vintages together, and everyone loved it. Then I understood that nothing was impossible in wine."**
>
> —Gilles Azzoni

Good-humored and loquacious, Gilles Azzoni is one of the two avuncular chieftains of southern Ardèche natural winemaking (the other being Gérald Oustric) known for a range of zero-zero wines incomparable for their value, purity, and idiosyncrasy.

A native of the Paris suburbs, Azzoni studied winemaking at Mâcon-Davayé in 1976, and worked for conventional estates in Pommard and Provence before establishing his estate in Ardèche in 1983. Little by little he eliminated synthetic treatments, adopting organic viticulture in 1997. Around the same time, he tasted Gérald Oustric's first experiments with sulfite-free vinification, and they struck a chord. So Azzoni followed Oustric in converting his production to natural vinification. The wines of southern Ardèche would never be the same.

Azzoni's wines were assemblages, vinified in steel, sometimes aged for short periods in used oak. When they were great—like the 2015 Homage à Robert—they were keen and swift as arrows. Azzoni retired after 2015, ceding 5 hectares (12.3 acres) to young natural winemaker Samuel Boulay and 1 hectare (2.5 acres) to his son Antonin, while keeping another hectare for himself. Azzoni's estate wines are now rarities for those who can discern the sculpture in volatile acidity. Antonin Azzoni runs Le Raisin et l'Ange as a négociant house, specializing in organic grape purchases from southern Ardèche.

THE WINE TO TRY

Vin de France–Nedjma

A blend of varying proportions of chardonnay, grenache blanc, and sauvignon, often with some skin maceration, Nedjma shows shimmering, tropical fruit and raw acid tang.

Since: 1983. Gilles began vinifying naturally in 2000. Since 2015, his son Antonin has run the business as a négociant house.

Winemaker: Antonin Azzoni. Gilles still produces a tiny quanity of wine from 1ha (2.5 acres) of his own vines.

Vineyard surface: 2ha (4.9 acres). Until Gilles's retirement, he farmed 7ha (17.3 acres).

Viticulture: Organic since 1997

Négociant work: All wines are négociant wines from organic grape purchases.

Grapes: Viognier, syrah, merlot, and more

Appellations: All wines released as Vin de France

Filtration/Fining: None since 2007

Sulfites: None added since 2000

ANDRÉA CALEK

Andréa Calek and Stefana Nicolescu

ALBA-LA-ROMAINE

Andréa Calek (*right*) with his friend Gérald Oustric of Domaine du Mazel outside Paris natural wine shop Crus et Découvertes in 2013

Idealists abound in the world of natural wine, as do their opposites, the pragmatists. The Czech émigré turned Ardèche winemaker Andréa Calek is a rarer type: a vigneron-nihilist, cackling at the pieties of the culture surrounding him.

For years after his 2007 installation in Ardèche, Calek lived in a trailer amid his vines. Combined with his maniacal sense of humor and provocative fashion sense, this caused journalists to paint him as an archetypal punk outsider. Yet Calek's résumé is longer than that image would imply. For decades he worked as an agent for barrels, representing the cooperage François Frères in Eastern Europe. He was already involved in biodynamics circles in 1997, when Provence vigneron Dominique Hauvette hired him to help her install biodynamics. Later, pursuing winemaking studies in Carpentras, Calek encountered the enologist Yann Rohel. He began working with Rohel as a hired-gun natural winemaker, and met Rohel's friends in the Beaujolais, notably Morgon vigneron Guy Breton. This was the era in which Ardèche vigneron Gérald Oustric drew influence from Rohel and the Beaujolais natural wine circle. Oustric offered to help set Calek up as a vigneron in Ardèche.

Today Calek still works alongside Oustric during harvest. Gone is the trailer; in its place is a splendid, bioclimatic wooden home and winery, in which Calek and his girlfriend, Stefana Nicolescu (who joined him in 2016), plan to host a seasonal pop-up restaurant someday. Calek's bristly whites and savory reds have always been partly intended to shock. But they retain a wholesome, digestible aspect, a testament to the sincerity that underpins his work.

THE WINE TO TRY

Chatons de Garde

Chatons de Garde is a semi-carbonic-macerated, barrel-aged syrah, often intensely reductive. With aeration, it broadens out to show a baritone class that belies its reactionary cuvée name ("guard kittens," apparently a jab at militant feminism).

Since: 2007

Winemaker: Andréa Calek. Stefana Nicolescu has taken the lead since 2016.

Vineyard surface: 5ha (12.3 acres)

Viticulture: Practicing biodynamic

Négociant work: Some grapes are purchased from Domaine le Mazel.

Grapes: Grenache, syrah, merlot, and more

Appellations: All wines released as Vin de France

Filtration/Fining: None

Sulfites: None added

NATURAL HISTORY

LA CAVE DES VIGNERONS D'ESTÉZARGUES

Founded east of Avignon in 1965, La Cave des Vignerons d'Estézargues was a cave cooperative like any other until the arrival, in 1988, of winemaker Jean-François Nicq (see page 293). A native of Lille, Nicq had just completed winemaking studies in Mâcon and Libourne alongside his friend Thierry Puzelat (see page 150). Within the next few years, Puzelat would introduce Nicq to Marcel Lapierre and the band of innovative natural vignerons in Villié-Morgon. Nicq soon convinced the ten winegrowers of Estézargues to give him a freer hand in vinifying and marketing the cooperative's wines.

Beginning in 1995, Nicq took the unusual step of creating loose "parcel" cuvées for each individual winegrower, as a way to motivate them to improve their farming methods. In 1997, Nicq ceased commercial yeast inoculation on all tanks. By the time he left to begin his own estate in the Roussillon in 2001, Nicq had converted the entirety of the cooperative's output to natural vinification, conditioning all wines without fining or filtration.

Today the cooperative is run by Nicq's successor, Denis Deschamps, who continues to naturally vinify the fruit of its members' 420 hectares (1,038 acres).

"The only way to make things advance—to stop filtering, to stop yeasting, and then to stop adding sulfites—was to show the winegrowers, little by little, that the wines sold better that way."

—Roussillon vigneron Jean-François Nicq, former cellar master at La Cave des Vignerons d'Estézargues

---

La Cave des Vignerons d'Estézargues, during construction in March 2021

DOMAINE L'ANGLORE

Eric, Marie, Thibault, and Joris Pfifferling

TAVEL

A beekeeper by trade, Eric Pfifferling began growing wine for the local cave cooperative when he inherited 3 hectares (7.4 acres) of family vines in Tavel in 1988. Through his wife, Marie, Eric met Jean-François Nicq (see page 293), who introduced the Pfifferlings to natural wine and encouraged them to begin their own estate. The Pfifferlings left the cave cooperative in 2002, an inauspicious year: freak floods destroyed most of their harvest. On Nicq's recommendation, future Languedoc vigneron Axel Prüfer came to help the Pfifferlings with vinification that year.

Eric Pfifferling soon proved himself a talented winemaker in his own right, possessed of uncommonly great terroir. The heart of the estate's production are hundred-plus-year-old grenache vines on the Tavel appellation's historic sandy soils. For the Pfifferlings, true Tavel has a spiritual kinship with the ancient archetype of Châteauneuf-du-Pape; theirs is a "ruby" wine far darker than most present-day rosés, yet lighter than most red wines. It's produced using a variant of Lapierre-style cool-carbonic vinification, with a maceration of just a few days.

"Natural wine carried an innovative message: the idea of making wine in total simplicity, with modest means, with a certain authenticity, true fermented grape juice."

—Eric Pfifferling

The couple's sons, Thibault and Joris, joined the estate in the late 2010s. In addition to their famed rosés, the Pfifferlings produce a wide range of silken, graceful single-vineyard reds. All reds are vinified whole-cluster, and the harvest is vatted at around 15°C (59°F) after a short refrigeration. There is also a Vin de France direct-press rosé, Chemin de la Brune, and a lone white, Sels d'Argent, from grenache blanc and bourboulenc.

From left: Eric, Marie, Thibault, and Joris Pfifferling

THE WINE TO TRY

See Wines to Know: The Rhône, *page 229*.

Since: 1988. Wines have been bottled since 2002.

Winemaker: Eric Pfifferling. Thibault and Joris Pfifferling joined the estate in 2017.

Vineyard surface: 20ha (49.4 acres)

Viticulture: Organic

Négociant work: None

Grapes: Grenache, mourvèdre, carignan, and more

Appellations: Tavel for rosé. Lirac. All other wines released as Vin de France.

Filtration/Fining: None

Sulfites: Generally 10g/L at assemblage before bottling

Natural Wine Dining

The length of the Rhône River between Lyon and Avignon abounds with local delicacies. Among the most irresistible are *caillettes ardéchoises*, fist-size spiced pork meatballs laced with chard and held together with pork caul. Apricot orchards are plentiful, though it's regrettably rare to find organic ones. Renowned cheeses of the region include the savory Picodon, from raw goat milk, and the sheep's-milk-based Saint-Félicien, which is delightfully runny when ripe.

Unfortunately, this wealth of native country specialties is not reflected in the region's dining scene, which is quite barren, particularly outside of the summer tourist season. Book in advance for these places, and just in case, load up on caillettes at the local butcher shops.

La Cachette

VALENCE

Since 2006, chef Masachi Ijichi has won plaudits for his refined and ambitious reimagining of classic French cuisine at this one-Michelin-star restaurant in Valence. He deserves equal praise for his consistent support of natural wines, which occupy half the impressive wine list. Refined and conventionally chic, La Cachette is a favorite of Hervé Souhaut, Thierry Allemand, René-Jean Dard, François Ribo, Jean Delobre—everybody who's anybody in the northern Rhône.

La Tour Cassée

VALVIGNIÈRES

Valvignières bistro and bed-and-breakfast La Tour Cassée illustrates the lackadaisical country charms of southern Ardèche. Longtime owners Claire and Jean-Claude Bouveron run it seasonally, closing and opening largely when they feel like it. The cuisine is homespun and satisfying, ranging from gazpachos to beef carpaccios and tajines. Along with the courtyard dining and the foosball table, the big draw is the wine list: a dream lineup of well-aged bottles from veteran natural vignerons throughout France.

La Courtille

TAVEL

Opened in 2018, La Courtille is a seasonal bistro in Tavel run by Natalia Crozon with the support of her boyfriend, Thibault Pfifferling, and his family. Sited in the walled courtyard of a former silk factory, it is a countryside homage to their former workplace, the historic natural wine bistro Le Baratin (see page 403). Local grenache-based rosés glimmer from tabletops beside steak tartare, tarama, and poached veal brain.

The interior courtyard of La Courtille in Tavel

Legends in the Making

In the northern Rhône, astronomic real estate prices and limited land availability and planting rights make it nearly impossible for young natural winemakers to gain a foothold. While vines aren't plentifully available in southern Ardèche, either, far lower real estate prices and the beneficent influence of Gérald Oustric and Gilles Azzoni have brought a boom of younger natural winemakers to the area. Similarly, because real estate prices around Avignon and Orange are forbidding for newcomers, young natural winemakers without family estates have clustered farther west in the Gard, closer to the picturesque town of Uzès, where vineyards and cellar space are more accessible.

The best of this new generation in the Rhône combine the revolutionary spirit of Oustric and Azzoni with the technical mastery of Dard et Ribo and Domaine Gramenon.

Ad Vinum

Sébastien Chatillon

VALLABRIX

Normandy native Sébastien Chatillon gained renown as the longtime sommelier at Paris restaurant Le Chateaubriand (2009–2016) before his friend Valentin Vallès encouraged him to make wine. His prior experience working for Anjou's René Mosse gave no indication of the wildly experimental tack he would take as a winemaker. His one-off micro-cuvées, from grapes purchased throughout the Gard and beyond, often include direct-press red juice co-fermented with whole-cluster fruit, and occasionally see oxidative aging. The results are wiry, intensely fruited, unclassifiable wines. In 2020, he became a vigneron-négociant with the purchase of 3 hectares (7.4 acres) of grenache, syrah, and sauvignon near Vallabrix.

Domaine des Miquettes

Paul Estève and Chrystelle Vareille

CHEMINAS

Paul Estève and Chrystelle Vareille began producing wine in Saint-Joseph in 2003, taking over the estate of the cave cooperative vigneron for whom they'd been working upon his retirement. They left the cave cooperative immediately and installed organic and biodynamic agriculture. After visiting Georgia in 2011, Estève and Vareille became France's foremost proponents of amphora winemaking. Since 2012, all wines are vinified without filtration or additives including sulfites. The couple use freestanding Spanish tinajas (see page 104) for red wines and buried tinajas for orange wines. Their wines are at the forefront of a beautiful culture clash, resulting in powerful, unprecedented wines like their Madloba Saint-Josephs.

Paul Estève in his cellar

Le Clos des Grillons

Nicolas Renaud

ROCHEFORT-DU-GARD

Nicolas Renaud was working as a history teacher when he tasted a 2003 wine by Eric Pfifferling and had an epiphany. He dedicated himself to founding a wine estate, which he began in his garage in 2006 after work experience with the biodynamic Châteauneuf-du-Pape estate Domaine de la Vieille Julienne. Today Renaud farms 12 hectares (29.6 acres) of vines in a practicing-biodynamic way, producing diaphanous cool-carbonic reds and energetic whites that rival the work of his mentor at Domaine l'Anglore.

Daniel Sage

RIOTORD

Former Lyon wine dealer Daniel Sage began making unsulfited wine in 2011 in an abandoned factory on Mont Pilat. (He has since moved to a converted farmhouse even farther up the mountain.) He farms 1.8 hectares (4.4 acres) of vines in and around northern Saint-Joseph, but most of his work consists of négociant fruit purchased in southern Ardèche. Sage has developed an innovative vinification style, submerging whole-cluster grapes in direct-press juice in an exotic variation on carbonic maceration. The striking, pale results are never confused for simple Ardèche country wines. Their arresting labels are actual works of art by the late Belgian surrealist Jean Raine.

Les Frères Soulier

Charles and Guillaume Soulier

SAINT-HILAIRE-D'OZILHAN

Youthful brothers Charles and Guillaume Soulier began producing wine in their native Saint-Hilaire-d'Ozilhan in the northern Gard in 2015, recovering vines their father used to farm. Guillaume, a former landscaper, takes the lead in vineyard work, while Charles, who produced corporate events before interning for Mas de Daumas Gassac, takes the lead in vinification. They farm 10 hectares (25 acres) organically, with an emphasis on permaculture and non-plowing. Charles's cellar work is equally radical, featuring quite a few oxidative direct-press whites from red grapes. Wines are pressed manually and never pumped, and they never see filtration or additives. Seek out red and rosé versions of La Clastre, a syrah of surpassing delicacy and herbaceous, olive-toned finesse.

Charles Soulier at the door to his cellar

Valentin Valles

SAINT-QUENTIN-LA-POTERIE

What began in 2011 as ten barrels of négociant wine in a garage is now a 7-hectare (17-acre) estate on siliceous sands at Vallabrix for Valentin Valles, a charismatic, gravel-voiced Saint-Quentin native whose entry into natural wine circles came while working as a server at erstwhile Sanilhac natural wine bistro Le Tracteur. There he met his future employer Eric Pfifferling, for whom he worked in the vines until the mid 2010s. Today Valles's impactful, boldly pure winemaking is characterized by a slow, gentle vertical press and used-barrel aging. Widely admired for his generosity and easygoing mentorship, Valles has comes to influence recent generations of Gard vignerons no less than Pfifferling himself.

Christian Binner retrieving grape samples from his rows of Grand Cru Wineck Schlossberg

10. Alsace

The fairy-tale half-timber architecture of Alsatian villages is enough to make you forget the region is essentially a long, thin battlefield between France and Germany. Its vineyards span the eastern foothills of the Vosges mountains and encompass a kaleidoscope of terroirs: limestone, granite, gneiss, schist, sandstone, and more.

There are four key pioneers of natural winemaking in Alsace, most of whom began experimenting with unsulfited vinification in the 1990s.

Bruno Schueller began vinifying at his estate in 1982 at just seventeen years old. Right from the start he was sparing with sulfites, thanks to a traumatic sulfur overdose experience that occurred while he was wicking foudres for a neighboring estate.

Jean-Pierre Frick, meanwhile, was among the first of France's winegrowers to embrace biodynamics back in 1981. He began vinifying certain wines without added sulfites in 1999.

Christian Binner and his father, Joseph, were introduced to natural wine circles by the Paris agent Jean-Christophe Piquet-Boisson (see page 36), who brought Marcel Lapierre to Alsace to meet them in 1998.

Rounding out the quartet is Nothalten's Patrick Meyer, who produced certain noncommercialized natural wines as far back as the late 1990s.

According to Binner, the four vignerons were only vaguely aware of one another's efforts in natural vinification until they were introduced by the wine journalist Sylvie Augereau (see page 168) in the early 2000s.

Since then, natural winemaking in Alsace has achieved critical mass, aided by efforts like Christian Binner's Les Vins Pirouettes, a distribution structure intended to help estates converting to natural winemaking find a market for their first natural cuvées. Alsace-based natural wine enologists Pierre Sanchez and Xavier Couturier, who practice their craft as DUO Oenologie, also deserve credit for supporting the nascent natural wine community in the region (the pair are also involved in Les Vins Pirouettes).

Total vineyard surface:
15,700ha (38,796 acres)[1]

Rate of organic agriculture:
32 percent[2]

Key grapes:
Riesling, pinot gris, pinot blanc, gewürztraminer, muscat, sylvaner, auxerrois, pinot noir

1 France Agrimer 2019
2 Dernières Nouvelles d'Alsace 2021

Wines to Know

Alsace's natural white wines have all the exotic personality of their conventional counterparts. But they tend to be drier (an obligation that comes with nonfiltration), riper, bursting with vitality. The region is also the most promising zone of France for orange wines, with pinot gris, muscat, and gewürztraminer all yielding consistently positive results with skin maceration.

Domaine Christian Binner | Alsace–Si Rose

Christian Binner professes to having long been skeptical of orange wines—until 2012, when experiments made with his intern, the future California winemaker Evan Lewandowski, changed his mind. The result is Binner's flaming-orange Si Rose, a majestically ripe, two-vintage blend of heavily-foot-trodden pinot gris and gewürztraminer. Disregard the questionable taste of its name (a pun on liver disease) and focus on its handsome structure and its long, ginger-inflected apricot fruit.

Domaine Gérard Schueller | Alsace Pinot Noir–LN012

Husseren-les-Châteaux vigneron Bruno Schueller creates intense, perfumed pinot noirs without parallel in Alsace and beyond. The cryptic name LN012 comes from Schueller's experiments with unsulfited vinification: it refers to the analysis of this wine's inaugural unsulfited 1997 vintage, which showed 12mg/L of total sulfites. The wine comes from a 0.3ha (0.75 acre) parcel within the Grand Cru Eichberg, and sees whole-cluster maceration of two to three weeks, followed by pressing and aging in old barrels for two years without topping up. Neither as floral nor as ethereal as Schueller's Bildstoeckle and Le Chant des Oiseaux pinot noirs, LN012 instead marries muscular, ferrous fruit with his signature structural volatile acidity, born of oxidative élevage along with a kiss of botrytis at harvest.

Domaine Pierre Frick | Alsace Grand Cru Vorbourg Pinot Gris–Macération

At a time when skin-macerated white wines are rarely produced within renowned French appellations, it's significant that Pfaffenheim biodynamic vigneron Frick produces several grand cru wines with this method. His pinot gris from Vorbourg sees a maceration of just over a week, before aging in old oak foudres. Usually quite dry and rendered slender by pad filtration, it shows a balance of purity and precision that is rare in its genre, evoking quince, clementine, and incense.

DOMAINE CHRISTIAN BINNER

Christian, Béatrice, Monique, and Joseph Binner

AMMERSCHWIHR

Christian Binner at the Salon des Vins Libres tasting in Mittelbergheim

The Binner estate was already ahead of the curve under patriarch Joseph Binner, who estate-bottled his wines and never used herbicides or other chemicals in farming. Around the time his son Christian took over in 1998, natural wine agent Jean-Christophe Piquet-Boisson introduced the Binners to Morgon vigneron Marcel Lapierre, who would inspire their transition to unsulfited vinification in the 2000s. During this time Christian and his then wife, Audrey, also introduced biodynamic agriculture.

Several events heralded a new chapter at the estate in the early 2010s. In 2012, the Binners completed construction of a stunning new bioclimatic cellar, naturally insulated with a prairie roof. The same year, Christian's cellar intern, future California winemaker Evan Lewandowski, finally convinced him of the merits of orange winemaking. Then, in 2013, Christian and Audrey split up. These changes seem to have inaugurated a renaissance at the estate. A wine range that was promising but stylistically uneven throughout the 2000s became, from about 2012, a triumphant parade of lush, pure, unfiltered and unsulfited whites and reds, supplemented by a panoply of splendid orange wines.

"The more we take care in the vines, the less we have to do in the cellar. It's like a child that we educate with care and respect—he lives his own life. The more you stress him in an unhealthy way, the more he'll bust your balls as an adolescent or an adult."

—Christian Binner

In the cellar, Christian practices long élevage in ancient foudres, accompanying his boldly ripe white wines for years until he deems them ready to bottle. If residual sugar remains at that time, he'll append the ingenious suffix *qui gazouille* ("that babbles") to the wine name to indicate that the wine is liable to be lightly effervescent.

THE WINE TO TRY

See Wines to Know: Alsace, *page 247.*

Since: 1770. Estate bottling began before World War II. Joseph Binner took responsibility in 1965. Christian Binner took responsibility in 1998.

Winemaker: Christian Binner

Vineyard surface: 10ha (25 acres)

Viticulture: Organic since founding. Biodynamic since the early 2000s.

Négociant work: None, but Binner sells the wines of Les Vins Pirouettes.

Grapes: Riesling, pinot gris, pinot noir, and more

Appellations: Alsace Grand Cru Schlossberg, Alsace Grand Cru Wineck-Schlossberg, Alsace Grand Cru Kaefferkopf, Alsace, Crémant d'Alsace

Filtration/Fining: None in recent years

Sulfites: None added since 2012

DOMAINE PIERRE FRICK

Jean-Pierre and Chantal Frick

PFAFFENHEIM

Pfaffenheim biodynamic and natural wine pioneer Jean-Pierre Frick cuts a Tolkienesque figure, with his tall grace, elfin features, and eerily measured tones in conversation. Frick seems to think and work at a different rhythm to the rest of the world. Unlike many early Rudolf Steiner enthusiasts, he continues to refine his vinification approach, growing closer to the natural style of the younger generations who have apprenticed themselves to him to learn from his vast biodynamic experience. Frick credits his friend Pierre Overnoy and Strasbourg wine merchant Michel Legris with inspiring his transition to natural vinification.

One of the first things you see when arriving chez Frick are rows of spongelike squares, about the size of letter paper, leaning against his winery. They are cellulose wine filtration pads, which Frick dries in the sun after use and then burns to heat his house and winery. Since his first experiments with natural vinification in 1999, Frick has increased the proportion of his range vinified naturally each year. He has also embraced skin maceration for white grapes, largely to combat the sluggish direct-press white fermentations that have afflicted natural vignerons in recent years. Combined with his light touch in filtration, this yields nuanced, delicate macerated whites for the tannin-averse, like his Grand Cru Vorbourg pinot gris. (Frick avoids the term "orange wine.") Wines age in a range of old oak foudres.

Frick is as renowned for his intellectual approach to viticulture as he is for his willingness to share and explain his methods. He doesn't trim vine apexes, instead braiding them together, because he finds it helps the plants better resist temperature fluctuation.

THE WINE TO TRY

See Wines to Know: Alsace, *page 247.*

Jean-Pierre Frick beside the oven he uses to heat his home and winery

---

Since: The estate has been in the Frick family for twelve generations. Jean-Pierre and Chantal Frick took over from Pierre Frick in 1968.

Winemaker: Jean-Pierre Frick

Vineyard surface: 12ha (29.6 acres)

Viticulture: Biodynamic since 1980

Négociant work: None

Grapes: Pinot noir, auxerrois, pinot blanc, and more

Appellations: Alsace, Alsace Grand Cru Steinert, Alsace Grand Cru Vorbourg, and more

Filtration/Fining: Most wines see pad filtration.

Sulfites: None added on certain cuvées since 1999. Others see a small dose at bottling.

DOMAINE JULIEN MEYER

Patrick and Mireille Meyer

NOTHALTEN

Instinctive and self-taught, Nothalten vigneron Patrick Meyer credits an early tragedy, the death of his father when Patrick was five, with giving him the freedom to innovate in wine growing and winemaking when he took the reins of his family estate in 1981 at age nineteen. After an early test run of the chemical agriculture he'd been taught in school produced lamentable results, Meyer took a different approach, eventually embracing organic and later biodynamic certification. He first began experimenting with natural vinification in the mid-1990s, though it would be many years before he adopted it for the majority of his range.

Today, alongside Jean-Pierre Frick, Meyer ranks among the most passionate biodynamics enthusiasts in Alsace, and often hosts training sessions for other vignerons. His peers appreciate his willingness to go out on a limb in biodynamic thought. One example is his decadelong collaboration with local natural wine enologists DUO Oenologie, in which they've experimented with homeopathic dilutions of sulur (along with other elemental additives) in winemaking. In conversation, he can be quite out-there, preferring to speak in Steinerian terms of cosmic energies rather than of maceration temperatures or racking techniques.

His wine range in recent years has included both boldly natural higher-end cuvées, which enjoy significant cult acclaim, and simpler, more accessible cuvées, of which the whites see filtration.

THE WINE TO TRY

Alsace Grand Cru Muenchberg Riesling

Muenchberg is an almost 18-hectare (44.5-acre) cru beside Nothalthen, renowned for the volcanic influence in its soils. Meyer's chef d'oeuvre from it sees extended aging in barrels, often up to four years, with infrequent topping up. Redolent of tamarind and Szechuan cuisine, it shows an underexplored dark side of riesling.

Patrick Meyer (*right*) during a game of pétanque at the Salon des Vins Libres d'Alsace in Mittelbergheim

---

Since: The estate has been in the Meyer family for generations. Patrick Meyer took responsibility for the estate in 1981.

Winemaker: Patrick Meyer

Vineyard surface: 8ha (19.7 acres)

Viticulture: Practicing organic since 1993. Practicing biodynamic since 1998. Certified in both since 2012.

Négociant work: None

Grapes: Sylvaner, riesling, gewürztraminer, and more

Appellations: Alsace Grand Cru Muenchberg, Alsace, Crémant d'Alsace

Filtration/Fining: Entry-level whites are often filtered.

Sulfites: None added in recent years

TASTING DESTINATIONS
ALSACE NATURAL WINE SALONS

The history of natural wine in Alsace was written by four estates, so it should come as no surprise that those same four are behind most of Alsace's natural wine tastings. As L'Association des Vins Libres d'Alsace, a vigneron collective comprising Christian Binner, Jean-Pierre and Chantal Frick, Bruno and Elena Schueller, and Patrick and Mireille Meyer, they organize two of the three most important tasting salons. The third is co-organized by Paul Gillet, who worked for the Schuellers before beginning his own career as a vigneron in Touraine.

Far from being a closed or insular clique, however, these founding natural vignerons—and the salons they organize—are very welcoming, devoted to the betterment of the region and the standing of Alsatian wine worldwide.

Salon des Vins Libres d'Alsace

When: Early May

Who: Organized by L'Association des Vins Libres d'Alsace

For: Open to the public

Where: The location shifts each year. Previous editions have been held in the villages of Obermorschwihr, Strasbourg, and Molsheim.

Alsace's first natural wine salon, Salon des Vins Libres d'Alsace has been held every other year since 2008. (The 2020 salon was pushed back a year due to COVID.) This two-day event unites sixty-plus natural vignerons from throughout France and from as far afield as Georgia, Italy, and Spain.

D'Summer Fascht

When: Mid-July

Who: Organized by L'Association des Vins Libres d'Alsace, D'Summer Fascht is a summer festival featuring the natural vignerons of Alsace.

For: Open to the public

Where: The location shifts. Previous editions have been held at Domaine Christian Binner in Ammerschwihr and the Parc du Natala in Colmar.

D'Summer Fascht is a midsummer natural wine fair usually organized in years when the Salon des Vins Libres d'Alsace *isn't* held.

Salon Brut(es)

When: Early November

Who: The salon is organized by four Alsatian friends and longtime natural wine lovers, Bruno Schaller, Jean-François Hurth, Eric Bazard, and Paul Gillet. Gillet is presently also a biodynamic vigneron in Touraine, and formerly had a wine shop in Mulhouse.

For: Open to the public

Where: Motoco, an art space in Mulhouse

Among the newest and most well run of France's natural wine salons, Brut(es) draws a majority of the great natural vignerons of Alsace, along with an astutely balanced mix of zero-zero veterans (Clos Fantine, Daniel Sage, Patrick Bouju) and newcomers to natural vinification. It notably also includes vignerons from the Côtes de Toul and Germany's Mosel Valley.

A toast at the Salon des Vins Libres in Mittelbergheim

DOMAINE GÉRARD SCHUELLER

Bruno and Elena Schueller

HUSSEREN-LES-CHÂTEAUX

Plainspoken and down-to-earth, Bruno Schueller is a visionary in vinification whose puzzling wines confound and delight the most discerning palates.

His father, Gérard, never employed herbicides, and today Bruno avoids all vineyard treatments for a third of the estate. Schueller harvests early (despite possessing high-sited vines in Husseren-les-Châteaux), and claims to avoid botrytis, although it's often perceptible in his pinot noirs. He conducts a long, very hard pneumatic press, sufficient to lend a certain tannicity to even his direct-press whites. Yields are among the lowest one encounters in Alsace, around 40 hectoliters per hectare in most years.

Schueller gradually stopped using sulfites in vinification after taking over winemaking in 1982, and 1997 saw his first zero-sulfites-added cuvée. He welcomes the influence of oxygen in vinification: vats and small foudres are not topped up whatsoever during long aging periods. This is a regime that, by popular understanding, should destroy most wines. It renders Schueller's almost immortal. Schueller always bottles by individual tank or foudre, resulting in a head-spinning range of up to thirty-five cuvées per vintage.

> **"When you make wines without sulfur, the most difficult thing is to learn how to sleep. How to find serenity. The rest just needs time."**
>
> —Bruno Schueller

Schueller was the first in Alsace to experiment with skin maceration on white grapes, back in 2002, influenced by the Italian natural wine culture he discovered early on alongside his wife, Elena, who hails from Tuscany. In recent years, Schueller's fame has grown alongside that of his many renowned former interns and cellarhands, whose ranks include Savoie's Jean-Yves Péron (see page 263), Auvergne's Vincent Marie, Lazio's Gianmarco Antonuzzi of Le Coste (see page 351), and the Jura's Kenjiro Kagami.

THE WINE TO TRY

See Wines to Know: Alsace, *page 247*.

Bruno Schueller in his cellar

Since: Gérard Schueller began estate bottling at the end of the 1950s.

Winemaker: Bruno Schueller since 1982

Vineyard surface: 10ha (25 acres)

Viticulture: Organic certified and practicing biodynamic

Négociant work: None

Grapes: Pinot noir, riesling, pinot gris, and more

Appellations: Alsace Grand Cru Pfersigberg, Alsace Grand Cru Eichberg

Filtration/Fining: None, although he's not opposed to it in the future for certain wines

Sulfites: None added since 2013

Natural Wine Dining

Au Pont Corbeau in Strasbourg

Whether you will enjoy the Alsace dining scene depends on your affection for the rich, pungent local cuisine. Ubiquitous at informal occasions, *flammekueche* (or *tarte flambée* in French) is a roasted flatbread slathered in cream and topped with onions and bacon chunks. Tasty meat-studded *choucroute* (sauerkraut) is available practically by the bale at most restaurants. Aged versions of the acclaimed local Munster cheese are fragrant enough to empty football stadiums. For a taste of the sublime, however, check out these Alsatian natural wine destinations.

Restaurant Thierry Schwartz

OBERNAI

Chef Thierry Schwartz worked for Joël Robuchon in Paris before creating his own tasteful and ambitious dining destination in 2002 in the picturesque town of Obernai. Boasting a wood-fired oven, an adjacent bakery, and a commitment to biodynamic ingredients, Restaurant Thierry Schwartz is also one of just a few Michelin-starred restaurants in France to prominently foreground natural vignerons on its immense, value-laden wine list. In a dazzling flourish of tableside theater, servers bake bread and churn fresh butter for each table.

Au Pont Corbeau

STRASBOURG

Longtime proprietor Christophe Andt has ceded the reins of this cozy, wood-paneled riverside bistro to his daughter Coralie in recent years. He still makes regular appearances, however, and is only too happy to discuss the fruits of his more than twenty years collecting natural wine. The cuisine is well-executed Alsatian fare, sincere and hearty.

L'Esprit Libre

HORBOURG-WIHR

Among the more recent additions to the Haut-Rhin dining scene, L'Esprit Libre, opened in 2019, offers a lighter "*bistronomie*" take on Alsatian cuisine, dispensing with the choucroute and incorporating more seafood options in its spacious, terraced new space in Horbourg-Wihr.

ALSACE

Legends in the Making

Alsace today is the site of a sea change in winemaking. While the region's high vineyard prices make it difficult for outsiders to establish themselves, their absence is more than compensated for by established estates converting wholly or partly to natural vinification.

A striking aspect of Alsace's natural wine scene today is its harmoniousness. It is among the few regions where the natural wine elders are just as radical as their followers, and delight in sharing their knowledge and experience.

Jean-Marc Dreyer

ROSHEIM

Rosheim's Jean-Marc Dreyer took charge of his family estate in 2004 after an internship with Patrick Meyer. He began producing certain unsulfited wines in 2011, but it took a moment of personal crisis in 2013 for him to renounce sulfite addition for his entire production. In 2014, he obtained biodynamic certification. The majority of his 6 hectares (14.8 acres) are devoted to sylvaner and auxerrois. From these modest grapes—along with smaller surfaces of the other Alsatian varieties—Dreyer coaxes some of France's most wholesome and sublime orange wines. Worth checking out, too, is the cuvée of pinot noir he produces with a little help from his friend Raphaël Beysang, the southern Beaujolais vigneron.

Jean-Marc Dreyer in his cellar

Jean-François Ginglinger

PFAFFENHEIM

Jean-François Ginglinger is a longtime organic and (uncertified) biodynamic vigneron and an acolyte of both Jean-Pierre Frick and Bruno Schueller (Schueller is also Ginglinger's cousin). Ginglinger took charge of his family estate (which dates back to 1610) in 1999. Encouraged by Frick and Schueller, he began experimenting with unsulfited natural vinification as early as 2003, and eventually adopted the approach for his entire output. Ginglinger's wines, upon release, are often oily, volatile, overly gassy, or all three at once, like revolting teenagers. But wait the better part of a year—or longer, as the case may be—and they straighten out beautifully. His limber sylvaners, in particular, are high-water marks for complexity for this oft-neglected variety.

Christophe Lindenlaub

DORLISHEIM

Dorlisheim's Christophe Lindenlaub is an energetic mixed-agricultural farmer, who farms an equal surface of grains and vegetables alongside his 10 hectares (25 acres) of vines. His father, Jacques, raised cattle alongside wine growing until 1986. Christophe joined the estate in 1999 and initiated its conversion to organic certification in 2012, influenced by his friend Patrick Meyer. Since 2014, all vinification has been additive-free. Lindenlaub employs mostly steel tanks in his cellar, with little barrel aging. In his natural vinification, he reveals his own poetic streak, avoiding lieu-dit names in favor of evocative fantasy names like Matin Fou ("crazy morning"), a crystalline direct-press sylvaner.

Christophe Lindenlaub in his cellar

Yannick Meckert

ROSHEIM

With a résumé that includes stints working for Philippe Pacalet, Christian Binner, Patrick Meyer, Le Coste, and Claus Preisinger, expectations are high for this charismatic half-Alsatian, half-Burgundian renegade. None more so than his own: Meckert's first vintage was intended to be 2019, but, unhappy with the quality of his wines that year, he sold them off wholesale rather than bottle something disappointing. His true debut came in 2020, with a range that includes négociant wines from throughout Alsace as well as wines from his own parcels in Bernardswiller and Heiligenstein and one plot of pinot noir near Dijon that he co-farms with Burgundy-based friends. Meckert's emerging style is marked by slow, soft presses, whole-cluster maceration, and a connoisseur's choice in aging containers (high-quality oak foudres, barrels, and amphorae, many secondhand from his previous employers).

Jean-Pierre Rietsch

MITTELBERGHEIM

Jean-Pierre Rietsch's painstaking, intellectual approach to natural vinification has won acclaim since he first began producing low-sulfite wines in 2006. His diverse range of wines from around Mittelbergheim is recognizable for its handsome labels by local artist Marie Dréa, whose drawings depict subjects ranging from a pair of wolves to John Lennon and Yoko Ono, and for the wines' abiding tense, vertical profiles. Wines are harvested early and produced in a very dry style, far from the ripe opulence of Christian Binner or the exotic oxidative style of Bruno Schueller. The wines are thus well placed to win over palates fearful of Alsace's reputation for bigness and residual sugar. Skin-macerated whites like the muscat Murmure retain an angular profile, too.

Catherine Riss

BERNARDVILLÉ

The daughter of restaurateurs, Catherine Riss was born without family vines, but has nonetheless made a name for herself as one of Alsace's most virtuosic young vignerons. Before establishing her estate in 2012, she studied winemaking in Beaune and enology in Dijon, interning for Gevrey-Chambertin's Domaine Trapet. Her career back in Alsace began with work installing an estate for biodynamic Rhône négociant Michel Chapoutier. Her wines have made great strides since 2015, when she acquired her own cellar in Bernardvillé and obtained organic certification. Today they are perhaps the best values in the region, particularly her luminous, barrel-aged riesling De Grès ou de Force, from a sandstone site in Reischfeld.

Sons of Wine

BEBLENHEIM

Farid Yahimi began producing his first négociant wines in Christian Binner's former cellar in 2016. By 2020, he was based in his own cellar in Beblenheim, and technically became a vigneron with the purchase of a small vineyard two hours' drive away in the Côtes de Toul in Lorraine. But Yahimi's experiences within the world of natural wine go back much further. A longtime aficionado, he helped administrate the early natural winemaking group L'Association des Vins Naturels (see page 116) in the early 2000s. Yahimi's diverse and experimental range of négociant wines feels like the release of decades of pent-up imagination. It includes everything from skin-macerated cuvées of sylvaner, pinot gris, riesling, muscat, and gewürztraminer, to cross-regional blends.

Roussanne vines in Chignin-Bergeron at Domaine Partage

11. Savoie and Bugey

Natural wine can seem an afterthought in the epic French alpine landscape bordering Switzerland and Italy's Valle d'Aosta. The region's ski tourism and its proximity to the economic powerhouse of Geneva mean that for the natives, almost any other pursuit is more profitable than winemaking on its steep, difficult-to-farm terroirs.

Natural winemaking in Savoie has long been in tacit conflict with the regional focus on filtered, off-dry, often chaptalized sparkling wines. (This wine style is partly the heritage of a pre–climate change era when grapes on these mountain slopes rarely attained full ripeness.)

There is nonetheless a thread of natural wine history tying the Bugey region to neighbors to the north in the Jura and to the east in the Beaujolais. Following encounters in 1985 with Marcel Lapierre and Pierre Overnoy, the small-scale, self-made Mérignat vigneron Raphaël Bartucci embraced organic agriculture (initially without certification) and began a contemporaneous collaboration with natural wine guru Jacques Néauport. Néauport would help Bartucci refine a low-sulfite variation on the local Bugey-Cerdon recipe for filtered, sweet, low-alcohol sparkling wines.

A decade passed before the arrival of Villebois vigneron François Grinand, whose unsulfited wines as La Vigne du Perron would make him a modest icon for a more radical natural wine vanguard. Grinand did viticultural studies at Mâcon-Davayé in 1993 alongside Auvergne natural wine pioneer Jean Maupertuis and Rennes wine retailer Éric Macé, who introduced Grinand to Marcel Lapierre's circle. Grinand's production would become the most doggedly natural of the Bugey region. Joining Grinand on the radical, unsulfited front of French alpine winemaking in 2004 was Jean-Yves Péron, an acolyte of Cornas's Thierry Allemand and Alsace's Bruno Schueller, who found success bringing masterful skin-macerated white winemaking to Savoie.

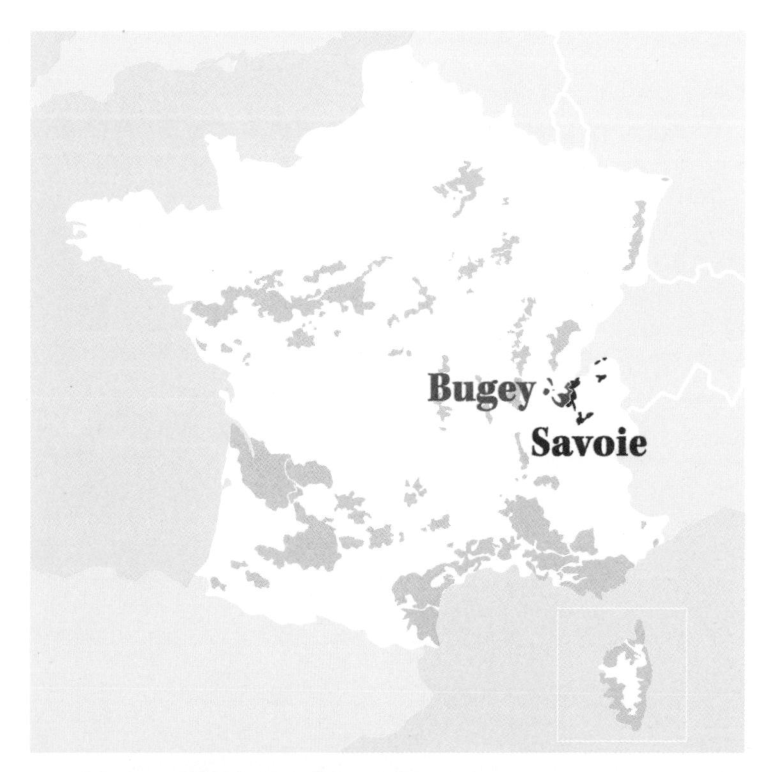

SAVOIE

Total vineyard surface: 1,800ha (4,448 acres)[1]

Rate of organic agriculture: 23 percent[2]

Key grapes: Jacquère, altesse, roussanne, gringet, mondeuse, pinot noir

BUGEY

Total vineyard surface: 800ha (1,977 acres)[3]

Rate of organic agriculture: 33 percent[4]

Key grapes: Gamay, poulsard

---

1 France Agrimer 2019
2 Agence Bio 2020 figures from Savoie and Haut-Savoie departments
3 Auvergne-Rhône Alpes Chambre d'Agriculture 2018 figures for Ain department
4 Agence Bio 2020. Ain department figures from Vitisphere.

SAVOIE AND BUGEY

Wines to Know

From Bugey, expect invigorating, bright gamay or pinot, as fresh as a lungful of mountain air—and, with the exception of the influential Bugey-Cerdon mentioned below, expect it to be dry, heretofore a rare style in a region known for sparkling sweet red. Savoie is a more mixed bag, containing everything from dark, wiry mondeuse reds to chirpy whites from jacquère to more ample whites from roussette, roussanne, and the rare gringet. Jean-Yves Péron's sinuous, zero-zero orange wines remain stylistic outliers for the region.

Domaine Raphaël Bartucci | Bugey-Cerdon

Raphaël Bartucci's brilliant Bugey-Cerdon is a platypus in the evolution of natural winemaking: an exotic creation that got left behind by what came after. First perfected in the 1980s by Bartucci and Jacques Néauport, the luminous, sparkling semisweet red wine is as bracing as a pomegranate and as succulent as a poached cherry. It remains rare and sought after by natural wine aficionados, despite being both filtered (after secondary fermentation) and lightly sulfited (at the start of fermentation), practices that have been stigmatized by later generations of natural wine drinkers. The wine's continued cult status shows just how rare such sensitive organic farming and vinification mastery are within the niche realm of sparkling semisweet reds.

Jean-Yves Péron | Vin de France–Les Barrieux

Jean-Yves Péron's white wines are iconoclastic in Savoie both for their radical, unsulfited, unfiltered purity, and for his practice of extended skin maceration, unheard of in the region before his arrival. They stand out for their succulent lees-iness and their filigreed tannins, often evoking the sweetly saline complexity of a fine Belgian lambic. Les Barrieux, a blend of roussanne, jacquère, and altesse that sees a full two-and-a-half-month maceration, is a fine introduction to the work of this quietly visionary vigneron.

La Vigne du Perron | Vin de France–Les Etapes

From vines he planted with his father on a limestone scree slope overlooking his village, François Grinand makes a luminous, cherry-toned pinot noir of rare exuberance. Vinified whole-cluster without pigéage and aged in old oak barrels, it proves that the terroir of his corner of Bugey can rival any site in the nearby Jura for complexity.

RAPHAËL BARTUCCI

MÉRIGNAT

Raised in the Moselle by parents who fled fascism in Italy, Raphaël Bartucci became the unlikely link between Marcel Lapierre's circle of early natural vignerons and the modern winemaking Bugey vignerons embraced in the 1980s. An electrician by trade, Bartucci began farming small parcels of rented vines in 1981, and began selling his wine in 1983. Over the years, he acquired land and planted his own vines.

Initially, Bartucci practiced the chemical viticulture prevalent at the time. But in 1985, friends in Mérignat introduced him to Marcel Lapierre, through whom he met Pierre Overnoy. With their encouragement, Bartucci began farming organically in 1986 and began collaborating with natural wine consultant Jacques Néauport. Together, they refined a low-sulfite version of the area's semisweet sparkling red.

"In every village, there are conflicts among families. So when people left, they didn't want to sell vineyards to their neighbors. That's why they sold the parcels to me, a foreigner! To piss off their neighbors!"

Bartucci makes a few hundred bottles of chardonnay each year, and has experimented with wine styles ranging from carbonic-macerated red gamay (a failure) to destemmed poulsard (more successful) to a *vin de paille*–style sweet wine ("truly formidable," to hear him tell it). His lone wine produced on a commercial scale is his Bugey-Cerdon. Typically 85 percent gamay and 15 percent poulsard, it sees a direct press, light sulfitage, and cold clarification. Bartucci enjoys powerful native yeasts—a testament to his sensitive farming—which work busily despite a regimen of intensive chilling. The wine sees just one stage of filtration, after disgorgement, and remains quite textured and intact. Today Bartucci expresses regret that a natural wine scene he witnessed from the start has become dogmatically opposed to practices he considers vital in producing his Bugey-Cerdon. He needn't worry: his small production sells out by summertime every year to a loyal following of connoisseurs.

THE WINE TO TRY

See Wines to Know: Savoie and Bugey, *page 259.*

Since: 1983

Winemaker: Raphaël Bartucci

Vineyard surface: 2.1ha (5.2 acres)

Viticulture: Practicing organic since 1986. Certified organic since 1997.

Négociant work: None

Grapes: Gamay, poulsard, chardonnay

Appellations: Bugey-Cerdon

Filtration/Fining: His Bugey-Cerdon is filtered after disgorgement.

Sulfites: 30mg/L at débourbage, after pressing

LA VIGNE DU PERRON

François Grinand

VILLEBOIS

A piano teacher by trade, Bugey native François Grinand enrolled in winemaking school at Mâcon-Davayé at age twenty-seven. There he encountered Rennes natural wine retailer Éric Macé and future Auvergne vigneron Jean Maupertuis, who put him in touch with Marcel Lapierre and the natural wine band of Villié-Morgon. Grinand sought to apply their ethos to vines he and his father had planted on limestone-scree terrain in the village of Villebois (with limited success, he says, until the early 2000s). After dallying with contemporary enology for the 1996 vintage, he returned to producing wines without filtration or added sulfites on the advice of Jacques Néauport and Pierre Overnoy. Grinand converted to organics in 1998, following the example of his nearest natural wine neighbor, Raphaël Bartucci.

"The Bugey will explode soon. There are many young vignerons setting up. But when I first started, I asked myself, besides a few crazies like me, who would come here to make wine?"

Like Auvergne's Domaine de Peyra, Grinand was ahead of his time in promoting zero-zero wines from a little-known region. After a series of disastrous yields in the mid-2000s, he went bankrupt in 2007—only to relaunch his estate in 2009 with the investment of some farsighted Belgian supporters. Today Grinand farms with the aid of his son Louis, supplementing his estate production with organic grape purchases in Savoie and the Mâconnais. Reds see two- to three-week whole-cluster carbonic maceration (except mondeuse, which is destemmed) and aging in old oak barrels. Whites age in old oak, too. In recent years, these vivid, delicate Bugey wines have attained well-deserved cult status in a younger generation of radical natural wine establishments.

THE WINE TO TRY

See Wines to Know: Savoie and Bugey, *page 259.*

Since: 1993–2007. The estate was refinanced and restarted in 2009.

Winemaker: François Grinand

Vineyard surface: 2.6ha (6.4 acres)

Viticulture: Practicing organic since 1998

Négociant work: Yes, since 2009, from organic grapes in Savoie and the Mâconnais

Grapes: Jacquère, altesse, mondeuse, and more

Appellations: All wines released as Vin de France since 2000

Filtration/Fining: None

Sulfites: None added since 2009

DOMAINE BELLUARD

Dominique Belluard

AYSE

Until his suicide in 2021, Dominique Belluard was for many years the renowned, practicing-biodynamic custodian of about half the remaining 20 hectares (49.4 acres) of the rare, indigenous Savoyard grape gringet at the estate his father founded in the Arve Valley. Belluard's exotic, tasteful whites and Champagne-method sparklers from the grape won the admiration of natural and conventional palates alike.

Belluard's path to wine stardom was slow. He studied winemaking in Beaune and in 1988 joined his family estate, which at the time practiced chemical agriculture and specialized in *méthode ancestrale* wines from gringet. Fellow Savoyard vigneron Michel Grisard introduced Belluard into biodynamic circles in 1995. Belluard adopted many biodynamic practices over the years, but resisted certification because he continued using fungal sprays not permitted under biodynamics. His vinification, too, was marked by conservatism: elegant whites like the gringet-based Les Alpes and Le Feu saw Kieselguhr filtration and light sulfitage. In the best vintages, he produced a fascinating skin-macerated, unsulfited gringet entitled Pur Jus. Belluard also produced small quantities of an intense unsulfited mondeuse, aged in concrete egg tanks.

Since: 1947. Dominique Belluard officially joined the estate in 1988.

Winemaker: Dominique Belluard until 2021. The future of the estate is uncertain.

Vineyard surface: 10.5ha (25.9 acres)

Viticulture: Mostly practicing biodynamic since 2001

Négociant work: None

Grapes: Gringet, mondeuse, altesse

Appellations: Vin de Savoie, Ayse

Filtration/Fining: Whites see Kieselguhr filtration before bottling, as does Les Perles du Mont Blanc.

Sulfites: All wines except the Pur Jus gringet and mondeuse see a light sulfitage before bottling.

DOMAINE PARTAGÉ

Gilles and Christine Berlioz

CHIGNIN

A fanatic for vineyard work with immense sideburns and a permanent suntan, Gilles Berlioz began with under a hectare (2.5 acres) of family vines in 1990, and increased his surface to a peak of 7 hectares (17.3 acres) that decade. Since embracing organics in 1999, Berlioz and his wife, Christine, have come to produce a sought-after range of stirring Chignin jacquères and sumptuous Chignin-Bergeron roussannes.

The Berliozes' quest for purity is perceptible in their evolving cellar procedures. They ceased oak-aging starting in 2006, and as of 2018 all their white wines pass winter in horizontal fiberglass egg containers to promote a continual *bâtonnage* on lees. Crisp reds from mondeuse (and a touch of persan) see whole-cluster, semi-carbonic maceration, without a starter ferment or pigéage. Pumping is banned from their cellar; all movements are done by gravity, or by hand with buckets. The wines remain conservative in terms of bottle conditioning. Low sulfite addition, occasional fining, and filtration are the norm for whites, while reds see just a low sulfite addition.

In 2016, the Berliozes took the curious step of changing the name of their estate to Domaine Partagé ("The Shared Estate") to honor the cooperative input of all their employees and interns.

Since: 1990

Winemakers: Gilles and Christine Berlioz

Vineyard surface: 4.5ha (11.1 acres)

Viticulture: Certified organic since 2002. Practicing biodynamic.

Négociant work: None

Grapes: Jacquère, altesse, roussanne, and more

Appellations: Chignin-Bergeron, Chignin, Roussette de Savoie, Vin de Savoie

Filtration/Fining: Whites occasionally see bentonite fining and are usually filtered. Reds are not filtered.

Sulfites: Usually 20–30mg/L at bottling

JEAN-YVES PÉRON

CONFLANS

After enology school in Bordeaux, Parisian-born biochemistry major Jean-Yves Péron worked for two natural wine legends: Cornas's Thierry Allemand (as did classmate Hirotake Ooka) and Alsace's Bruno Schueller. The natural winemaking project Péron initiated in 2004 in his mother's native Savoie proved his appetite for a challenge equaled that of his mentors. He took on old vines of jacquère and mondeuse on steep granitic schist soils near the village of Conflans, later adding parcels in nearby Fréterive. Péron soon became France's highest-profile proponent of skin maceration of white grapes (aka orange wines), a style he discovered while working for Schueller. In 2017, Péron debuted a new winery high on the hillside near his vineyards in Conflans.

"With macerated whites, you have structure, tannins, and bitterness—and yet they're almost easier to drink than direct-press whites, where you just have the volume of the fruit and the acidity. It's bizarre."

Péron's whites see minimal or no clarification after pressing, and are mostly skin-macerated, whole-cluster, for anywhere from two weeks to a year. In riper vintages, they are beguiling, savory brews: salty, yellowed, yet full of light.

Péron's reds are no less surprising. Far from the grape's dark, tannic archetype, Péron's whole-cluster mondeuses can seem like hardy mountain cousins to Jurassien trousseau. They are light reds for mildly masochistic tastes, low on alcohol, low on fruit, yet full of character. His négociant wines from Piedmont, meanwhile, are stylistically revelatory, showing what that region's native grapes can become when liberated from outmoded local vinification styles (which tend to hinge upon destemming).

Such is Péron's wizardry that he can make jacquère taste like cantillon and grignolino taste like grenache. Neither cheap nor crowd-pleasing, his wines reveal their ingenious construction slowly, in glimpses, like Alexander Calder mobiles.

THE WINE TO TRY

See Wines to Know: Savoie and Bugey, *page 259.*

Since: 2004

Winemaker: Jean-Yves Péron

Vineyard surface: 3ha (7.4 acres)

Viticulture: Practicing biodynamic

Négociant work: Yes, since 2011, from grapes in Alsace, Savoie, and Piedmont

Grapes: Jacquère, roussanne, mondeuse, and more

Appellations: All wines released as Vin de France

Filtration/Fining: None

Sulfites: None added

SAVOIE AND BUGEY

Natural Wine Dining

A boudin noir by charcutier Alain Grèzes at Annecy cave à manger Midget

The last half decade has seen natural wine dining spots appear in force in Savoie. These are mostly caves à manger: wine shops equipped with small kitchens that allow them to double as low-key restaurants (see page 39). (Bugey, with far less tourism, remains bereft of a dining scene.) The region's gastronomic restaurants, reliant on older tourism, have been less receptive to natural wine than its wine bars and caves à manger have been.

Most Savoyard classics involve some combination of melted cheese, potatoes, and pork. The local specialty *tartiflette*, for example, highlights Reblochon, a rich, mildly nutty raw cow's-milk cheese. You'll also find overlap with Swiss cuisine in raclette cheese, traditionally served using a special device that slowly sears one exposed face of a cheese disc the size of a scooter's front wheel, allowing diners to rake (*racler* in French) a glistening wave of melted cheese onto a plate of steaming potatoes and charcuterie.

L'Auberge Sur-les-Bois

ANNECY

Chef Daniel Baratier led a dashing career in Paris at Le Sergeant Recruteur and Les Déserteurs before opening his tony L'Auberge-sur-les-Bois in his native Savoie in 2018. Avoid the ready-made conserves—the real show begins with main courses for two persons, like a well-aged Blonde d'Aquitaine steak with bone marrow, artichokes, and olives. The wine list leans conservative but contains a solid selection of natural wines from Savoie and beyond.

Midget

ANNECY

Carine Paris and Benjamin Béranger, the partners behind Midget, met while working in restaurants in Australia, where their colleagues apparently referred to them both by the nickname they'd later adopt for this adorable Annecy cave à manger, opened in 2020. Paris and Béranger alternate chef and sommelier duties week to week, proposing well-informed and radical wine selection alongside a menu where artfully adorned charcuterie shares space with simple fresh preparations.

Kamouraska

ANNECY

Opened in 2016, Kamouraska is the intimate and wildly popular chef's table of chef Jérôme Bigot and Canadian sommelier Marie-Hélène Tardif, who offer exquisite small plates from land and sea accompanied by a radical and savvy natural wine selection. Book well in advance.

Avicenne

THONON-LES-BAINS

Avicenne owner-operator Thomas Aragonès worked as a server in gastronomic restaurants in the US before embracing sommellerie after a stint at Annecy two-star restaurant Clos des Sens. He opened his cozy cave à manger in the Lake Geneva town of Thonon in 2017, emphasizing natural wines and simple, well-sourced share plates and bistrot cuisine.

SAVOIE AND BUGEY

Legends in the Making

In the last decade, Bugey has finally seen an influx of ambitious young natural vignerons. Savoie, with much higher real estate prices—and greater distances separating the few estates working naturally—remains comparatively conservative. Many of both regions' most promising younger vignerons trained in the natural wine communities of nearby Jura and the Beaujolais before establishing their estates.

Corentin Houillon

CHAUTAGNE

After years of training in Australia and Oregon and four years working as winemaker at Domaine de la Ville de Morges in the Vaud region of Switzerland, young Jura native Corentin Houillon took over from the retiring vigneron at Domaine Veronnet in Chautagne in 2019. Under Houillon the estate comprises 4 hectares (10 acres) of gamay, gamay teinturier, pinot noir, mondeuse, jacquère, and altesse. From the start, Houillon's work has attracted major interest—thanks both to his impressive winemaking lineage (he is the nephew of Maison Overnoy-Houillon's Emmanuel Houillon) and his revelatory unsulfited and unfiltered takes on the native Savoyard grapes. From his sandstone-scree soils, he coaxes a pinot noir of surpassing rosiness and finesse.

La Combe aux Rêves

Grégoire Perron

JOURNANS

Grégoire Perron had almost a decade of vineyard experience under his belt, including work for the Jura's Jean-François Ganevat, when he established his tiny Bugey estate in 2010. His location in Journans places him at a crossroads, encompassing the chardonnay, gamay, poulsard and pinot of the southern Jura, as well as the jacquère, altesse, and mondeuse of Savoie. It can be hard to identify the labels of his ever-shifting cuvées as the work of the same winemaker (each features a unique graphic design by his partner, Judicaëlle), and indeed, Perron still seems to be finding his voice in vinification. There are misses, but the hits can strike hard, like a gamay-mondeuse pét-nat entitled La Grenadine. Perron deserves credit for sticking to his guns about zero sulfite addition and zero filtration, even through difficult vintages that saw him expand to négociant work from purchased grapes in the Mâconnais and Savoie.

Louis Terral

MÉRIGNAT

Former industrial engineer Louis Terral began working in viticulture in 2013, and put in time at a series of celebrated Beaujolais estates, including Julie Balagny, Michel Guignier, and Jean-Louis Dutraive—as well as with Mâconnais maven Philippe Valette—before founding his own estate in his girlfriend's native Bugey in 2016. In just a handful of hail-strewn vintages, he's achieved biodynamic certification and managed to confirm what many have long suspected about the local Bugey terroir: that it's apt for producing keen and ethereal natural, still, dry red wines. Working with a vertical press, zero sulfite addition, and no filtration, Terral is producing pure, glimmery gamays sure to make his Beaujolais mentors proud.

Corentin Houillon in the vines above his cellar

Carignan vines, in January, at Fontedicto in the Herault

12. The Languedoc

It's not easy to pinpoint a heyday in the history of winemaking in the Languedoc, the vast swath of Mediterranean plains, mountains, and plateaus spanning Carcassonne to the suburbs of Nîmes. Historically poor, the region embraced cave cooperatives in the late nineteenth century as a way of bypassing the négociants who had long exploited them.

But no sooner had railroads and cave cooperatives made the Languedoc a source of cheap red wine throughout France than even cheaper wine from Algeria began to supplant it. In the early 1900s, this led to a series of protests by Languedoc winegrowers in favor of, yes, "natural wine"—by which they meant wine produced without excessive chaptalization or blending with Algerian wine. This was the last we'd hear about natural wine in the Languedoc for many decades.

With few exceptions (such as Mas Jullien's Olivier Jullien and Louis Julian in the Gard), it wasn't until the 1990s, when an economic crisis rendered cave cooperatives and négociant sales definitively unprofitable, that the region's more farsighted, naturally inclined winegrowers began bottling their own wine. Among them were Jean-François Coutelou near Béziers, Didier and Jean-Luc Barral in Faugères, Bernard Bellahsen in Caux, and, a few years later, their neighbors at Clos Fantine, and Anne-Marie Lavaysse in Minervois.

These early natural vignerons of the Languedoc were initially unaware of one another's work. Natural wine in the Languedoc has only become a community since around 2004, thanks to the involvement of Roussillon vignerons Jean-François Nicq and Edouard Laffitte, Tavel vigneron Eric Pfifferling, and Ardèche vigneron Gérald Oustric, who organized the first La Remise tasting in Marie Pfifferling's parents' storage hangar in Nîmes (see page 281).

Partly thanks to the network surrounding La Remise, the 2000s saw many new and newly converted natural vignerons appear in the Languedoc. Today the region is to French natural wine what Southern rap was to US hip-hop in the 1990s: its talented underdog, whose vast, varied potential is impossible to ignore.

Total vineyard surface:
199,000ha (491,740 acres)[1]

Rate of organic agriculture:
16 percent[2]

Key grapes:
Grenache blanc, macabeo, terret, grenache, carignan, cinsault, mourvèdre, aramon

1 France Agrimer 2019
2 Agence Bio 2019

THE LANGUEDOC

Wines to Know

Wine lovers tend to look to the Languedoc for simple, rich, good-value reds. Natural wines from the region can indeed hew to that profile, but they offer much greater range: from the gleaming, keenly fruity Gard reds of Alain Allier and Louis Julian to the structured, baritone Faugères of Domaine Léon Barral to the concentrated, nervy creations of Clos Fantine and Bernard Bellahsen. The best evoke the wild herbal aromas of the dense Occitan garrigue.

Clos Fantine | Faugères–Tradition

Made from very low yields of carignan, cinsault, aramon, syrah, and grenache (sometimes with a touch of mourvèdre), about 80 percent destemmed, the Andrieu siblings' Tradition cuvée sees a vatting in large concrete tank for about two weeks. Few red wines in its undervalued price range offer its age-worthiness, its china-bark complexity, its stirring vitality. It is a tantalizing taste of a dream-vision of Languedoc viticulture in an Edenic state.

Domaine Mouressipe | Vin de France–Càcous

Typically a blend of 70 percent grenache with 30 percent syrah, Càcous exemplifies the smiling, airborne bonbon fruit that Alain Allier musters from his limestone soils in Saint-Côme-et-Maruéjols. A carefree cousin to the work of Eric Pfifferling and Nicolas Renaud farther north, it often shows a tad less extraction, a touch more CO_2, and a toothsome dash of volatility.

Fontedicto | Vin de France–Promise

Radical natural winemakers from throughout France often recall how in the past—before the advent of contemporary enology, and in an era when more wine was purchased to be aged before consumption—high volatile acidity was not an obstacle to producing a great wine. In southern regions like Bordeaux and the Languedoc, as often as not, volatile acidity (combined with low yields) was a *prerequisite* of a great, age-worthy wine. Bernard Bellahsen's Promise evolves like the career of Cat Power: atonal and shrill in the early years, but with time, it becomes suave, confident, evocative of classics.

DOMAINE MOURESSIPE

Alain Allier

SAINT-CÔME-ET-MARUÉJOLS

Alain Allier's older brother took over his family's small wine-growing estate in the Gard, so Alain led a long, happy career as a car mechanic. But an innate love for vines impelled him to begin purchasing vines in his hometown, starting at the age of twenty-five. His vines are arrayed around a local hill called Mouressipe, the site of a Roman settlement dating to 500 BCE. For years, Allier sold his grapes to the local cave cooperative, until a mutual friend introduced him to Eric Pfifferling, whose early natural wines inspired Allier to become a vigneron. In 2004, well into his forties, Allier left the cave cooperative and began producing his own bristling, cool-carbonic Gard wines, which found a devoted following among natural wine bars in Paris.

"It's true that in the Parisian market, the demand is for *glou-glou* wines with low alcohol and low prices. But we can't do without ripeness. It has to be there."

At first he vinified in his garage and stored the bottles at his parents' house, but in 2016, he moved production into a purpose-built cellar. He practices a variant of the cool-carbonic vinification style he learned from Pfifferling. The harvest is refrigerated overnight before either a short, whole-cluster maceration or a direct press that lasts a full twelve hours. Certain wines have seen light sulfite addition in recent vintages to guard against mouse (see page 378); these cuvées do not mention natural vinification on the back label.

Following the precise characteristics of Allier's cuvées can be challenging. For one thing, their names are almost all in the Occitan dialect. For another, the wines, with one or two exceptions, all bear Allier's signature smiling, lucent fruit.

THE WINE TO TRY

See Wines to Know: The Languedoc, *page 269.*

Since: 2004

Winemaker: Alain Allier

Vineyard surface: 8.5ha (21 acres)

Viticulture: Organic

Négociant work: None. He even sells some of his grapes.

Grapes: Chardonnay, chasselas, clairette, and many more

Appellations: All wines are released as Vin de France.

Filtration/Fining: None

Sulfites: Certain tanks see 10mg/L at bottling, but the majority of wines remain unsulfited.

Louis Julian

Louis Julian is a thirteenth-generation Gard vigneron whose simple table wines exist in a genre all their own. Practicing organic since 1979 and certified since 1984, Julian's estate in Ribaute-les-Tavernes produces the kind of very inexpensive wines that used to be called *vins de consommation courant*, or wines to be drunk within the year. A liter of Julian's light red, beloved by such natural wine luminaries as Bruno Schueller and Philippe Jambon, costs €3.50 at the estate.

If it's possible to make natural wine so cheaply, why isn't everyone doing it?

The secret is, Julian's wines are not quite natural. He adds a powerful yeast to ensure a swift, clean fermentation. Such is Julian's confidence that total sulfites will remain below the 10mg/L threshold of detection in his finished wines that he alone in the wine world includes the phrase "wine without sulfites" on his labels (rather than the more common phrase "no added sulfites").

Yeasting is not Julian's only departure from natural wine orthodoxy. The press juice for his "rich" red, which clocks in at around 12.5 percent alcohol, undergoes tangential filtration. And the press itself is a high-volume continuous press from 1934, a model rarely encountered in quality winemaking. His wines ferment outdoors in massive, 600-hectoliter vertical steel tanks.

Julian's viticulture is as singular as his winemaking. His estate comprises, in addition to 22 hectares (54.3 acres) of vines, 30 hectares (74 acres) of grains, which are used to feed a flock of three hundred chickens that roam among the vines. No other fertilizers are used in his very wide-planted, hugely productive vines, which include no less than forty-one grape varieties, including many obscure hybrids.

"When I stopped using chemical fertilizers, for fifteen years my yields dropped. For ten years they stayed the same. I said, 'Am I sure I'm on the right path?' But when I saw the earth no longer eroded in my vines, I knew I was on the right path. And since then, yields have climbed."

Julian's wines are nourishing rather than intoxicating. They're miniature sunbursts of invigorating, spritzy Languedoc fruit, laced with the salinity of the ancient seabed that undergirds the vineyards. Are they natural wines? Does it matter?

Julian exports only a tiny quantity to longtime clients in the Netherlands. The rest is bottled on-demand in refillable plastic containers and liter bottles and sold at a small organic grocery store adjoining Julian's cellar.

LE TEMPS DES CERISES

Axel Prüfer

LA TOUR-SUR-ORB

Originally from East Germany, Axel Prüfer decamped to France with his girlfriend to avoid national service in 1998, only to fall headlong into natural wine when he began doing vineyard work in Pic Saint-Loup. At Mas Foulaquier, he met natural wine enologist Yann Rohel (see page 323), who introduced him to the wider natural wine scene. He befriended Jean-François Nicq (see page 293) and Eric Pfifferling as both were establishing their respective estates in the Roussillon and Tavel. (When Pfifferling's first vintage ran into flood trouble at harvest, Nicq sent Prüfer to help with vinification.)

Nicq's practices of early picking and strict carbonic maceration provided inspiration, but after founding his estate in 2004, Prüfer would develop a loopier, more abstract style all his own. He abandoned sulfite addition in 2009, and today experiments with extra-long carbonic macerations (up to eleven months). His main red cuvées are parcel bottlings that see shorter, strict carbonic maceration in fiberglass tanks. Élevage also occurs in tank. Though they're just a twenty-minute drive over the hills north of Faugères's schist soils, Prüfer's vineyards lie chiefly on limestone at high altitudes, which contributes to the abiding angular lightness of his wines.

"I don't make wines with the consumer in mind. I try to do the opposite, and yet I always find someone willing to drink them."

THE WINE TO TRY

Vin de France–La Peur du Rouge

La Peur du Rouge is a mind-bending, high-altitude chardonnay made with a carbonic maceration of underripe fruit with overripe fruit, harvested weeks apart. It offers soft, sustained tropical fruit and a pleasant, leesy grip on the tongue.

Since: 2004

Winemaker: Axel Prüfer

Vineyard surface: 7.5ha (18.5 acres)

Viticulture: Organic since 2005

Négociant work: In low-yield vintages, Prüfer purchases certified organic grapes.

Grapes: Grenache, merlot, carignan, and more

Appellations: All wines released as Vin de France

Filtration/Fining: None

Sulfites: None added since 2009

CLOS FANTINE

Carole, Corine, and Olivier Andrieu

CABREROLLES

Siblings Carole, Corine, and Olivier Andrieu inherited their father Jacques's organically farmed Faugères vines upon his death in 1997, just a year after he left the cave cooperative and began estate bottling. In the quarter century since, they've transformed Clos Fantine into one of the world's most expansive testing grounds for a bold vision of low-yield permaculture and zero-zero vinification.

Olivier takes the lead in the vines, while Carole handles administration and Corine manages vinification. A former inorganic chemist with a poetic streak, Corine abandoned sulfite addition and filtration for all wines beginning in 2005. Most impressively, the Andrieus eliminated copper treatments on their entire estate for eight years starting in 2011. (Disaster struck in 2018, when black rot obliterated their entire harvest; they resumed minimal copper treatments in subsequent years.) The family remain dedicated to permaculture. They have forsworn plowing in favor of mowing and sheep pasturage, and have applied natural fertilizer just once in the past two decades.

> **"Yield is not a goal for me. For me, the goal is that nature be beautiful and everyone be happy. I think that nature gives us what we need."**
>
> —Corine Andrieu

Their cellar, built in 2000, contains large concrete tanks along with a few fiberglass vats. Reds are destemmed around 80 percent and vinified and aged in cement. The principal cuvée is the Faugères Tradition, a blend of all red varieties save aramon (which is not permitted within the Faugères appellation). Each year, they also release an inkier, selection-level monovarietal Faugères Courtiol, made from whichever variety the siblings feel was the best that year (mourvèdre in recent vintages). Rounding out the range are the Lanterne Rouge, a nervy blend of old-vine aramon and cinsault, and an extraordinary skin-macerated terret.

THE WINE TO TRY

See Wines to Know: The Languedoc, *page 269.*

Corine Andrieu in the cellar at Clos Fantine

Since: Estate bottling began in 1996.

Winemaker: Corine Andrieu

Vineyard surface: 30ha (74.1 acres)

Viticulture: Organic permaculture

Négociant work: Yes, since 2018

Grapes: Carignan, mourvèdre, syrah, and more

Appellations: Faugères

Filtration/Fining: None

Sulfites: None added since 2005

DOMAINE LÉON BARRAL

Didier and Jean-Luc Barral

LENTHÉRIC

Didier Barral lassoing a wounded bull

Brothers Didier and Jean-Luc Barral left the local cave cooperative in 1991 and began to bottle their wine in 1993 as Domaine Léon Barral, an homage to their grandfather. Didier credits an early visit from Domaine de la Pinte founder Roger Martin with awakening his aesthetic interest in natural winemaking. Today the estate is a model of biodiversity: it also comprises a herd of black pigs, several horses, and sixty cows. In addition to their immense, long-aged Faugères wines, the brothers Barral are known for pioneering direct seeding under cover crop in viticulture (see page 62), as well as cow pasturage, which they began in 2000. They stopped plowing entirely in 2005, reasoning that creating a mulch with rolled grass would better trap moisture.

The Barrals produce three Faugères reds, vinifying the various grape varieties separately and assembling them after aging. Macerations last from three weeks to a month, and combine a blend of destemmed and whole-cluster fruit, with pigéage. They employ a small, old vertical press. The selection-level Jadis and Valinière cuvées see a little over two years' aging in barrels, with very little new oak, while the basic Faugères ages in a mixture of cement and steel. These are broad, muscular reds, redolent of wild thyme, rippling with black fruit. Since 1993, the Barrals have also produced one of France's earliest orange wines, a field blend of terret blanc, terret gris, viognier, and roussanne that sees a few hours' maceration in the press.

"At the beginning, when we started, we didn't know that natural wine existed. What we wanted to do was make wine without adding products."

—Didier Barral

THE WINE TO TRY

Faugères

Didier Barral's flagship Faugères is a blend of carignan, grenache, and cinsault. Mostly destemmed, aged in cement and steel, the wine preserves a mild tension to its deep, chocolate-dusted cassis fruit.

Since: Estate bottling began in 1993.

Winemaker: Didier Barral

Vineyard surface: 35ha (86.4 acres)

Viticulture: Practicing biodynamic

Négociant work: None

Grapes: Carignan, cinsault, mourvèdre, and more

Appellations: Faugères, IGP Pays d'Hérault

Filtration/Fining: None

Sulfites: Depends on bacterial analysis—if there's a lot of bacteria in the finished wine, he adds 20mg/L.

MAS COUTELOU

Jean-François Coutelou

PUIMISSON

Jean-François Coutelou is the dynamic and passionate vigneron behind Mas Coutelou, a fourth-generation estate in the backwoods Béziers suburb of Puimoisson. His father obtained Nature & Progrès certification back in 1987, setting the stage for what would become Coutelou's lifelong work of viticultural land art. Each year, Coutelou plants trees and creates hedges around each of his parcels in an effort to restore the natural Languedoc biodiversity that has been erased in his area by productivist chemical agriculture.

Coutelou's visionary work in the vines isn't for nothing, either, for he's a zero-zero sorcerer in the cellar. His neighbors and peers at Domaine Léon Barral and Clos Fantine might have the Faugères appellation on their side, but Coutelou more than compensates with his imaginative, freethinking work with a wealth of grape varieties. His Le Blanc displays supreme balance despite its imperceptibly high alcohol. His multitude of red cuvées tend to see long macerations with partial or total destemming; the results are gripping and athletic reds without a trace of glycerolic flab. If you get the chance to visit his cellar, be sure to set aside extra time to taste through his uncommercialized solera cave, which contains forty-seven soleras of otherworldly sweet oxidative wines dating back to his grandfather's era. It ranks among the wonders of the wine world.

THE WINE TO TRY

Vin de France–OW

From eight white grapes, including a healthy dose of muscats, Coutelou produces OW, one of France's most mesmerizing orange wines.

Since: Jean-François is the fourth generation to make wine at the estate, founded in 1876. Estate bottling began in 1993; wine was previously sold to an organic négociant.

Winemaker: Jean-François Coutelou

Vineyard surface: 13ha (32.1 acres)

Viticulture: Certified organic since 1987

Négociant work: None

Grapes: Syrah, grenache, carignan, and more

Appellations: All wines released as Vin de France

Filtration/Fining: None

Sulfites: None added since the mid-2000s

DOMAINE FONTEDICTO

Bernard and Cécile Bellahsen

CAUX

The patron saint of Languedoc organic, biodynamic, and natural wine, Bernard Bellahsen has followed a long and intuitive path to natural wine acclaim since arriving in France from Tunisia with his parents at age ten. After agricultural studies in Montpellier, he settled in the plains between Béziers and Narbonne to grow grapes to produce grape juice, not wine. Animated by a desire to work with animals, in 1980 he became probably France's only grower to use horse plowing for the production of grape juice. He discovered and adopted biodynamics soon after.

It wasn't until 1994, when he moved his farm to superior thin-soiled, limestone-and-basalt terroir in Caux, that he began fermenting his grape juice on a commercial scale. His estate sits near the site of a former Roman fountain—*Fontedicto* is a Latin translation of Fontarèche, which means "fountain that we hear." Working without filtration or additives, including sulfites, appears to have come as naturally as breathing to Bellahsen. His profound, long-aged reds found a cult following in early natural wine circles, and Japan soon became an important export market. Aged in old barrels followed by extended periods in steel tanks, his grenache, carignan, and syrah blends, like Coulisses and Promise, are compacted and spiky in youth, with a lancing volatility, not unlike the Beaujolais of Bellahsen's early biodynamic peer Christian Ducroux. Yet with age, the wines smooth out, like mountain ranges, becoming panoramic and serene. Until 2011, Bellahsen also produced a few barrels of terret from very old vines above his house.

Today Bellahsen is semiretired, having ceded most responsibility for the estate to his wife, Cécile. The couple also grow grain and mill it on-site. Bellahsen uses it to bake bread, which he delivers to friends and sells at the market in nearby Pézenas.

Bernard Bellahsen in his cellar during harvest

THE WINE TO TRY

See Wines to Know: The Languedoc, *page 269.*

Since: 1977, in a different part of the Languedoc. Bellahsen settled in Caux and began producing wine in 1994.

Winemaker: Bernard Bellahsen

Vineyard surface: 1.5ha (3.7 acres)

Viticulture: Practicing organic since 1977. Practicing biodynamic since about 1980. Certified biodynamic since 1993.

Négociant work: None

Grapes: Carignan, syrah

Appellations: All wines released as Vin de France since 2006. Formerly Coteaux du Languedoc.

Filtration/Fining: None

Sulfites: None added

LA SORGA

Antony Tortul

MONTAGNAC

Radical natural wine négociant Antony Tortul established La Sorga in 2008 in his father's garage, after six years working for conventional estates throughout the South of France and the Rhône. His wines soon became the toast of young natural wine bars in Paris and beyond, thanks to his insistence on zero-sulfite vinification, an impressive grape-scouting ethic, and memorably garish wine labels. The label art for cuvées like Sorgasme and Noir Métal—along with the forty-odd other micro-cuvées he releases each year—caught the zeitgeist, giving ample ammunition to critics who consider natural wine fundamentally unserious.

"Our wines age well because we make wines that are a bit extracted. We don't make wines to please the Paris market. We make wines that truly resemble us."

Tortul is, in fact, a very reflective winemaker, whose methods have evolved over the past decade. (He changed cellars twice, moving first to Béziers and later, in 2016, to Montagnac.) His grape sources, all certified organic or biodynamic, span the South of France and the Rhône, with an emphasis on the Languedoc. Like his friend Alice Bouvot in the Jura, he prefers to harvest very early, seeking to avoid high-alcohol wines, sometimes at the expense of mature fruit. Lately Tortul favors partially destemmed maceration in fiberglass tanks and sandstone jars, often including proportions of direct-press juice at the start of fermentation. In 2019, he began farming a small parcel of vines in Aspiran.

Keep an eye out for his rare, long-aged oxidative wines, which show the symphonic unctuousness of great rancio (a term used in Spain and southern France to denote wines produced with intentional oxidation, often combined with exposure to sunlight).

THE WINE TO TRY

Vin de France–Cuvée A Freux

Generally a one-and-a-half-month maceration of hand-destemmed grenache blanc, sometimes with a dash of marsanne, Cuvée A Freux ("Frightful Cuvée") is a bristly, oxidative, and lightly volatile orange wine, low on extraction and surprisingly refreshing, with a savory patina to its kumquat-flecked fruit.

Since: 2008

Winemaker: Antony Tortul

Vineyard surface: 0.8ha (2 acres)

Viticulture: Practicing organic. All grape purchases are certified organic.

Négociant work: Yes, it's almost all négociant work.

Grapes: Grenache, mourvèdre, carignan, and more

Appellations: All wines released as Vin de France

Filtration/Fining: None

Sulfites: None added

BRUTAL!!!

Emblazoned on countless wines from a plethora of natural vignerons, from France and Spain to Austria and beyond, "Brutal!!!" or "Brutal Wine Corporation" (as it is sometimes styled) is the first runaway-success meme of the natural wine world.

Its origins lay in a misunderstanding that arose in 2010. Danish natural wine importers Sune Rosforth and Henrik Sehested of Rosforth & Rosforth (see page 404) had brought Catalan natural vignerons Laureano Serres and Joan Ramón Escoda (see pages 348–349) to taste with Hérault vigneron Remi Poujol and Antony Tortul of La Sorga. The French and Danish were mystified by the Catalans' habit of exclaiming "Brutal!" after tasting things. Over dinner in Béziers, Poujol and Tortul, abashed at having their wines described as "brutal" all day, finally asked their new Catalan friends what, exactly, they meant by the phrase, which has negative connotations in French. In fact Serres and Escoda used it as a term of superlative praise, the way Americans say "awesome."

The group decided to employ this novel phrase for a communal wine brand, to be employed on experimental, one-off wines by vignerons who vinified the entirety of their production without additives. Axel Prüfer of Le Temps des Cerises, among the initial band of five vignerons to use the label, defines a "brutal" bottling as "an experiment that you can monetize." The group decided the Brutal Wine Corporation label should be disseminated via co-option, i.e., anyone who wanted it could use it.

The original significance of the "Brutal Wine Corporation" has been slightly lost as the notion has grown in popularity and unrelated estates have borrowed the phrase and the iconograpy. As Tortul wrote in a kind of on-demand press release in 2020, "This whim of drunken buddies has been transformed into a sales pitch for mediocre wines, sometimes not even natural ones!"

In response, in 2021 Tortul and his friends debuted a new "Brutal" wine label featuring gold lettering, the use of which they intend to limit to the vignerons originally associated with the "Brutal!!!" phenomenon.

The original five:
Antony Tortul, Remi Poujol, Axel Prüfer, Laureano Serres, and Joan Ramón Escoda

Those who followed shortly after, with the blessing of the original five:
Antonio Vilchez, Jean-Louis Pinto, Sébastien Dervieux, Patrick Bouju, Charles Dagand, Alice Bouvot, Gianmarco Antonuzzi and Clémentine Bouveron (Le Coste), Joseph Jefferies, Damien Bureau, Alain Castex, Wim Wagemans, Antonella Gerosa and Massimo Marchiori (Partida Creus), Pep Torres, and Norbert Dutranoy

A range of BRUTAL!!! bottles, from Rémi Poujol, Sébastien Dervieux, Tom Lubbe, and more

LE TEMPS FAIT TOUT

Remi Poujol

ADISSAN

Behind the gentle demeanor of a record store clerk, Remi Poujol is among the most uncompromising vignerons of the Languedoc. His profound, paysan wines represent the yang to the yin of his friend (and grape client) Antony Tortul's more inventive and media-friendly work at La Sorga.

Poujol converted to organics in 2002, later embracing horse plowing, following the example of his friend Bernard Bellahsen. Poujol adopted zero-zero natural vinification in 2008 upon completing construction of his heavily insulated cellar.

"I wanted to do a red and white, and that's it. To assemble the terroir of Adisan and make an ideal wine each time. To put terroir above the grape variety."

Poujol destems his red grapes entirely. Most see two- to three-week macerations in steel and fiberglass tanks, while his carignan Vieilles Vignes macerates for fully nine months. Only the Vieilles Vignes sees aging in used oak barrels (from Domaine d'Auvenay, no less!); other reds age in steel and fiberglass. With time, early notes of volatility and brettanomyces tend to integrate into wholesome, sanguineous Hérault reds of a quiet finesse. His lone white wine is a refreshing blend of ugni blanc, clairette, bourboulenc, and viognier, of which a portion undergoes light skin maceration and aging in sandstone amphorae.

THE WINE TO TRY

Le Temps Fait Tout Rouge

From a blend of carignan, grenache, and syrah, Poujol's red Le Temps Fait Tout ("Time Does Everything") can be spiny and reduced in youth; a few years' aging does wonders for its fine, leather-toned cassis fruit and herbal aromas.

Since: 1997. Estate bottling began in 2003.

Winemaker: Remi Poujol

Vineyard surface: 10ha (25 acres)

Viticulture: Certified organic since 2002

Négociant work: None

Grapes: Grenache, syrah, carignan, and more

Appellations: All wines released as Vin de France

Filtration/Fining: None

Sulfites: None added since 2008

LA MAZIÈRE

Fabrice and Momoko Monnin

PADERN

Fabrice Monnin in his cellar

A geologist by training, Fabrice Monnin discovered natural wine while harvesting with Marcel Richaud in Cairanne. He worked brief spells for Jean-Pierre Robinot at L'Ange Vin and for Cyril Bordarier at Le Verre Volé in Paris before returning to his native Besançon in 2001 to found Les Zinzins du Vin, an influential natural wine cave à manger that he ran for the next fifteen years. During this time, he became a client and friend of original La Mazière vigneron Jean-Michel Labouygues.

After Labouygues died in 2006, Monnin purchased ten barrels of mystery wine remaining in Labouygues's ramshackle cellar in Padern, not yet sure what he would do with them. In 2012, having decided to become a vigneron, Monnin purchased Labouygues's entire estate, which had fallen into disrepair.

Monnin makes wine in the sorcerous manner of Labouygues: tiny yields of 10 to 15 hectoliters per hectare, late harvests (often at 16 to 18 percent potential alcohol), extended oxidative aging in old oak for at least two years, and a free hand in cross-vintage assemblage. In fact, all of Monnin's wines contain some portion of wine from Labouygues's ten remaining barrels, which are replenished with new wine each year. Early reds from carignan and syrah were intense, thrillingly volatile, often off-dry; in recent years, Monnin has shortened maceration times and begun practicing a post-press infusion on macabeu marc, yielding a more slender wine style. Whites, from old-vine macabeu and grenache blanc, see skin maceration and oxidative aging; they are monuments of a nearly extinct, hugely rewarding oxidative Languedoc wine style.

"You don't make extraordinary wines, wines that make people shiver, without flaws. It's not possible. The magic installs itself where there are flaws."

—Fabrice Monnin

THE WINE TO TRY

Vin de France–Macabeu

Monnin's mostly macabeu white is labyrinthine in complexity, a ripe Languedoc take on Vin Jaune, a nonvintage puzzle without a solution.

Since: Fabrice Monnin purchased the estate in 2012.

Winemaker: Fabrice Monnin

Vineyard surface: 8ha (19.8 acres)

Viticulture: Practicing organic

Négociant work: None

Grapes: Carignan, grenache, macabeu, and more

Appellations: All wines released as Vin de France

Filtration/Fining: None

Sulfites: None added

PRECURSORS

Jean-Michel Labouygues

Little is known about the late Padern vigneron Jean-Michel Labouygues, who began producing masterpieces of oxidative, long-aged macabeu and carignan in the 1970s. When clients visited his legendarily messy cellar before his death in 2006, Labouygues would claim not to know precisely what was in his bottles, labeled merely "Vin de Table Français." To early natural wine collectors, Labouygues's wines offered a rare glimpse into historical modes of oxidative natural winemaking in the Aude. Fabrice Monnin, Labouygues's former client and successor at La Mazière, suggests the wines be tasted in the manner of old whiskey or rum.

Monnin has maintained Labouygues's cellar the way he found it: walls black with mold, unlabeled bottles jammed in crevices everywhere, and a logic-defying, M. C. Escher–esque agglomeration of ancient barrels, illuminated only by daylight from the street, for there is no electricity.

TASTING DESTINATIONS
LA REMISE

When: The end of March or early April, usually

Who: The natural vignerons of the South of France, from the southern Rhône to the Languedoc and the Roussillon

For: Intended for professionals, but open to the public

Where: The Palais de Congrès in Arles

No other tasting salon has promoted and united the natural wine scene of southern France like La Remise, first organized in 2004 by Roussillon vignerons Jean-François Nicq and Edouard Laffitte, Tavel vignerons Eric and Marie Pfifferling, and Ardèche vigneron Gérald Oustric. The first edition was held in Marie's parents' storage hangar, a site known colloquially as *une remise.*

Since 2012, the salon has been held in Arles (save for a break during COVID), and the role of organizer has changed hands several times. Under Le Raisin et l'Ange's Antonin Azzoni and fellow Ardèche vigneron Grégory Guillaume, La Remise remains as welcoming and youthful as ever. It's followed by a popular vigneron dinner and a party featuring live music.

"At the start, La Remise wasn't necessarily intended as a salon. At first it was a place for us and our friends to meet, to exchange points of view on the difficulties that we encountered in working naturally."

—Tavel vigneron Eric Pfifferling

LE PETIT GIMIOS

Anne-Marie and Pierre Lavaysse

SAINT-JEAN-DE-MINERVOIS

On a quest for self-sufficiency, Anne-Marie Lavaysse and her son Pierre established their polycultural farm in the hamlet of Gimios in 1993. The Lavaysses delivered their small grape production to the local cave cooperative until 1999, when, encouraged by fellow vignerons active in biodynamic circles like Didier Barral and Thierry Navarre, they began to bottle their own wines. In later years, they constructed their own two-story cellar insulated with lambswool.

Muscat à petits grains forms the basis for the majority of the Petit Gimios white and rosés wines. Reds are parcel assemblages in which carignan and grenache are complemented by obscure varieties like terret rose and oeillade. They see foot crushing, then macerations of around ten days, before élevage in steel tank. The intense Rouge de Causse derives from a co-planted ungrafted parcel dating to 1840!

"I'd found a place to live with my four cows. And there was a vineyard in ruin. I like plants a lot, and I said, 'Well, I'll try to save them.'"

—Anne-Marie Lavaysse

Adepts at agroforestry and rainwater collection, the Lavaysses lavish as much attention on their vegetable garden and their cow as they do on their vines. Pierre Lavaysse is an avid surfer and rock climber and a fluent Anglophone; in recent years, he's been aided at the estate by his American girlfriend, Heather Schultz.

THE WINE TO TRY

Vin de France–Muscat Sec Des Roumanis

The Lavaysses' gauzy dry muscat is an aromatic cloudburst of wildflowers and lavender with mild oxidative inflections.

Pierre (*left*) and Anne-Marie Lavaysse outside their cellar

Since: 1993. Anne-Marie began estate bottling in 1999.

Winemaker: Pierre Lavaysse

Vineyard surface: 5.7ha (14 acres)

Viticulture: Certified biodynamic

Négociant work: None

Grapes: Grenache, carignan, alicante, and more

Appellations: All wines released as Vin de France

Filtration/Fining: None

Sulfites: None added since 2002

TASTING DESTINATIONS
MONTPELLIER NATURAL WINE SALONS

The last weekend of January—before the more famous natural wine salons in Angers and Saumur—Montpellier is the site of a separate circuit of salons. Geography and timing give them a different flavor from those of the Loire. Among international buyers, only the diehards bother with the detour south; conversely, French professionals seem more inclined to take a winter weekend by the Mediterranean than one in the freezing, wet Loire.

Le Vin de Mes Amis

When: The last weekend of January in Montpellier, and toward the end of November in Paris

Who: Organized by Charlotte Sénat of Domaine Sénat

For: Intended for professionals, but in practice, anyone can get in

Where: The Montpellier salon is held at Domaine de Verchant in the suburb of Castelnau-le-Lez. The Paris salon is held at the Maison de l'Amérique Latine in the 7th arrondissement.

Minervois vigneron Charlotte Sénat has hosted this swank natural wine salon in some form since 2004. It unites many of the most industry-approved practitioners of the loose ideology of natural wine, along with a scattering of their younger, more radical protégés.

Les Affranchis

When: The last weekend of January in Montpellier, and at the end of March in Paris

Who: Organized by sommelier and wine writer Laetitia Laure

For: Intended for professionals, but in practice, anyone can get in

Where: The locations shift. Recent Montpellier editions have been held at the Château de Flaugergues. The Paris edition is usually held in the 11th arrondissement.

Blois-based sommelier and wine writer Laetitia Laure has organized a roving series of natural wine salons called Les Affranchis ("The Made Men" in mob lingo) since 2010. Anchoring the salons are genuine "made men" of natural wine, like Thierry Puzelat and Nicolas Carmarans—but Laure includes many younger natural vignerons just getting started in natural vinification as well.

Les Vignerons de l'Irréel

When: The last weekend of January

Who: Organized by Hérault vignerons Julie Brosselin and Ivo Ferreira

For: Open to the public

Where: An event space behind a bowling alley at 188 avenue du Marché Gare, a long walk from the Montpellier train station

Ivo Ferreira and Julie Brosselin organize the most lively and radically zero-zero of the Montpellier salons, which by the evening turns into a party with live music, craft beer, and hundreds of unsulfited natural wines.

Beaujolais vigneron Cyrille Vuillod pouring at Les Vignerons de l'Irréel

Natural Wine Dining

Chai Christine Cannac in Bédarieux

The cuisine of the Languedoc varies so much between subregions that "Languedoc cuisine" is an almost meaningless term. Lamb, pork, and beef are staples in the mountains, while saffron-scented fish soup, brandade, and other pan-Mediterranean specialties typify its coastal cuisine.

But no one goes to the Languedoc for the food. Wealthy vacationers tend to prefer Provence's Côte d'Azur, with the result that there are very few good restaurants in the region, let alone many serving natural wine.

The few restaurants animating the Languedoc natural wine scene today reveal a very recent infusion of youthful energy from outside the region.

Les Canons

MONTPELLIER

Upstairs is a terraced wine bar and wine shop and downstairs is a fancy-free bistro at this semi-subterranean Montpellier natural wine destination, opened in 2018 by Anne Lapierre (youngest daughter of Marcel) and her boyfriend, Gwenaël Boisrame. Perhaps the only restaurant in the region where gamays outnumber grenaches, it is packed to bursting during the Montpellier natural wine salons.

Chai Christine Cannac

BÉDARIEUX

Sommelier and wine agent Christine Cannac has run this eponymous cave à manger on a sunny square beside the town hall in the hardscrabble Haut-Languedoc town of Bédarieux since 2007. A former student of Jean-François Coutelou (back when he taught at a local hospitality school), Cannac goes way back in natural wine circles, and has the gem-stuffed cellar to show for it. Visiting wine geeks and locals alike sip back vintages from Bruno Schueller and Didier Barral while snacking on platters of tinned fish and charcuterie sourced by gourmet food dealer (and fellow natural wine veteran) Betty Le Coq.

Picamandil

PUISSALICON

The spirit of Raymond Lecoq's La Cave Saint-Martin lives on in Frédéric Lamboeuf's inviting and unpretentious café-slash-épicerie-slash-wine bar Picamandil. He offers local natural wines, cheeses, charcuterie, and conserves, as well as organic bread his wife, Rebecca, bakes in her adjacent bakery.

THE LANGUEDOC

Legends in the Making

Throughout the 2000s, many Languedoc natural vignerons drew inspiration from the masterful Faugères of Didier Barral and the cool-carbonic macerations of Eric Pfifferling and Jean-François Nicq. Nowadays, the upstart natural vignerons of the Languedoc (like their counterparts in the Roussillon) increasingly take their cues from zero-zero vignerons like Bernard Bellahsen, the Andrieu siblings of Clos Fantine, and Jean-François Coutelou. The most promising include many newcomers to the region.

La Graine Sauvage
Sybil Baldassarre

CAUSSINIOJOULS

Charismatic young Italian enologist Sybil Baldassarre set up her 2-hectare (5-acre) estate in Faugères in 2015 with the help of crowdfunding company Terra Hominis. She made a name for herself with her old-vine, tank-aged Faugères blanc Rocalhas, a swooningly pretty, white-floral blend of grenache blanc, marsanne, and roussanne from vines in Cabrerolles. Baldassarre works alongside her partner, Bergerac native Alexandre Durand, who makes his own range of energetic unsulfited reds as Pèira Levada.

Olivier Cohen

ARGELLIERS

Nice native Olivier Cohen went from studying law into a charmed career in natural wine, first at Olivier Labarde's renowned Nice wine bar La Part des Anges (see page 326), then with internships with Maison Valette in the Mâconnais, Domaine Arena in Corsica (see page 320), Domaine Rivaton in the Roussillon, and Thierry Allemand in Cornas. In 2014, he and his wife purchased their 7-hectare (17.3-acre) estate in Argelliers, just a half hour west of Montpellier. Today he farms his merlot, syrah, cinsault, grenache, carignan, and cabernet sauvignon organically, with a regimen of direct seeding under cover crop and minimal plowing. His estate wines, emblazoned with cryptic motifs illustrating their names—such as Ronds Noirs ("Black Circles") and Ronds Rouges ("Red Circles")—are produced with patience and inventiveness.

Les Bories Jefferies
Joseph and Amandine Jefferies

CAUX

A thoughtful, bespectacled Brit from Warwickshire, Joseph Jefferies decided to become a vigneron after falling in love with his wife, Amandine, whose father owned a patch of vines on the ancient lava flow terroir near Caux in the Hérault. He began in 2003 and was self-taught for several years, until he encountered Bernard Bellahsen and Remi Poujol. His wines have been made without filtration or sulfites since 2015. His greatest achievement is La Pierre de Sisyphe, a blend of 70 percent terret with grenache and marsanne aged on *gross lees* that brims with volcanic minerality.

Sybil Baldassarre in her Faugères vines

Vineyards at Domaine Gauby in Calce

13. The Roussillon

Among the grossest oversimplifications of the wine world is the way the Roussillon is habitually considered an appendage of the Languedoc. The Roussillon is distinct both geologically and culturally—not least because its southerly Pyrénées-Orientales department was a part of Spanish Catalonia until 1659.

Wine culture has always been quite particular in this warm, windy corner of the Mediterranean, historically renowned for fortified wines, oxidative or rancio wines, and sweet wines. The Roussillon's natural wine culture is equally distinctive.

As in the Languedoc, natural winemaking took root on a commercial scale only with the advent of estate bottling, which began with the slow collapse of cave cooperative culture. Calce vigneron Gérard Gauby, who left his local cave cooperative in 1985, is considered the forerunner of organic viticulture and great dry winemaking in the region. It wasn't until ten years later that former Corbières vigneron Alain Castex and his erstwhile companion Ghislaine Magnier would establish a beachhead for natural wine in Banyuls-sur-Mer. They presaged what would become a first wave of outsiders establishing estates in and around Latour-de-France in 2002 and 2003, including the influential former Gard winemaker Jean-François Nicq, Burgundy-trained former mushroom dealer Bruno Duchêne, Domaine Gauby alumnae Jean-Louis Tribouley and Tom Lubbe, the Rhône-trained Loïc Roure, and Saumur-Champigny émigré Cyril Fhal. They were followed by a second wave who worked in much the same spirit: vignerons like Stéphane Morin, Laurence Krief, Antony Guix, and Edouard Laffitte.

These vignerons came to the Roussillon for the manifest grandeur of its still-affordable terroirs. With the exceptions of Bruno Duchêne and Laurence Krief, who obtained vines in Banyuls, these vignerons all purchased vines in the gneiss and granitic sands of the villages surrounding Latour-de-France. There, co-planted old-vine grenache and carignan are abundant; what's lacking are vignerons willing to farm them. The paysan winemaking culture of the Roussillon largely went bust with the decline in popularity of the local sweet and fortified wines, ruined by a dependency on chaptalization and filtration.

Today, thanks to the efforts of its natural vignerons and its proximity to the vibrant natural wine scene in and around Barcelona in Catalonia, the Roussillon is among France's most dynamic wine communities—a cosmopolitan, cross-cultural frontier for natural wine.

Total vineyard surface:
20,500ha (50,657 acres)[1]

Rate of organic agriculture:
29 percent[2]

Key grapes:
Macabeo, grenache blanc, carignan blanc, carignan, cinsault, grenache, syrah, mourvèdre

1 France Agrimer 2019
2 Agence Bio 2019

Wines to Know

With few exceptions, the Roussillon was until the early 2000s principally known for boozy sweet wines and the occasional weighty red. The natural wine renaissance that has occurred since then has brought the spotlight to its racy, light Mediterranean reds and orange wines, like those produced by Jean-François Nicq, Tom Lubbe, and Alain Castex, as well as the potential of its Banyuls schist terroirs to yield complex reds of surpassing finesse, as shown in the work of vignerons like Bruno Duchêne and Laurence Krief of Domaine Yoyo.

Bruno Duchêne | Collioure–Val Pompo

Bruno Duchêne's Val Pompo is an impressive demonstration of how low yields and careful farming on Banyul's elevated schist terroir can transform even a workmanlike, low-acid grape like grenache blanc into something subtle and majestic. The wine sees a direct press and is barreled down for a season's élevage two to three days after fermentation begins in tank. Its panoramic tropical fruit represents one of the heights of southern French white winemaking.

Les Foulards Rouges | Vin de France–Octobre

Octobre is not Jean-François Nicq's grandest wine. But it is the most emblematic of his abiding affection for Lapierre/Néauport-style cold-carbonic vinification, which—for better or for worse, depending on whom you ask—he and his friends have since spread to many natural estates in the Rhône and the Roussillon. Critics might complain that Octobre, a primeur of syrah with a touch of grenache that sees ten-day maceration before one-month élevage in *demi-muid*, barely evokes syrah, let alone its Albères gneiss terroir. Supporters say, who cares, when the result is a diaphanous, shimmery, deliriously drinkable wine like Octobre?

Les Vins du Cabanon | Vin de France–Canta Mañana

Alain Castex's cuvées are linked by his peculiar winemaking process, in which the whites, harvested first, make up the bottom layer of a tank, upon which, after a light pressing and revatting of the white marc, red grapes are layered over top. Canta Mañana, a kicking, electric rosé, is what happens when free-run red juice from merlot, syrah, grenache, and mourvèdre passes through barely pressed marc of macabeu and grenache blanc, absorbing the white grapes' perfume. It's a featherlight, dry, aerial rosé from a region whose conventional wines are more associated with power, fruitiness, and weight.

LES VINS DU CABANON

Alain Castex

FOURQUES

The genial grandaddy of radical natural winemaking in the Roussillon, Alain Castex today is at the quiet close of a long and eventful career. From 1981 to 1994, he produced wine with his first wife in Corbières, and began to practice organics in 1989. He started experimenting with natural vinification in 1990 or 1991 after meeting the famed enologist Max Léglise at a conference in Béziers. After the dissolution of his marriage, he and his new companion, Ghislaine Magnier, moved to Banyuls and became the first natural vignerons in the region, farming incredibly steep, rocky, low-yield sites manually and vinifying without added sulfites or filtration.

Castex says he truly fell into the world of natural wine in 1997. His Belgian importer introduced him to Hérault vigneron Bernard Bellahsen, and soon after Castex met natural wine agent Jean-Christophe Piquet-Boisson, who introduced him to Marcel Lapierre, Pierre Overnoy, and others. The delicate, handcrafted wines he made with Magnier during their twenty-year career in Banyuls would make the couple's fame. In 2015, the two split and sold the estate to young vigneron Jordi Perez (see page 301). Castex retains a small surface of old rented vines in Fourques, where, now in semiretirement, he rents a cellar ironically larger than the one he'd been working in for the previous two decades. From his co-planted vines, he does several passes of harvest, layering the grapes in his small tanks and producing several different cuvées as the tanks are filled.

"The fact that there are now people who sell reverse osmosis machines—saying that, with them, you can make a natural wine without additives—means that our original idea of natural wine was good. It corresponds to a demand."

Ever curious to try new things, Castex has ceased barrel aging since moving to Fourques and instead uses Italian and Spanish amphorae.

THE WINE TO TRY

See Wines to Know: The Roussillon, *page 289.*

Since: From 1981–1994, he made wine in Corbières. From 1995–2015, he made wine with ex-companion Ghislaine Magnier in Banyuls. He's made wine in Fourques since 2016.

Winemaker: Alain Castex

Vineyard surface: 2.5ha (6.2 acres)

Viticulture: Organic since 1989

Négociant work: Yes, he buys some organic syrah and merlot.

Grapes: Macabeu, oeillade, carignan, and more

Appellations: All wines released as Vin de France

Filtration/Fining: None

Sulfites: None added

BRUNO DUCHÊNE

BANYULS

Already steeped in natural wine culture from his native Blois (where he was a neighbor of the Puzelats) and his decadelong career as a Burgundy-based wholesaler of wild mushrooms, Bruno Duchêne moved to Banyuls in 2002 after a year working for Frédéric Cossard in Saint-Romain. In Banyuls, Duchêne found himself at the front line of the Roussillon's nascent natural wine movement. He acquired his first vines—the ones that yield his cuvée La Pascole—from Alain Castex, whom he befriended upon arrival.

The old, co-planted grenache, grenache blanc, grenache gris, and carignan vines that yield his estate cuvées are all planted on Banyuls's steep schists at around 300 meters (984 feet) altitude. Duchêne takes the unusual approach of differentiating his parcels by farming strategy: some are cable plowed, others are grassed over, and the parcel that yields his wine Anodine is entirely worked by hand with hoes. He avoids copper treatments, and supplements his very low yields (often under 10 hectoliters per hectare) with a baroque, ever-changing range of négociant purchases, often from his many renowned vigneron friends (collaborators include Dominique Derain, Bertrand Jousset, and Nicolas Carmarans). In 2012, after a decade of vinifying in a tiny cellar in Banyuls's Cap d'Ona neighborhood, Duchêne moved across town into Les 9 Caves (page 300), a collaborative winery (and restaurant) project for which he'd organized investment capital.

"Pascole, Anodine, Val Pompo, and La Lune are all my main cuvées. All the rest are little cuvées to amuse myself. Because it's good to travel and meet people and continue to live."

For red fermentations, Duchêne layers whole-cluster and destemmed grapes in steel and wooden vats, before aging in Burgundy barrels. His reds are sumptuous, riper in style than those of early peers like Castex or Jean-François Nicq, yet never overly marked by the Roussillon heat.

THE WINE TO TRY

See Wines to Know: The Roussillon, *page 289.*

Since: 2002. Moved to Les 9 Caves in 2012.

Winemaker: Bruno Duchêne

Vineyard surface: 4ha (10 acres)

Viticulture: Practicing organic

Négociant work: Yes, it comprises over half his production each year.

Grapes: Grenache blanc, grenache gris, grenache, carignan

Appellations: Collioure, IGP Côte Vermeille

Filtration/Fining: None

Sulfites: Occasionally a dose of 10–20mg at bottling, but wines are more often unsulfited.

DOMAINE YOYO

Laurence Krief

MONTESQUIEU-DES-ALBÈRES

Laurence Krief's banker still admonishes her for her unserious estate name, which comes from a nickname she earned for being hyperactive in childhood. A former head of visual merchandising for French prêt-à-porter brand Naf Naf, Krief sought a career change and did viticultural studies in Rivesaltes, where she befriended an influential generation of fellow students: Edouard Laffitte, Stéphane Morin, and Véronique Souloy of Domaine Matin Calme. Laffitte introduced her to the revelatory natural wines of Marcel Lapierre, as well as those of her future companion Jean-François Nicq.

Today Krief and Nicq share a cellar in Montesquieu-des-Albères, where Krief also farms 3 hectares (7.4 acres) of vines. Krief divides her range between wines made from these soils, mostly sandy granite and gneiss, and wines from 4 hectares (10 acres) of steep, unmechanizable schist within the Banyuls appellation. Overall yields are very low, between 10 and 20 hectoliters per hectare. Like Nicq, Krief harvests very early, often within the first ten days of August. Her wines from Banyuls, like KM31 and Akoibon, are distinguished by greater structure. But Krief also has a different touch in vinification than her companion. Her wines possess a wholesome vitality all their own.

"The first two bottles of natural wine I tasted were Marcel Lapierre and a bottle from Jean-François [Nicq]. When I tasted them, I felt I'd been searching for that. I dreamed of making that."

THE WINE TO TRY

Vin de France–KM31

From century-old vines of grenache gris with some carignan and grenache, KM31 shows an agile, finely spiced, licorice profile, persistent and profound.

Since: 2005. Wines were first estate bottled in 2007.

Winemaker: Laurence Krief

Vineyard surface: 7ha (17.4 acres)

Viticulture: Organic, certified since 2008

Négociant work: None

Grapes: Carignan, grenache, mourvèdre, and more

Appellations: All wines released as Vin de France

Filtration/Fining: None

Sulfites: Generally zero added, though Krief is not opposed to small doses when necessary.

LES FOULARDS ROUGES

Jean-François Nicq

MONTESQUIEU-DES-ALBÈRES

Lille native Jean-François Nicq befriended Cheverny vigneron Thierry Puzelat while the two were at viticultural school together in the late 1980s, first at Mâcon-Davayé and later in Libourne (where they were sent in an exchange of poor students). The mild-mannered yin to Puzelat's pugilistic yang, Nicq would discover the early natural wines of the Beaujolais alongside his friend, and go on to play a similarly critical role in spreading the gospel of natural vinification throughout France.

Nicq's first post upon leaving school was as cellar master of La Cave des Vignerons d'Estézargues, where he spent the 1990s putting in place the first natural vinification program in a cave cooperative. While there, he befriended Eric and Marie Pfifferling, who at the time deposited their grapes at a different cooperative. He introduced them to the wines of Lapierre and friends, and encouraged them to begin estate bottling. They did so in 2002, the same year Nicq established his own Roussillon wine estate, Les Foulards Rouges, with the aid of his childhood friend Bijan Mohamadi.

Nicq made a splash as the first to apply Lapierre-style cold-carbonic maceration to organically farmed Roussillon grenache, carignan, and syrah. Very early harvested cuvées like Les Vilains are low-alcohol, grenadine-toned thirst-quenchers. Others, like Frida and Les Glaneurs, see partial destemming, yielding a fuller, more opulent profile, although stylistically they adhere to the same theme. Nicq's wines seem to defy the heat of their region; he credits low yields—15 to 30 hectoliters per hectare—with permitting an early harvest without acquiring green or unripe notes.

"Phenolic maturity doesn't mean much. Grenache here is phenolically ripe between 14 and 15 percent alcohol, which makes the kind of wines we don't drink anymore. If you want freshness and acidity, you need to harvest earlier."

THE WINE TO TRY

See Wines to Know: The Roussillon, *page 289.*

Since: 2002. From 1989–2001, Nicq ran La Cave des Vignerons d'Estézargues.

Winemaker: Jean-François Nicq

Vineyard surface: 15ha (37 acres)

Viticulture: Certified organic

Négociant work: None

Grapes: Grenache, syrah, muscat à petits grains, and more

Appellations: All wines released as Vin de France

Filtration/Fining: None

Sulfites: Generally zero added, though Nicq is not opposed to small doses when necessary.

LE BOUT DU MONDE

Edouard Laffitte

LANSAC

Quiet and meticulous, Montélimar native Edouard Laffitte has the profile of an ideal cellarhand—which is what he was for seven and a half years, working alongside Jean-François Nicq at La Cave des Vignerons d'Estézargues. Laffitte took over running the cooperative when Nicq departed to found Les Foulards Rouges, only to follow Nicq in the same direction in 2004. He did viticultural studies in Rivesaltes, where he helped foment a second wave of natural vignerons in the Roussillon. In 2005, he acquired vines and began sharing a former cave cooperative with Loïc Roure (see opposite).

Belying his innate reserve, Laffitte has surprisingly strong ideas about vinification. Unlike Roure and Nicq, he adapts Lapierre-style long, cold-carbonic maceration to the high temperature of the Roussillon by going *even colder*, bringing his harvests to 4°C (39°F) before vatting into steel and fiberglass tanks. He also practices longer macerations than many of his peers, usually from three weeks to a month. Vinification occurs entirely without pumps. Laffitte's vines in Lansac, Rasiguères, and Cassagnes are predominantly north-facing, with more syrah than his peers. The result of his very cold, long carbonic macerations? Dark wines, mildly tannic, full-fruited but not silky. His lone white, Brave Margot, is a delicious and unusual roussanne.

Since: 2005

Winemaker: Edouard Laffitte

Vineyard surface: 7.7ha (19 acres)

Viticulture: Organic certified

Négociant work: None

Grapes: Carignan, grenache, grenache gris, and more

Appellations: All wines released as Vin de France

Filtration/Fining: None

Sulfites: Generally zero added, though Laffitte is not opposed to small doses when necessary.

DOMAINE LÉONINE

Stéphane Morin

SAINT-ANDRÉ

A former photographer, Stéphane Morin moved to the Roussillon for winemaking studies in Rivesaltes, where he found himself in the same class as Laurence Krief, Véronique Souloy, and Edouard Laffitte. Laffitte introduced him to the natural wines of Jean-François Nicq, whose cool-carbonic-macerated, old-vine Roussillon wines would become a stylistic template for Morin.

Morin purchased his estate in Saint-André from a retiring vigneron who still farmed in traditional ways, without herbicides. He constructed a striking winery with a prairie roof for insulation; the structure is built to permit him to avoid pumping at all stages of winemaking.

Morin's blues-inspired wine labels (Chuck Barrick, Rock Deluxe, etc.) can resemble a parade of forgettable vins de soif. Yet his early-harvested aerial blends of grenache, carignan, and syrah punch above their weight (and low price), offering admirably pure, fruit-forward thrills. He presses his low-yield harvests very gently, extracting very little lees. All estate wine and most négociant wines age in used barrels, where they clarify to a pristine limpidity. Few take such obsessive care in crafting simple sunny thirst-quenchers.

Since: 2005

Winemaker: Stéphane Morin

Vineyard surface: 10ha (25 acres)

Viticulture: Certified organic

Négociant work: Yes, since 2019

Grapes: Grenache, syrah, grenache gris, and more

Appellations: All wines released as Vin de France

Filtration/Fining: None

Sulfites: 10mg/L at bottling

DOMAINE DU POSSIBLE

Loïc Roure

LANSAC

Having decided on a whim to work in wine, Loïc Roure took a sommelier course in 1999, then sought an internship at the Paris bistro of natural wine pioneer Jean-Pierre Robinot purely because Robinot looked like an amusing guy. Roure's instinct proved correct. Roure crisscrossed France and worked for influential natural winemakers in Anjou, Alsace, the Rhône, and Ardèche before founding his estate in 2003.

Drawn to the Roussillon because he'd vacationed there as a youth, he found himself encouraged by recently installed vignerons like Cyril Fhal and Jean-Louis Tribouley. Roure purchased a huge former cave cooperative in the remote hamlet of Lansac, and soon began sharing it with Edouard Laffitte. In recent years, Roure has been divesting himself of less-productive old vines in the Agly valley and acquiring more vineyards at higher altitudes in the Caudiès-de-Fenouillèdes.

Some reds, like Charivari, undergo cool-carbonic maceration, while others, like the syrah Couma Acò, are entirely destemmed and foot-treaded. Roure is equipped with the same refrigeration as Laffitte but uses it far less, preferring to vat at ambient temperatures.

"I had just arrived for an internship with Thierry Allemand, and Gérald Oustric and Yvon Métras came to visit. And two days later it was Hervé Souhaut and René-Jean Dard. After spending two nights in the company of these guys, I knew I wanted to become a vigneron."

THE WINE TO TRY

Côtes du Roussillon–C'est Pas la Mer à Boire

This structured, vinous assemblage of Roure's two primary terroir types and his three red varieties (mostly grenache, with some syrah and carignan, all vinified separately) yields a more age-worthy form of his signature fleet-footed Roussillon reds.

Since: 2003

Winemaker: Loïc Roure

Vineyard surface: 11ha (27.2 acres)

Viticulture: Organic

Négociant work: Yes, since 2008, under the label "En Attendant la Pluie"

Grapes: Macabeu, carignan gris, grenache gris, and more

Appellations: Côtes du Roussillon for certain wines, otherwise Vin de France

Filtration/Fining: None

Sulfites: Some wines see at 7mg/L at bottling.

CLOS DU ROUGE GORGE

Cyril Fhal

LATOUR-DE-FRANCE

Parisian-born winemaker Cyril Fhal was initiated into a culture of great wine while running the Saumur-Champigny estate Château de Targé in the 1990s, a position where he found himself neighbors (and soon friends) with Clos Rougeard legend Charly Foucault. At the Latour-de-France estate he founded in 2002, Fhal brings formidable savoir faire to natural winemaking on the schist and gneiss hills of the Roussillon. Masterful, transparent whites from macabeu and carignan blanc and atypically elegant reds from grenache and carignan place Fhal in the stylistic lineage of Domaine Gauby, another early influence.

Fhal's yields are very low, generally in the range of 15 to 20 hectoliters per hectare. In vinification, he eschews most modern interventions, including temperature-control systems. Reds are macerated whole-cluster in 25-hectoliter vats at ambient temperatures, with moderate foulage at the bottom.

Fhal has an ambivalent relationship with the rest of the natural wine scene, preferring to stand slightly apart. He aims for a different audience from many of his peers; his wines are priced higher, and his highly ambitious farming spares little expense. Indeed, what earns respect from the Roussillon natural wine community—even more than his extremely pretty wines—is Fhal's manifest workaholism and mania for the nuances of plant and animal life in farming.

Since: 2002

Winemaker: Cyril Fhal

Vineyard surface: 7.5ha (18.5 acres)

Viticulture: Certified organic since 2003. Practicing biodynamic.

Négociant work: Yes, since 2016, under the label "Hors Champ"

Grapes: Carignan, grenache, cinsault, carignan blanc, and more

Appellations: IGP Côtes Catalanes

Filtration/Fining: None

Sulfites: Small doses of 10–20mg/L at bottling

DOMAINE TRIBOULEY

Jean-Louis Tribouley

LATOUR-DE-FRANCE

Latour-de-France vigneron Jean-Louis Tribouley was among the first wave of natural vignerons to set up in the Roussillon in the early 2000s, alongside Jean-François Nicq and Cyril Fhal. A former welder and roofer from the Franche-Comté, he followed winemaking studies in Beaune with six months of work experience for Gérard Gauby before setting up his estate in Latour-de-France in 2002.

True to his Burgundian training, Tribouley never went in for the carbonic maceration popular among many of his peers. He favors destemming for reds, before macerations ranging from one to three weeks. All reds age in steel, while whites from macabeau and grenache blanc age in old oak barrels with minimal racking. One-third of his vineyards is on schist, with the rest on a mixture of gneiss and granitic sands.

Tribouley, like most other Roussillon natural vignerons save Fhal, Duchêne, and Gauby, is preoccupied with accessibility and aims to produce natural wine in a frank and affordable style. As honest as he is pragmatic, he admits to adding yeast nutrients in vinification when he deems it necessary, and has in the past experimented with lysozyme use (see page 115), although today he rejects the practice. Unshowy, with a quiet, old-school grace, Tribouley's wines are workmanlike in the best possible sense of the term.

Since: 2002

Winemaker: Jean-Louis Tribouley

Vineyard surface: 10.5ha (26 acres)

Viticulture: Certified organic

Négociant work: None

Grapes: Carignan, grenache, grenache blanc, and more

Appellations: Côtes du Roussillon, IGP Côtes Catalanes

Filtration/Fining: Reds see plate filtration.

Sulfites: 5mg/L at assemblage for whites, 5mg/L before bottling for reds

DOMAINE MATIN CALME

Anthony Guix

MILLAS

Anthony Guix of Domaine Matin Calme might belong to the same second-wave generation of Roussillon natural wine estates as onetime friends like Laurence Krieff and Stephane Morin, but in truth his initiation to natural wine came earlier, when he studied at Mâcon-Davayé alongside Domaine de Peyra's Stéphane Majeune in the late 1990s. Guix would take to heart the zero-zero radicalism that was in the air at that time, later surpassing his Roussillon peers in his quiet commitment to nonintervention in viticulture and winemaking.

"In this profession, there aren't many of us, the purists."

A diminutive, half-Catalan native of Normandy, Guix worked as a consultant enologist in Bordeaux before establishing Matin Calme in 2006 with then wife Véronique Souloy (see page 300). Initially the couple were close with their peers in the Roussillon. But Guix's noninterventionist vision of winemaking proved divergent from the dominant, more pragmatic school espoused by Jean-François Nicq, and the two vignerons cut contact in the late 2000s.

Guix's organic farming, on his high-sited, old-vine vineyards on granite and schist around Belesta, is marked by the non-use of fertilizers. Typical yields are jaw-droppingly low, under 10 hectoliters per hectare. He harvests right at the start of ripeness, with the result that wines are typically between 12 and 13 percent alcohol. On sappy, granular reds from grenache and carignan he conducts monthlong carbonic macerations, while his small quantity of whites see a brief, whole-cluster maceration in a fridge before pressing. All wines are vinified and aged in steel. Soft-spoken and affecting, Guix combines the technical mastery of an accomplished enologist with a monk's do-nothing determination.

THE WINE TO TRY

Vin de France–Ose

A blend (usually) of grenache gris, maccabeu, carignan blanc, and muscat, Ose is a transporting, vital white, with a lancing volatility and a wiry tropical fruit that can recall the work of Laureanno Serres in Spain's Terra Alta.

Since: 2006

Winemaker: Anthony Guix

vineyard surface: Just under 7ha (17.3 acres)

Viticulture: Organic

Négociant work: Occasional small-scale organic grape purchases to supplement low yields

Grapes: Grenache, grenache gris, maccabeu, and more

Appellations: All wines are released as Vin de France.

Filtration/Fining: None

Sulfites: None

PRECURSORS

Domaine Gauby

In 1985, two decades before his region became a hotbed of natural winemaking, Calce vigneron Gérard Gauby left his cave cooperative and put the Roussillon on the map for wine aficionados worldwide. Beginning with 5 hectares (12.3 acres) of family vines, he expanded the estate to 45 hectares (111 acres) and produced a streamlined range of masterfully constructed whites and reds from local grape varieties that won international acclaim throughout the 1990s and 2000s.

A perceptive farmer, Gauby began transitioning away from chemical viticulture as far back as 1989 and became certified organic in 1996, validating the chemical-free approach for generations who followed. Today he is renowned as a pioneer of agroforestry in viticulture, having literally changed the landscape of his vineyards in Calce by interspersing his vineyards with more than five thousand trees.

While he never emphasized naturalness in vinification, Gauby helped to establish many Roussillon natural vignerons, including Tom Lubbe and Cyril Fhal.

Gauby was always restrained with sulfite addition and eschewed other corrective techniques, along with filtration. But Domaine Gauby's wines made a leap further into natural vinification when Gauby's son Lionel succeeded him as winemaker in 2009.

Ghislaine (*left*) and Gérard Gauby in their cellar

TASTING DESTINATIONS
INDIGÈNES

When: The last weekend of April

Who: Organized by Slow Food Roussillon president Jean Lhéritier, Mataburro vignerons Laurent Roger and Melissa Ingrand, micro-négociant-distributor Laurent Pujol, chef Marc Meya, and Via del Vi proprietor Romain Margueritte

For: Open to the public

Where: Église des Dominicains in Perpignan

Billed as a tasting salon of wines from Catalonia and Occitanie—that is, the Roussillon, northeastern Spain, and the Languedoc—Indigènes is named in homage to Perpignan's first natural wine bar, whose founder, Christophe Albero, passed away in 2016. Since 2017, the salon has been the premier showcase for the resplendent cross-cultural exchange happening among natural winemakers in southern France and northern Spain.

MATASSA

Tom Lubbe

ESPIRA-DE-L'AGLY

Tom Lubbe is a singular figure in Roussillon natural winemaking: a foreign outsider (a South African raised in New Zealand) turned consummate insider, whose success has only encouraged him toward more radical natural winemaking. After interning for South African estate Welgemeend in the late 1990s, Lubbe worked three years for Roussillon pioneer Gérard Gauby—a life-enriching experience, in that Lubbe wound up marrying Gauby's sister Nathalie.

"The life of the soil is very important in getting a lot of flavor with low alcohol. If there's no microbial life, the flavor doesn't come."

Lubbe began making his own wines as Matassa in 2003, initially working in partnership with his friend Sam Harrop, with whom he'd purchased the parcel in Coteaux du Fenouillèdes that gave the project its name. The first vintage was produced in his and Nathalie's living room, after which Gérard Gauby loaned Lubbe the former Gauby winery. In 2019, Lubbe recovered the former Jolly Ferriol cellar and estate in Espira-de-l'Agly.

Harvest at Matassa begins very early, at the end of July in recent years. Many red cuvées are co-fermented macerations of red and white grapes, while most white grapes see skin maceration. His underripe, whole-cluster aesthetic, in orange wines as in reds, yields grippy, bright, low alcohol elixirs, as vital and trebly as early White Stripes singles.

Lubbe's eloquence in English—and his self-evident winemaking genius—has helped make his estate a magnet for interns from non-Francophone nations, many of whom have gone on to become influential natural winemakers in their own right (particularly in Austria).

THE WINE TO TRY

Vin de France–Mambo Sun

A co-fermented blend of 80 percent old-vine grenache and 20 percent macabeu Mambo Sun shimmies with long, melodic tannins and crisp dark-cherry fruit.

Since: 2003

Winemaker: Tom Lubbe

Vineyard surface: 25ha (61.8 acres)

Viticulture: Certified organic, practicing biodynamic

Négociant work: None

Grapes: Macabeu, grenache gris, muscat d'Alexandrie, and more

Appellations: All wines released as Vin de France since 2014

Filtration/Fining: None

Sulfites: None added since 2007

THE ROUSSILLON

Natural Wine Dining

A confluence of Roman, Arab, Jewish, Spanish, and French influences, Roussillonnaise cuisine heavily features beans and tomatoes, along with seafood like anchovies and monkfish around the Mediterranean coast. (The town of Collioure even has its own appellation for local anchovies.) The Roussillon's double identity—at once French and Catalan—has helped preserve its distinctive local culinary traditions to a greater extent than those of many other regions.

Les 3 Journées in Perpignan

Les 3 Journées

PERPIGNAN

Founded in 2015 by former Domaine Matin Calme vigneron Véronique Souloy on a shady backstreet square in Perpignan, Les 3 Journées is an adorable Franco-Korean lunch canteen with a splendid natural wine list featuring the wines of Souloy's vigneron friends.

Sagí Taverna

PERPIGNAN

Chef Yannick Ferez runs this locavore natural wine cave à manger on rue Fabriqués d'en Nadals, emphasizing house-made vermouth along with razor-sharp Catalan small plates that rival anything on offer in Barcelona.

Les 9 Caves

BANYULS-SUR-MER

Anchoring the Banyuls natural wine scene, Les 9 Caves is the shared cellar space and vacation apartments founded by Banyuls vigneron Bruno Duchêne in 2012. Its spacious bar and restaurant is operated by Dutch couple Natasja and Jan Paul Delhaas, who formerly owned Wink restaurant in Amsterdam. Chef Natasja has a field day with the bounty of the Catalan coast, while sommelier Jan Paul ably serves up bottles from the site's vast wine selection.

El Xadic del Mar

BANYULS-SUR-MER

At El Xadic del Mar, a block from the beach in Banyuls-sur-Mer, ex–Le Verre Volé manager Manu Desclaux has created an endearing, homespun, one-man-show table d'hôte.

Le Coq à l'Âne

LATOUR-DE-FRANCE

Pocket-size wine bistro Le Coq à l'Âne is run by Morgan van der Horst, brother of local vigneron Saskia van der Horst. He offers a panoply of rough-and-ready natural wines from debutant vignerons of the region, alongside fresh and sincere bistro cuisine.

THE ROUSSILLON

Legends in the Making

Within the past twenty years, the Roussillon has transformed from a backwater of the wine world to a flourishing natural wine community. Today certain generational differences are visible in the approaches of its vignerons. If the down-to-earth, egalitarian winemaking of Edouard Laffitte, Loïc Roure, and Jean-Louis Tribouley set the tone for the 2000s, it's the no-expense-spared high artisanship (and high prices) pioneered by Cyril Fhal that typifies more recently installed vignerons in the region. The savviest of the new generation of Roussillon vignerons set their sights high, choosing difficult, high-altitude terroirs and old, low-yield vines, despite the work they require.

Le Temps Retrouvé

Michaël Georget

LAROQUE-DES-ALBÈRES

A native of Chinon in the Loire, Michaël Georget followed an interest in biodynamics to work experience in Alsace and the Roussillon before establishing his own estate on 4.5 hectares (11.1 acres) of very old vines in 2012. An advocate of horse plowing, Georget vinifies unsulfited, unfiltered, low-yield wines separately according to grape variety in small fiberglass tanks. Sometimes spritzy, occasionally marked by high volatility, his wines are often more profound than the natural vins de soif the region is known for.

Casot des Mailloles

Jordi Perez

BANYULS-SUR-MER

Energetic young Catalan vigneron Jordi Perez has one of the hardest jobs in wine. In 2015, he took over the estate of (mostly) retiring vignerons Alain Castex and Ghislaine Magnier, confronting the simultaneous challenges of unmechanizeable low-yield vines on Banyuls's steep schists, a tiny cellar smack in the center of town, and, of course, the expectations that come with stepping into the shoes of natural wine legends. Perez, whose previous wine experience was at conventional estates, is also learning natural vinification as he goes. Casot des Mailloles wines are expensive.

Jordi Perez in his cellar in Banyuls

Yet the prices reflect the value of Perez's superb terroir, proven by his predecessors.

Pedres Blanques

Rié and Hirofumi Shoji

BANYULS-SUR-MER

A couple from Japan who met at the viticultural school in Beaune, Rié and Hirofumi did work experience with two legendary Freds of Burgundy—Mugnier and Cossard—before establishing their 3.5-hectare (8.6-acre) estate in Collioure in 2017. Their intense, profound first vintage of whole-cluster, old-vine grenache raised eyebrows with its price tag: €20 ex-cellar, a figure more in the spirit of Burgundy than the Roussillon. The Shojis became a cause célébre in wine circles in 2018, when their local prefecture deemed their business economically unviable and refused to renew their work visa. A surge of support from the wine community helped overturn the ruling, allowing the Shojis to remain among the most highly touted young vignerons of the South of France.

Canon-Fronsac vines at Château Moulin-Pey-Labrie

14. Bordeaux and the Southwest

Little more than a high-speed rail line links the historically aristocratic, internationalist, and prosperous wine culture of Bordeaux with the paysan winemaking of the French Southwest.

Bordeaux is isolated by its historical prosperity, a legacy of hundreds of years of profitable trade with England and the Netherlands.

Stratospheric vineyard prices in Bordeaux mean ownership has long been divorced from the actual work of viticulture and vinification. These are not circumstances that induce much reflection or willingness to resist mercantilism among Bordeaux producers. Natural winemaking in Bordeaux has been limited to a few proud eccentrics and outsiders.

The story of natural wine is drastically different elsewhere in the Southwest, where the local wines remain unknown to most consumers. Here you find mostly dark, characterful reds; what historical wine merchants called "black wine." The Southwest is also the ancestral home of the pétillant naturel (see page 112), while Gaillac also has a long, overlooked tradition of sweet and oxidative winemaking.

Winemaking in the Southwest retains its paysan smallholder spirit. But the few natural vignerons working in appellations like Cahors, Buzet, and Aveyron find themselves isolated geographically, separated by great distances from the more populous natural winemaking communities of the Loire and the Beaujolais. Natural winemaking in the more remote areas of the Southwest persists thanks to traditionalist holdouts like Gilles Bley of Clos Siguier, as well as thoughtful neo-vignerons like Aveyron's Nicolas Carmarans, an acolyte of Marcel Lapierre.

The most significant natural winemaking community in the Southwest is found in and around Gaillac, north of Toulouse, where figures such as Patrice Lescarret of Causse Marines, Bernard Plageoles, and Laurent Cazottes have found shared cause in promoting organic agriculture and natural vinification.

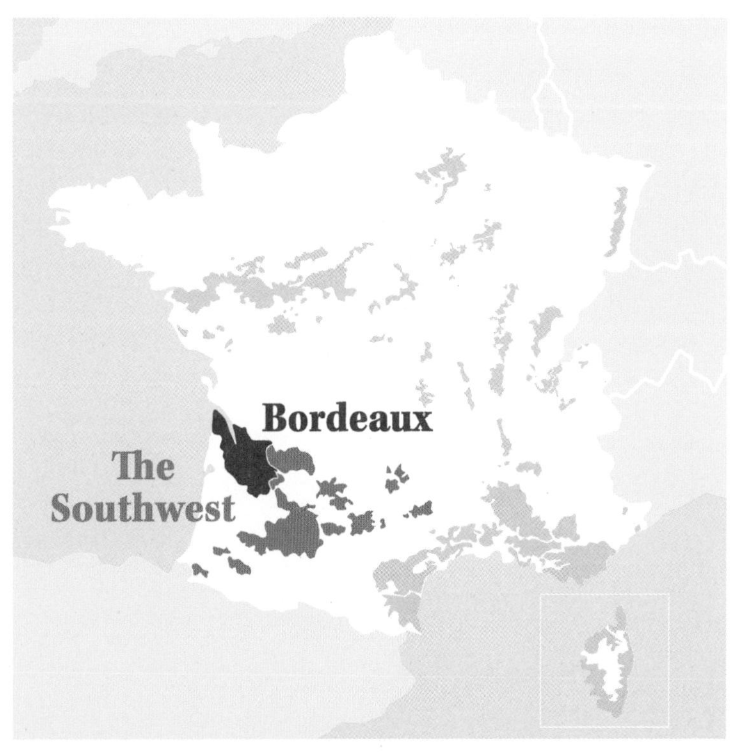

BORDEAUX

Total vineyard surface: 115,300ha (284,913 acres)[1]

Rate of organic agriculture: 11.6 percent[2]

Key grapes: Merlot, cabernet franc, cabernet sauvignon, malbec, semillon, sauvignon

THE SOUTHWEST

Total vineyard surface: 52,900ha (130,719 acres)[3]

Rate of organic agriculture: 19 percent[4]

Key grapes: Fer servadou, duras, prunelart, mauzac, merlot, malbec

---

1 Figures for Gironde department, France Agrimer 2019
2 Agence Bio 2019
3 France Agrimer 2019 figures for Dordogne, Landes, Lot et Garonne, Gers, Lot, Tarn, Tarn et Garonne, and Haute Garonne departments
4 Agence Bio 2020

Wines to Know

Bordeaux connoisseurs, prepare to be disappointed. There simply is very little natural Bordeaux being made, and what's out there tends not to aim for much in the way of grandeur, like the cheerful Bordelais thirst-quenchers of Château Moulin Pey-Labrie (Le Puy's peculiar Barthélemy and the Saint-Émilion of Château Meylet are notable exceptions). The rest of the Southwest is more stylistically varied and rewarding. It encompasses everything from scrappy, middleweight reds from fer servadou to fulsome, appley pétillants naturels to muscular, peach-fuzzy orange wines from mauzac.

Domaine Plageoles | Gaillac–Mauzac Nature

The ancestral method sparklers of the French Southwest are the original pétillants naturels: bottle-fermented sparkling wines produced without added sugar or yeast. Domaine Plageoles employs a fascinating historical method of reusable canvas cloth filtration to pause fermentation before bottling their Mauzac Nature, which generally finishes with 5 to 10 grams of residual sugar. The result is a buoyant, unpretentious apéritif sparkler, evoking fresh green apples and orgeat.

Nicolas Carmarans | IGP Aveyron–Maximus

Given the diversity of his grape sources, it's hard to associate Carmarans's red wines with the magnificent splendor of his local mountain terroir. (For that, look to his rare Selves chenin.) In Maximus, we find instead Carmarans's signature expression of cool-carbonic fer servadou, an homage to great natural Beaujolais accented with the wild, sanguineous character of the Aveyronnais grape. It's a landmark style, transplanted to a new frontier.

Le Puy | Vin de France–Barthélemy

Before shelling out for Le Puy's pricey parcel-selection cuvée Barthélemy, it's important to know what you're getting into. Most high-end Bordeaux wines try to offer the power, extraction, and tannins that mass-market consumers associate with age-worthiness. In this regard, Barthélemy brings a book of poems to a gunfight. It is a curious, fine-boned, intellectual wine, offering delicate floral and tobacco-toned pleasures that feel blasphemously Burgundian. Its unusual texture is the result of an out-there aging regimen that sees the wine stirred ("dynamized," in biodynamics talk) on its *gross lees* in used barrels several times a week for two years.

LE PUY

Jean-Pierre and Pascal Amoreau and Harold Langlais

SAINT-CIBARD

Patriarch Jean-Pierre Amoreau was among the first biodynamic winegrowers in Bordeaux, starting in the late 1980s. Amoreau's father, Pierre-Robert Amoreau, who began estate bottling in 1965, had never used herbicides; he also originated the estate's curious method of red vinification, which encourages gentle, recurrent oxygen exposure during maceration in cement tank. The estate's signature red, Emilien, sees two years' aging, racked often between foudre and barrel. All the estate's wines derive from the far end of the limestone plateau that underpins Saint-Émilion and Pomerol.

Le Puy's nebulous commercial structure extends to the Closerie Saint Roc, a separate organic estate, and the négociant label Duc des Nauves, which processes all grapes not yet certified organic within the family's acquisitions, and which was created to satisfy demand in the Chinese market. Since joining his father at the estate in 1996, Pascal Amoreau has quintupled Le Puy in size.

His Scottish brother-in-law, Steven Hewison, joined the estate in 2011, and today oversees winemaking. The estate was known as Château Le Puy until the 2017 vintage, when the Amoreaus seceded from the Francs Côtes de Bordeaux appellation, losing the right to use the term "Château" in the process.

Pascal Amoreau in the cellar at Le Puy

THE WINE TO TRY

See Wines to Know: Bordeaux and the Southwest, *page 305.*

Since: The estate has existed since 1610. Estate bottling began in 1965. Jean-Pierre Amoreau took over in 1990, and his son Pascal joined him at the estate in 1996.

Winemaker: Steven Hewison

Vineyard surface: 50ha (123 acres) at Château le Puy. The family also maintains the 16ha (39.5-acre) Closerie Saint Roc.

Viticulture: Practicing biodynamic since the late 1980s. Certified biodynamic since 2006.

Négociant work: Yes, under the label "Duc des Nauves," sold in developing markets

Grapes: Merlot, cabernet sauvignon, and more

Appellations: Formerly Francs Côtes de Bordeaux. All wines are released as Vin de France since 2017.

Filtration/Fining: None

Sulfites: The cuvée Emilien sees sulfite addition at devatting. Most other cuvées see none added.

CHÂTEAU MEYLET

Michel and David Favard

SAINT-ÉMILION

Château Meylet's tiny surface of Saint-Émilion on clay-limestone sands has been owned by the Auvergnat Favard family since 1875. Michel Favard, who took it over from his father in 1978 and began estate bottling, discovered biodynamics in 1985, and by 1987 had become its earliest practitioner in Bordeaux. Favard credits meetings with Jacques Néauport, Marcel Lapierre, Philippe Pacalet, and Alsace's Jean-Pierre Frick with leading him toward natural vinification.

Saint-Émilion appellation authorities routinely denied Favard the right to use the appellation. Seeking a break, Favard rented the estate to the enologist Stéphane Derenoncourt for five vintages from 2005, returning only in 2010. His son David Favard succeeded him 2015. The past decade has been marked by lower yields (in the range of 25 hectoliters per hectare) and experimentation with new types of aging and vinification containers.

> **"I cared for myself naturally, with homeopathy, and I asked myself, why shouldn't I take care of my earth and my vegetation naturally?"**
>
> —Michel Favard

Michel Favard (*left*), David Favard, and David's daughter

Meylet's Saint-Émilion is about 70 percent merlot. Almost entirely destemmed, the wine sees pigéage during its seven- to fifteen-day maceration in large wooden vats, beginning at cool temperatures thanks to the use of dry ice. The wine ages for one and a half to two years in barrels, many new.

THE WINE TO TRY

Saint-Émilion–Origine

Produced in 2015, Origine was the estate's first totally unsulfited Saint-Emilion. Its long, dazzling red fruit stays airborne on the palate despite impressive concentration.

Since: Michel Favard took over the family estate in 1978. His son David Favard rejoined the estate in 2012 and took over in 2015.

Winemaker: Since 2016, David Favard with the aid of Mauro Catarinella and Anne Calderoni

Vineyard surface: 1.6ha (3.9 acres)

Viticulture: Practicing biodynamic since 1987. Certified since 2010.

Négociant work: Yes, since 2019

Grapes: Merlot, cabernet sauvignon, cabernet franc, malbec

Appellations: Saint-Émilion Grand Cru Classé

Filtration/Fining: Ceased egg white fining in 1992. No filtration.

Sulfites: No sulfites in vinification since 1996. No sulfites in élevage since 1998. Today it depends on the cuvée. Some wines see no sulfite addition; others see a small dosage at bottling.

CHÂTEAU MOULIN PEY-LABRIE

Grégoire and Bénédicte Hubau

FRONSAC

Wine lovers Grégoire and Bénédicte Hubau knew only the basics of how to farm vines and make wine when they left careers as a pharmacist and a computer specialist, respectively, to purchase Château Moulin Pey-Labrie in 1988. Corsican vigneron Antoine Arena, who had the neighboring stand at the Salon des Caves Particuliers in Lyon, introduced the Hubaus to future friends like Catherine and Pierre Breton and Marcel Richaud.

In this way, the Hubaus joined the ranks of the more commercially savvy wing of early French natural wine. The couple maintained a second, neighboring Fronsac estate, Château Haut Lariveau, which produces a more accessible counterpart to their longer-aged Canon-Fronsac at Château Moulin Pey-Labrie. The Hubaus tiptoed to natural vinification at a gentle pace, forswearing yeast inoculation only in the early 2000s. Even today their Beaujolais-inspired low-sulfite cuvées are a separate range from their more classic wines. This realistic and intelligent couple take their winemaking for what it is: a second career meant, above all, to give pleasure to themselves and their clients.

Since: 1988. The Hubaus purchased Château Haut Lariveau in 1990.

Winemakers: Grégoire and Bénédicte Hubau

Vineyard surface: 7.6ha (18.8 acres) at Château Moulin Pey-Labrie; 7.9ha (19.5 acres) at Château Haut Lariveau

Viticulture: Certified organic since 2014. Practicing biodynamic.

Négociant work: None

Grapes: Merlot, malbec

Appellations: Canon-Fronsac, Fronsac

Filtration/Fining: Rosé and certain reds are filtered.

Sulfites: Depends on the cuvée. Many wines see sulfite dosage during élevage and at bottling.

CAUSSE MARINES

Patrice Lescarret and Virginie Maignien

VIEUX

Bordeaux-raised and -trained Patrice Lescarret was only a few years into a hired-gun winemaking career in 1993 when he threw caution to the wind and purchased Causse Marines, formerly the mixed-agriculture estate of a local cooperateur. The estate is home to some of the oldest parcels of massal-selection indigenous grape varieties in Gaillac, and Lescarret soon began working alongside his neighbor Bernard Plageoles to preserve and sustain these varieties in the face of Gaillac AOC authorities bent on aligning the region with Bordeaux varieties. Lescarret was arguably the first in the region to begin organic viticulture in 1997. His estate became certified organic in 2005, the same year Virginie Maignien, his intern turned collaborator and life partner, joined the estate.

With a couple of exceptions, Causse Marines wines are released as Vin de France, partly the result of a conflict between Lescarret, who enjoys varietal vinification, and the local AOC, which insists upon blends. In early years, the wines evinced a mild conservatism, including a freer hand in sulfite dosage and occasional fining and filtration. Lescarret's garrulous rebel nature led him further toward an ascendant natural wine aesthetic. Today many wines see zero sulfite addition, while fining and filtration are mostly a memory for the estate.

Since: 1993. Virginie Maignien arrived in 2005.

Winemakers: Patrice Lescarret and Virginie Maignien

Vineyard surface: 12ha (29.6 acres)

Viticulture: Practicing organic since 1997, certified since 2005

Négociant work: None

Grapes: Braucol, syrah, duras, and more

Appellations: Most wines released as Vin de France. Occasionally a few in the Gaillac AOC.

Filtration/Fining: None in recent years

Sulfites: Little or none added in recent years

DOMAINE PLAGEOLES

Famille Plageoles

CAHUZAC-SUR-VÈRE

Generations of the Plageoles family have helped cement the estate's status as a pillar of the natural wine community and the local Gaillac appellation.

"My grandfather planted two hundred ondenc vines and never saw the wine we made with them. As vignerons, we always work for the following generation."

—Bernard Plageoles

Since 2015, Domaine Plageoles has been run by brothers Romain and Florent Plageoles, along with their mother, Myriam, and their father, Bernard. The estate's groundbreaking work began with the brothers' great-grandfather, Marcel. In 1970, a few years before his untimely passing, Marcel Plageoles grafted plantations of two native grapes that had almost disappeared after phylloxera: the red duras and the white ondenc. His son Robert was the first to produce varietal wines from these grapes, and exhorted the local viticultural syndicate to emphasize native grapes rather than the cabernet sauvignon and syrah then in vogue.

While Domaine Plageoles always practiced native yeast vinification, it was only under Bernard Plageoles in the mid-1990s that the family stopped using herbicides. Bernard would bring the estate into organic certification in 2007 at the urging of his UK importer. Today Florent and Romain Plageoles maintain the estate's value-laden wine range, which includes everything from bright, sapid dry white verdanel to muscular red braucol.

THE WINE TO TRY

See Wines to Know: Bordeaux and the Southwest, *page 305.*

Bernard Plageoles in his tasting room

Since: 1805. Robert Plageoles ran the estate 1970–1983. Bernard Plageoles ran it from 1983 until his sons Romain and Florent took over in 2015.

Winemakers: Florent and Romain Plageoles

Vineyard surface: 30ha (74.1 acres)

Viticulture: Ceased herbicide use in 1995. Certified organic since 2007.

Négociant work: Yes, since 2016, under the label "Contre-Pied"

Grapes: Braucol, duras, prunelart, and more

Appellations: Gaillac, Côtes du Tarn

Filtration/Fining: White and sweet wines are filtered. Sparkling mauzac sees filtration *a là manche* (sleeve filtration), in which the wine is passed through reusable custom canvas sheets.

Sulfites: Most wines see a light dosage at bottling.

LAURENT CAZOTTES

VILLENEUVE-SUR-VÈRE

Laurent Cazottes (*right*) with his father, Jean

Paysan spirits and wine impresario Laurent Cazottes only began to bottle his wine in 2010, but he's been promoting organic agriculture and natural winemaking in the Tarn ever since he founded his artisanal distillery in 1999 at age twenty-three. His eaux-de-vie, produced on his father's 1950s alembic still, quickly captured the imaginations and the palates of France's famed chefs. Today all the fruit for Cazottes's brandies comes from his own farm, with the exception of some purchased organic pears for his bestselling Poire Williams.

The first to adopt organic certification in the Gaillac region, Cazottes produced grapes for his eaux-de-vie and for home winemaking throughout the 2000s. In 2009, he collaborated with chef Julien Bourdariès to organize investment among fellow organic vignerons for Gaillac bistro Vigne En Foule. Today, Cazottes also imports and distributes natural wines from peers throughout France and beyond.

"My father and I were very complementary, because I had my sense of gastronomy and finesse, and my father had the paysan sense."

Cazottes's own wine range consists of well-crafted, unbeatably priced bistro wines from local Gaillac varieties. Like his spirits, they are excellent, accessible ambassadors for Tarn gastronomy.

THE WINE TO TRY

Vin de France–Prunelart

Cazottes's prunelart sees destemming and just three days' maceration before aging in tank. It shows the vibrant, quaffable side of a deep, often broody grape.

Since: 1998. Estate bottling began in 2010.

Winemaker: Laurent Cazottes

Vineyard surface: 3ha (7.4 acres) of vines (18ha/44.5 acres total including orchards, sunflowers, grains, etc.)

Viticulture: Organic

Négociant work: Yes

Grapes: Mauzac rose, braucol, prunelart, and more

Appellations: All wines released as Vin de France

Filtration/Fining: None

Sulfites: Prunelart sees no added sulfites. Other wines see sulfite addition of 20–30mg/L.

CLOS SIGUIER

Gilles Bley

MONTCUQ

Clos Siguier vigneron Gilles Bley produces wines that feel like time capsules, a peek into the sort of patient, traditionalist winemaking championed by early *vin sans soufre* wine bars of Paris like Les Envierges and L'Ange Vin.

Bley never adopted organic certification, but uses no herbicides. His estate is situated away from the heavier soils of the Lot River, at 400 meters (1,312 feet) elevation on the red clay and limestone slopes of Montcuq. He attributes the lower, balanced alcohol levels of his wines with the cooler nights his vines enjoy on the hillsides.

At Clos Siguier, the destemmed harvest is vinified in large steel and cement tanks. Fermentation, with some pigéage, lasts just over a month on average. After the turbulent early stages, Bley takes great care to leave very little oxygen exposure in the tanks, using special chimney fittings to overfill them. (This is, interestingly, a practice that goes entirely against the grain of the modern micro-oxygenation technology that emerged from Bley's own region.) Two-year aging occurs primarily in tanks, with some large oak barrels. Bley's black, svelte, smoky Cahors is among the only wines in its bargain price point to demonstrate a seeming immortality: bottles even decades old rarely disappoint.

Since: Gilles Bley took over the estate from his father in 1985.

Winemaker: Gilles Bley

Vineyard surface: 15ha (37 acres)

Viticulture: Practicing organic

Négociant work: None

Grapes: Malbec, tannat

Appellations: Cahors

Filtration/Fining: None

Sulfites: 10mg/L is added after malolatic fermentation.

DOMAINE DU PECH

Ludovic Bonnelle and Magali Tissot

SAINTE-COLOMBE-EN-BRUILHOIS

Buzet vigneron Ludovic Bonnelle is a familiar face at Paris natural wine tastings—partly because his wide grin is unforgettable, and partly because it takes work to sell the deep, long-aged, low-yield biodynamic reds he produces with his partner, Magali Tissot (a cousin of renowned Jura vigneron Stéphane Tissot). Theirs are expressive, challenging wines, which would surely buck all expectations of the Buzet AOC, except that almost no one has any.

Bonnelle and Tissot took over the estate in 1997 upon the death of Tissot's father, Daniel, who farmed conventionally. The couple made a clean break with past practices in 2003 when they stopped machine harvesting, and soon after they began installing organic and biodynamic viticulture. They employ phytotherapeutic treatments, avoiding excessive copper, and forswear all fertilizers. Their wide-planted vines are primarily the red grapes of Bordeaux, along with a 1-hectare (2.5-acre) band of sauvignon blanc. The estate's soils are divided between gravelly clays and clay-limestone. Red wines ferment in wooden vats or steel, depending on the cuvée. Among the more remarkable aspects of the couple's work is their patience, aging certain cuvées for up to four years in foudre.

Since: 1978. Ludovic Bonnelle and Magali Tissot took over the estate in 1997.

Winemakers: Ludovic Bonnelle and Magali Tissot

Vineyard surface: 14ha (34.6 acres)

Viticulture: Practicing organic since 2005. Certified organic since 2008, certified biodynamic since 2009.

Négociant work: None

Grapes: Merlot, cabernet franc, cabernet sauvignon, sauvignon

Appellations: Buzet

Filtration/Fining: None

Sulfites: Small doses were added during vinification in 2010 and 2013; the 2014 wines saw some sulfite addition during élevage. Otherwise, none added.

NICOLAS CARMARANS

MONTÉZIC

An Aveyronnais raised in Paris, Nicolas Carmarans was born into natural wine. He drank his first glass of wine at age thirteen at the bar of family friend Bernard Pontonnier's Café de la Nouvelle Mairie in 1979. In 1994, Carmarans would purchase that bar, inaugurating its heyday as a Parisian natural wine destination. Inspired by his vigneron friends, Carmarans began planting gamay and chenin vines in the high hills of his native northern Aveyron in 2003. By 2007, he'd become a full-time vigneron and négociant winemaker, producing wines principally from fer servadou and chenin. (His gamay vines never flourished in the Aveyron climate.)

"Fer servadou is a rustic grape. Here the wines are black and thick, with astringence and tannins. I don't work like that. I was raised on Beaujolais."

Carmarans's vineyards include young vines on a lakeside site beside his winery, and another site with steep, restored vine terraces at 500 meters (1,640 feet) altitude. The rest of his production is from organic fer servadou grapes purchased from growers in his village and in nearby Marcillac. Carmarans practices what he learned from Marcel Lapierre: for reds, he conducts cool-carbonic maceration and old-oak élevage, preferring lightness over extraction. (The exception is his cuvée Mauvais Temps, which sees partial foulage.) By his own admission, Carmarans is still figuring out his Aveyronnais terroir: Maximus derives from red clay, Mauvais Temps from granite, and Fer de Sang from clay-limestone. His young-vine chenin, Selves, derives from granite. Savvy and urbane, a true gentleman farmer, Carmarans divides his time between Paris and Aveyron. Keep an eye out for his rare collaborative cuvées with his motorcycling pal, Banyuls vigneron Bruno Duchêne.

THE WINE TO TRY

See Wines to Know: Bordeaux and the Southwest, *page 305.*

Since: 2003

Winemaker: Nicolas Carmarans

Vineyard surface: 3ha (7.4 acres)

Viticulture: Organic

Négociant work: Yes

Grapes: Fer servadou, chenin, négret de banhars, and more

Appellations: IGP Aveyron

Filtration/Fining: None

Sulfites: Usually none added; occasionally a microdose at bottling

ES D'AQUI

Jean-Louis Pinto

MOULIN-NEUF

Patient, paysan, and preternaturally laid-back, Toulouse-trained enologist Jean-Louis Pinto spent the 2000s working for conventional estates in the Languedoc and the southern Rhône. Disillusioned with the conventional wine industry, he quit his job when he started a family, and went to work vinification season with his longtime friend Antony Tortul of La Sorga, who encouraged him to start his own négociant winemaking operation. Purchasing grapes from throughout the Southwest, the Roussillon, and the Languedoc, Pinto began making natural wine in his companion's native Gaillac in 2010, before moving to the village of Moulin-Neuf in his own native Ariège in 2015.

The name of his business translates to "it's from here," referring to his early labels, which featured old maps that indicated the source of his organic grape purchases.

Now over a decade in, Pinto's resolutely zero-added-sulfite winemaking has remained almost defiantly primitive. He employs a manual press; he has no means of temperature control; until as recently as 2017, he bottled his wines by hand, individually, with a funnel. He conducts whole-cluster maceration of both whites and reds, with very little pigéage or pumpovers, in small fiberglass tanks, before aging in tank, oak barrel, or sandstone amphora.

"Often people do négociant purchase to make big volumes. For me it was kind of the opposite. It wasn't to make cheap wine. It was to make pretty wines."

THE WINE TO TRY

Vin de France–Orange Mauzanic

A barrel-aged, long maceration of mauzac from limestone terroir in nearby Limoux, Pinto's Orange Mauzanic offers a fistful of passion fruit and crabapple, with a delightful quiver of volatility.

Since: 2010

Winemaker: Jean-Louis Pinto

Vineyard surface: 0.6ha (1.5 acres) planted in 2021

Viticulture: Organic

Négociant work: Yes, it's all négociant work, and will remain so until 2023 at earliest.

Grapes: Braucol, grenache, cinsault, and more

Appellations: All wines released as Vin de France

Filtration/Fining: None

Sulfites: None added

Natural Wine Dining

While France's Southwest contains a tapestry of individual regional cuisines, none loom as large in the imagination as that of Gascony, a historical province extending south and east of Bordeaux until the limits of Basque Country in the west and the hills of the Roussillon in the east. It is ground zero for foie gras; in its dishes, duck and goose fat are splashed as liberally as olive oil is in Italy. Toulouse and the Tarn are rarely in the hazy historical delimitation of Gascony, but they are excellent places to source much of the same heart-stoppingly memorable cuisine.

For vegetarians, or anyone wishing to moderate their consumption of duck fat, it will be a relief to discover that the Southwest's natural wine destinations run the gamut from robustly traditional to brightly modern.

La Légende

SAUVETERRE-DE-BÉARN

Located in a village of fifteen hundred souls east of Bayonne, this forward-thinking and ever-evolving locavore bistro and natural wine bar was founded in 2018 by ex-Parisian private chef Laura Schiffman and Le Verre Volé alum Alistair Bellahsen (a nephew of Bernard Bellahsen). Now run by Bellahsen, La Légende is known for its homespun bistro charm and its connoisseur's selection of natural wines.

Vigne en foule

GAILLAC

A collaboration between chef Julien Bourdariès and Gaillac vigneron-distiller Laurent Cazottes, with involvement from Cazottes's fellow organic vignerons Bernard Plageoles, Michel Issaly, and Patrice Lescarret, Vigne en foule is an upscale yet egalitarian town-center brasserie showcasing the best wine and ingredients the Tarn has to offer.

L'Épicurien

ALBI

Swedish chef Rikard Hult opened this classy yet easygoing brasserie in central Albi in 2005. It's a perfect pit stop for well-appointed burgers and steak tartare (along with more involved creations), accompanied by an astute selection of natural wines.

Soif

BORDEAUX

Self-taught chef Cécile Lambré and sommelier Nico Lefevre run this immaculately tasteful natural wine bistrot in central Bordeaux, where the city's in-the-know diners enjoy duck tartare and grilled squid alongside cult natural wines from throughout France.

Vigne en foule in Gaillac

BORDEAUX AND THE SOUTHWEST

Legends in the Making

Importers, sommeliers, and retailers the world over are eagerly waiting for a dynamic young natural wine talent to emerge in Bordeaux. They will have to keep waiting. Even the more acclaimed new estates in the region tend to be natural only by Bordeaux standards: biodynamic, perhaps, but invariably conservative in the cellar.

More engaged natural winemaking is occurring in the less-fussed-over zones of Gaillac, Cahors, and Jurançon, where inexpensive vineyard land and a heritage of old autochthonous vines offers ripe terrain for new generations of Southwest natural vignerons.

Château Lafitte

Antoine Arraou

MONEIN

Former photographer Antoine Arraou returned to his 5-hectare (12.3-acre) organic family estate in 2012, where he initiated a transition to biodynamics and agroforestry in the vineyards and launched a slim new range of natural wines that brought a flurry of attention to this overlooked French Basque region. Arraou's debut still whites from petit manseng, Argile and Orange, are both aged in clay amphorae; Orange sees a three-week maceration. Also of note is his gripping gros manseng pétillant naturel, Funambule.

Simon Busser

PRAYSSAC

A disciple of Anjou natural wine legend Olivier Cousin, Simon Busser returned to his native Cahors to begin producing courageously zero-zero natural wines in 2007. Today he farms 5 hectares (12.3 acres), primarily côt, with some merlot and tannat, and works all his soils with horse plowing. His titanic, vivid reds are the result of monthlong macerations, mostly destemmed, and aging in used barrels.

Simon Busser outside a wine tasting in Anjou at the winery of his friend Baptiste Cousin

L'Ostal Levant

Louis and Charlotte Pérot

PUY-L'EVÊQUE

Louis and Charlotte Pérot left careers in publishing and book sales in Paris in 2012 to produce wine in Cahors, first working alongside their friend Simon Busser. In 2015, they took on 2 hectares (5 acres) of vines, primarily côt, which they supplemented with plantings of their own, including certain hybrid varieties. Their initial vintages haven't been easy, but the dedicated and talented couple quickly earned the support of like-minded zero-zero vignerons like Jean-François Coutelou, who gave them grapes to help compensate for their frost shortfall in 2017. Their own wines, aged in an underground quarry without electricity, already possess a sap and vigor that rivals those of their mentors.

Vermentinu vineyards at Domaine Arena

15. Provence and Corsica

As any Corsican will tell you, the vociferously independent, nominally French island has little in common, culturally, with Provence. If the island is lumped in with the mainland French southeast in this chapter, it is only because neither region is home to many influential natural vignerons.

In both regions, high real estate demand promotes a conservatism in viticulture and vinification that is at odds with the risks involved in making wine naturally.

In much of Provence, estates are acquired not by idealistic young natural vignerons, but by investment groups, celebrities, and billionaires.

The traditional, synthetic-chemical-free farming and additive-free winemaking that would come to be known as "natural" were preserved in Provence by an actual nobleman, Château Sainte-Anne's François Dutheil de la Rochère, known as "Le Marquis." His neighbor and acolyte, the natural wine enologist Yann Rohel, would later introduce Dutheil to Marcel Lapierre and the early natural vignerons of Villié-Morgon.

Later, in 1996, the Alpilles vigneron Henri Milan took inspiration from encounters with soil expert Claude Bourguignon and Sologne vigneron Claude Courtois, deciding to embrace biodynamic farming and natural vinification. By the early 2000s, natural winemaking began to gather some small momentum in Provence. Beside Domaine Milan in the Alpilles, the Savoyard neo-vigneron Dominique Hauvette converted her own acclaimed estate to biodynamics in 2003, with the aid of future Ardèche vigneron Andréa Calek. A year later, former Parisian politician Jean-Christophe Comor founded his Domaine les Terres Promises estate in the Var, farming organically and vinifying without additives (save low doses of sulfites) from the start.

The story in Corsica is different. The island's sole link to mainland France's early natural wine communities was maintained by the lawyer turned Patrimonio vigneron Antoine Arena, who befriended peers like Marcel Lapierre and Guy Breton at tasting salons throughout France. Arena's influence looms large to this day in Corsica, where many credit the recent ban on herbicides within the Patrimonio AOC to his decades of advocacy.

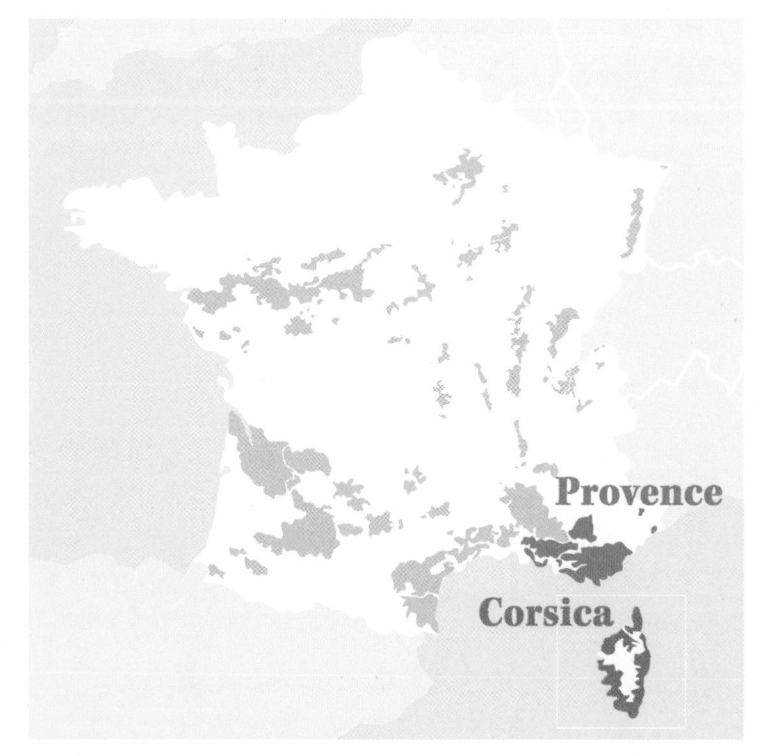

PROVENCE

Total vineyard surface:
85,700ha (211,769 acres)[1]

Rate of organic agriculture:
30 percent[2]

Key grapes:
Ugni blanc, rolle (vermentino), clairette, grenache, syrah, mourvèdre, cinsault, tibouren

CORSICA

Total vineyard surface:
6,000ha (14,826 acres)[3]

Rate of organic agriculture:
33 percent[4]

Key grapes:
Nielucciu (sangiovese), bianco gentile, vermentino, grenache

1 France Agrimer 2019
2 Agence Bio 2020
3 France Agrimer 2019
4 Agence Bio 2020

PROVENCE AND CORSICA

Wines to Know

Corsican natural wines are known for richness, and an appealing spice. Reds often scan like sangiovese that has attended a Swiss boarding school, retaining an Italian accent with a whole new set of manners. Climate change has rendered many of the island's whites a little ungainly in recent years.

Provence is lately known for its ubiquitous pale rosé, churned out by huge, investment-portfolio estates whose careless farming practices give a bad name to organics. Its natural estates have instead made their reputations with intense, structured reds, like the Bandols of Château Sainte-Anne or the Alpilles wines of Dominique Hauvette and Domaine Milan. These estates produce great rosé, too, but it's always deeper and more vinous than the supermarket version.

Château Sainte-Anne | Bandol Blanc

A hidden gem of a white in a landscape famous for rich reds, Château Sainte-Anne's saline and vertical Bandol blanc is a fifty-fifty blend of two unassuming varieties, ugni blanc and clairette. The Dutheils credit its surprising coastal-scrub complexity to veins of white sand deep in the undersoil of the parcels. Whatever the source, it is among the most refined whites of southern France.

Domaine Antoine Arena | Patrimonio–Morta Maio

Patrimonio patriarch Antoine Arena planted nielluccio (the local term for sangiovese) on the deep schist soils surrounding his house in 2001. Despite relatively severe pruning and debudding, the Arenas still practice about one-fifth green harvest on the vigorous young vines. In the wineglass, the wine shows firm tannins and sanguineous, spiced-cherry fruit.

Domaine Milan | Vin de France–S&X

Its seemingly lascivious title is an ode to the estate's longtime cellar master, Sébastien Xavier. But the wine itself does recall the perennial cliché of sex in a wineglass, offering surging and persistent dark red fruit, evocative of the work of Domaine Gramenon. A short but impressively extractive maceration of young grenache from stoney limestone and blue clay terroir, S&X ages for a year in used oak before release.

DOMAINE ARENA

Antoine, Antoine-Marie, and Jean-Baptiste Arena

PATRIMONIO

Antoine Arena in his cellar, with vegetables from his garden

Semiretired today, the charismatic and urbane former lawyer Antoine Arena built his family estate into the leading light of Corsica's northerly Patrimonio appellation, which, partly thanks to his longtime advocacy, banned herbicides in 2021. It is a handsome capstone to a winemaking career Arena began in 1980 after returning to Corsica from Nice. Arena expanded the estate to its present 14 hectares (34.6 acres); since 2014, production has been divided among himself and his two sons, Antoine-Marie and Jean-Baptiste, who've worked at the estate since the early 2000s.

> **"Here, in the Mediterranean, we plow three times a year, but the third is just aesthetic. Two passes suffices easily. It allows us to touch the soils less."**
>
> —Antoine-Marie Arena

The Arenas' key vineyard sites are the east-facing, clay-limestone Carco (including bianco gentile plantings on the Hauts de Carco), the south-facing, sunnier clay-limestone Grotte di Sole, and Morta Maio, on schist soils around the family house. Today Antoine-Marie works with the Carco and Hauts de Carco vineyards; his brother, Jean-Baptiste, farms Grotte di Sole; and their father farms Morta Maio along with a small planting of bianco gentile. While each vigneron is nominally independent, the wine ranges don't show much diversion from the Arena house style. Reds see destemming before moderately long maceration in steel or cement; some see aging in large oak barrel, with others remaining in tank. Whites, always sulfited and sometimes filtered, age in steel.

THE WINE TO TRY

See Wines to Know: Provence and Corsica, *page 319.*

Since: 1980. The mixed-agriculture estate has been in the Arena family since the eighteenth century, but Antoine Arena initiated viticultural specialization in 1977.

Winemakers: Antoine Arena, Jean-Baptiste Arena, and Antoine-Marie Arena

Vineyard surface: 14ha (34.6 acres)

Viticulture: Practicing organic since the start. Certified organic since 2007.

Négociant work: None

Grapes: Nieluccio, vermentinu, muscat à petits grains, and more

Appellations: Patrimonio, Muscat du Cap Corse

Filtration/Fining: White wines are occasionally filtered.

Sulfites: Light sulfite dosage before bottling on all wines

DOMAINE LES TERRES PROMISES

Jean-Christophe Comor

LA ROQUEBRUSSANNE

Neo-vigneron Jean-Christophe Comor cuts a peculiar figure in the sunbaked hills of the Var. Owl-eyed, eloquent, and slightly hunched, he's a former souverainiste politician and law professor who renounced politics in 2002 to make natural wine in the Var, a few hours from his native Aix.

Comor never attended winemaking school; instead, he appealed to vignerons he knew and admired for advice, including Marcel Richaud and Pierre Breton. As a winemaker he's both scrupulous and playfully experimental, displaying particular finesse with light, sun-kissed whites from native varieties like ugni blanc, clairette, and carignan blanc. His rosés, while filtered, are vastly more pure than the Provençal standard, produced with native yeast (of course) and without blocking malolactic fermentation.

"Enologists try to work on the grapes as soon as they're in the harvest case. Because everything that's alive is a source of anguish for an enologist. The alive is the unknown."

Today Comor's idiosyncratic range of wines—eleven regular cuvées, plus various micro-cuvées each year—includes highlights like the lightly macerated, foudre-aged carignan blanc Analepse and a suave, powerful Bandol (vinified at Château Pibarnon) called L'Amourvèdre.

Jean-Christophe Comor presenting his primeur wine at Paris wine shop Caves Augé, November 2013

THE WINE TO TRY

IGP Sainte Baume–Analepse

In literature, an "analepse" is a flashback—and such is Comor's carignan blanc, a grape once common in the Var that is no longer permitted in its AOC wines, even though it is manifestly well suited to the local climate. Comor practices a very light maceration on it, yielding a refreshing palate of fennel-toned fruit.

Since: 2004

Winemaker: Jean-Christophe Comor

Vineyard surface: 15ha (37 acres)

Viticulture: Certified organic since 2006

Négociant work: Yes

Grapes: Grenache, carignan, cinsault, and more

Appellations: Coteaux Varois en Provence, Bandol, IGP Sainte Baume

Filtration/Fining: Rosés and some whites are filtered.

Sulfites: Certain cuvées see zero added, but most see a small dosage at bottling.

CHÂTEAU SAINTE-ANNE

Françoise and Jean-Baptiste Dutheil

ÉVENOS

The late François Dutheil de la Rochère never embraced chemical farming, and made efforts to certify his estate as organic as far back as 1976. In 1987, Dutheil's neighbor introduced him to Yann Rohel, then a young enology student. Rohel would go on to introduce Dutheil to Marcel Lapierre and Jacques Néauport, sparking a friendship that lasted until Dutheil's death.

At that point, Rohel left the Beaujolais to manage Château Sainte-Anne for three years, assuring the transition to Dutheil's widow, Françoise, and her children, Marie and Jean-Baptiste. Today Jean-Baptiste and Françoise run the estate, where they've changed little, save for the installation of temperature-controlled steel tanks in 2016. Françoise, an agricultural engineer by training, prefers low yields in the range of 30 hectoliters per hectare, and refuses wire training for the estate's gobelet-trained vines. Ten of the estate's 14 hectares (35 acres) are within the Bandol appellation; the rest are old vines immediately adjacent in the Côtes de Provence appellation. The commune of Évenos benefits from its location at the end of a funnel of gorges that cools the warm sea breezes, making it cooler than the rest of Bandol. The estate's whites and rosés undergo malolactic fermentation without impediment from sulfite additon or filtration, a rarity in Provence. Château Sainte-Anne's long-lived Bandol reds, destemmed and fermented without pigéage in steel tanks before aging in foudre, display an underappreciated mineral side to mourvèdre.

"What we don't like in certain Bandol wines is to have the mouth attacked by tannins. With us, it's the Lapierre school, the finesse, the silky, integrated tannins."

—Françoise Dutheil

Jean-Baptiste (*left*) and Françoise Dutheil outside their tasting room

THE WINE TO TRY

See Wines to Know: Provence and Corsica, *page 319.*

Since: The sixteenth-century estate has been in the Dutheil family for five generations. François Dutheil de la Rochère began producing wine in the 1970s.

Winemaker: Jean-Baptiste Dutheil since the early 2000s

Vineyard surface: 14ha (34.6 acres)

Viticulture: Practicing organic since the start. Certified organic since 1976.

Négociant work: None

Grapes: Ugni blanc, clairette, grenache, and more

Appellations: Bandol, Côtes de Provence

Filtration/Fining: The rosé sometimes sees filtration.

Sulfites: 20mg/L at bottling

PRECURSORS

Yann Rohel

Alongside his former mentor Jacques Néauport and classmate Philippe Pacalet, Yann Rohel was among those most responsible for disseminating the early credo of natural vinification throughout France in the 1990s and 2000s. He discovered a love of vinification while working alongside conventional winemaker Cyrille Perzinski at Château Pibarnon, then attended enology school in Beaune in 1987, where his classmates included Philippe Pacalet. That same year, Rohel encountered natural wine in his native Bandol when a friend of his mother's introduced him to François Dutheil de la Rochère of Château Sainte-Anne (see opposite). The next year, Pacalet introduced Rohel to the Lapierre circle, including Jacques Néauport, with whom Rohel would travel as an assistant for two years.

Animated by a desire to know all stages of wine production firsthand—rare among enologists of his era—Rohel did vineyard work for Marcel Lapierre in 1990. The same year, on a whim, he offered to help harvest with Domaine de la Tour du Bon, where he met the new cellar master, a young Thierry Puzelat. He would soon put Puzelat in touch with Lapierre. Rohel went on to manage the Lapierre-Chamonard collaboration Château Cambon in 1995. Upon the death of his mentor Dutheil in 1996, Rohel returned to Bandol to manage Château Sainte-Anne for three years. In 1998, he provided vinification counsel to Mas Foulaquier's Pierre Jéquier. In 1999, Rohel began what would be over a decade's collaboration with Marcel Richaud, leading the southern Rhône estate toward natural vinification. The list of vignerons influenced by Rohel goes on and on, including Guy Breton, Andréa Calek, and Axel Prüfer.

After a breakdown in 2009, Rohel quit the world of wine for six years. He returned only in 2015 to help his friend Richaud, whose wife had just died.

Yann Rohel (*left*) with former vigneron Jean-Baptiste Selles in the garden of their friend Guy Breton in the Beaujolais

Rohel then followed a girlfriend to Mexico, where he planned to work in landscaping and gardening—only to run into fellow Lapierre alum Louis-Antoine Luyt, who involved him in Bichi, a Mexican natural wine estate. Today Rohel divides his time between the Beaujolais, Bandol, and the southern Rhône, where he consults for Domaine Richaud.

DOMAINE HAUVETTE

Dominique Hauvette

SAINT-RÉMY-DE-PROVENCE

There is a certain association between strong-willed characters and an affection for horses that Baux-de-Provence vigneron Dominique Hauvette personifies almost to the point of parody. Now in her seventies, she raises horses while continuing to produce landmark unfiltered and fairly low-sulfite Provençal wines. A sign on the door of her office flatly states "No tastings."

A Sorbonne-trained expert in rural law, Hauvette established her estate with the help of her retiring restaurateur father in 1988, intending only to farm grapes, but she soon apprenticed herself to the luminaries of the slim tradition of ambitious vinification in the French south: Eloi Dürrbach of Domaine de Trevallon and Laurent Vaillé of Domaine de la Grange des Pères. Like these estates, Hauvette produces long-macerated destemmed long-aged reds, and whites of surpassing finesse, all on native yeasts. Her work has always edged closer to the aesthetics of one of her early clients, the influential natural wine author and bistro owner François Morel: she doesn't use sulfites in vinification, and tends to degas less than her mentors, resulting in less polished, more soulful wines. Practicing organic from the start, Hauvette transitioned to biodynamic agriculture in 2003 on the insistence of one of her employees, future Ardèche vigneron Andréa Calek.

"I'm against thermal shocks in vinification. I'm against all that is brutal. Because yeasts and bacteria are living beings. It's not a chemical mechanism, it's a mechanism of metabolism and life."

Hauvette was an early adopter of concrete eggs, which she uses for vinifying and aging her old-vine cinsault-based Amethyste, her age-worthy rosé Petra, and her creamy clairette-marsanne-roussanne blend Dolia.

THE WINE TO TRY

Les Baux-de-Provence–Cornaline

Concentrated, tea-toned, with the ferrous note of a rare steak, Hauvette's Cornaline is a blend of grenache, syrah, and cabernet sauvignon that sees a year in wooden vats and a year (or longer) in foudre before release. It resembles its maker in its sternness, reserve, and almost bitter authority.

Since: 1988

Winemaker: Dominique Hauvette

Vineyard surface: 17ha (42 acres)

Viticulture: Practicing organic since 1988. Certified biodynamic since 2003.

Négociant work: Yes, some grenache for the Roucas cuvée

Grapes: Clairette, marsanne, roussanne, and more

Appellations: Les Baux-de-Provence, IGP Alpilles. Briefly Côtes du Roussillon.

Filtration/Fining: None

Sulfites: Low sulfite doses at bottling

DOMAINE MILAN

Henri, Emmanuelle, Théo, and Natalie Milan

SAINT-RÉMY-DE-PROVENCE

Natalie (*left*) and Théo Milan in the mourvèdre and grenache vineyard that yields the estate's primeur wine

Retired Alpilles vigneron Henri Milan credits encounters in 1996 with soil expert Claude Bourguignon and Sologne vigneron Claude Courtois with encouraging his quest to illustrate the potential of his patchwork of Baux-de-Provence terroirs—which include blue marne and limestone scree along with gravels and sands reminiscent of Châteauneuf-du-Pape—using natural vinification. Milan's own aesthetics, as evidenced by the rich, bold wines he produced before retiring in 2014, took inspiration from classic Bordeaux. (Today the similarity still shows in the family's impressive merlot Le Jardin.) It was during this early era that tasting Milan's wines would inspire the Beaujolais vigneron Jean-Claude Lapalu to embrace natural winemaking.

Milan first began producing certain red cuvées with zero added sulfites in 2000, identifying the expanding zero-zero range with a butterfly motif on the label. Many of these wines remain in circulation, and they always present a striking contrast between their powerfully built contents and their daffy label designs, whose eye-searing magenta also features heavily in the estate's signage and tasting-room décor.

The estate turned a page in 2014 when siblings Emmanuelle and Théo took over vinification. (Emmanuelle divides her time between Domaine Milan and the estate of her Alsatian husband, Mathieu Deiss, son of famed biodynamic vigneron Marcel Deiss.) Théo—and his American wife, Nathalie, former head sommelier at The Modern in New York—brings a more fluid and open-minded style to the estate's range, expanding it to include bright, short-carbonic wines like Le Vallon (a blend of grenache, syrah, cinsault, and nielluccio), as well as intense orange wines like Luna & Gaia, from grenache blanc, rolle, and roussanne.

THE WINE TO TRY

See Wines to Know: Provence and Corsica, *page 319.*

Since: 1956. Henri Milan took over from his father in 1986. He retired in 2014.

Winemakers: Emmanuelle and Théo Milan, along with cellar master Sébastien Xavier

Vineyard surface: 15ha (37 acres)

Viticulture: Practicing organic since 1988. Certified since 2002.

Négociant work: Yes, since 2017

Grapes: Grenache blanc, rolle, roussanne, and more

Appellations: All wines released as Vin de France since 2007. Previously wines were released as Les Baux-de-Provence AOC.

Filtration/Fining: None

Sulfites: First vintage without added sulfites was 2000. Wines with butterflies on the label see zero added sulfites, while others see 15–20mg/L at bottling.

Natural Wine Dining

If Provence isn't teeming with natural wine estates, the situation with natural wine restaurants is quite the opposite. The city of Nice in particular is home to the most dynamic community of natural wine lovers outside Paris. Its natural wine restaurants offer contemporary, product-driven market cuisine, often drawing inspiration (and ingredients) from the nearby Italian border.

The dining scene in Corsica is more conservative, partly due to its restaurants' heavy reliance on French family tourism. Corsican regional identity also tends to conspicuously trump any considerations of naturalness on the island's wine lists.

Le Canon

NICE

At Sébastien Perinetti's boxlike bistro on rue Meyerbeer, his baroquely overwritten menu highlights an impressive array of farm ingredients, many sourced from over the border in Italy. Don a sweater and ask to browse the walk-in wine fridge, which houses an enviable connoisseur's selection of radical natural wines.

La Part des Anges

NICE

Founded in 1998 by Olivier Labarde (see page 328), La Part des Anges offers a peerless array of France's greatest natural wines, along with superbly selected charcuterie, conserves, cheeses, and other small composed plates. In recent years, charismatic chef Brice Fortunato has taken the cuisine at La Part des Anges to another level. He's the type to go harvest vegetables from local farmers on his day off, and serve them the following day alongside veal raised by Faugères vigneron Didier Barral.

Lavomatique

NICE

Brothers Grégoire and Hugo Loubert make a world-class sommelier-chef duo at this maniacally fun small-plates destination built inside a former laundromat in Nice's old town. Grégoire, who fell

La Part des Anges in Nice

Jeanne in Antibes

into *vin sans soufre* circles while still in his teens, now has over two decades' experience at the cutting edge of natural wine. His dealer's-choice service style perfectly complements the exuberant street-food variations coming out of the kitchen.

Jeanne

ANTIBES

Opened in in the touristic old town in 2017 by the youthful and preternaturally savvy sister-act Elsa and Marine Gauthier, Jeanne is a cozy, two-story wine bistro at the forefront of the southern French dining scene. Marine studied winemaking and worked for wineries around the world before amassing Jeanne's splendid cellar, which includes sweetheart allocations from the likes of Alain Castex. Elsa, who formerly worked in fashion PR, divides her time between the dining room and the kitchen, where she shares duties with her boyfriend, Austrian chef and négociant winemaker Paul Schuster.

Les Buvards

MARSEILLE

Owners Laetitia Pantallaci and her partner, Fred Coachon, offer sincere market-menu bistro cuisine and a stellar selection of past, present, and future greats of French natural winemaking at this cozy, natural cave à manger two blocks north of Le Vieux Port.

Le Gibolin

ARLES

Of the Paris bistro owners who first supported the nascent idea of *vins sans soufre* in the 1980s, only two are still in business. One is the couple behind Paris's Le Baratin. The other is Luc Desrousseaux, who once ran L'Echanson in Paris's 14th arrondissement and now runs Le Gibolin, an enchanting natural wine bistro in the Camargue town of Arles. His partner Brigitte Cazalas dishes out the wines (mostly well aged and by the glass), while Desrousseaux's unpretentious menu contains, among more involved creations, France's greatest *frites*.

Olivier Labarde on Bringing Natural Wine to the Côte d'Azur

Few have done more to mentor successive generations of natural vignerons and natural wine restaurateurs than Nice wine retailer Olivier Labarde, a Provençal native who founded his roomy, welcoming cave à manger La Part des Anges in 1998. Like his friends Cyril Bordarier of Paris's Le Verre Volé (see page 403) and Jean-Michel Wilmès and Nicolas Vauthier of Troyes's Aux Crieurs de Vin (see page 342), Labarde got his start in wine working for conventional wine retail chain Repaire de Bacchus. In his quarter century at La Part des Anges, he's inspired a fervent Côte d'Azur natural wine clientele, and helped nurture the careers of vignerons ranging from Olivier Cohen to Julien Besson.

"At Repaire de Bacchus, we couldn't drink the wines we sold. They were too classic. In 1992, we tasted Domaine Gramenon, then we quickly discovered the others around them. It was a revelation for us. So as soon as we opened our own businesses, we started with them. The Beaujolais guys—Guy Breton, Jean Foillard, Yvon Métras. Gérald Oustric. Domaine de Peyra in Auvergne. Thierry Puzelat. Christian Chaussard."

"It was really complicated selling natural wine at the start. We delivered a message, but it wasn't understood."

"At La Part des Anges, we don't get too much into technical stuff with customers. We speak of wines that taste free, wines that are very *digeste*, that don't give you a headache. Wines that are easy to drink. They can be inexpensive thirst-quenchers, but also very complex wines."

"Provence is late to natural wine. Historically it's a region that sells its wine well. So estates don't have many problems with money. So they figure why take risks, why complicate things? But for a few years now, some of the most curious estates have started to move toward this type of wine."

Legends in the Making

It's hard to be rosy about the potential for new natural viticultural communities to arise in Provence. As an older generation retires, its land and estates are often snapped up by investment groups for the production of highly technical rosé. (For example, Domaine de l'Ile, whose former owner Sébastien Le Ber was a longtime proponent of native yeast fermentation and malolactic fermentation for its rosé from Porquerolles, was purchased in 2019 by Chanel, whose estate manager promptly installed a regime of cryovinification and blocked malolactic fermentation.)

Several bright spots nonetheless emerge from this scene of remorseless viticultural agglomeration. Below are a few instances where new managers or younger-generation owners are demonstrating the feasibility and the cultural value of natural winemaking.

Domaine de la Cavalière
Julien Besson
LOURMARIN

Nice native and Le Verre Volé alum Julien Besson runs this 47-hectare (116-acre) Luberon estate owned by San Francisco investment banker Jeffrey Ubben. Besson produces vivid, young-vine natural wines from the estate's 8 hectares (19.7 acres) of organically farmed grenache, bourboulenc, syrah, and merlot, along with two marvelous olive oils. The unfiltered red primeur is a treat.

Domaine de Sulauze
Karina and Guillaume Lefèvre
MIRAMAS

At their 30-plus-hectare (74-acre) biodynamic mixed-agricultural farm near the Étang de Berre, Karina and Guillaume Lefèvre produce a stylistically broad spectrum of wines, ranging from pleasant filtered rosé to balanced and expressive low-sulfite cinsault and grenache. Also of interest is their charming paysan beer production.

Cadavre Exquis
Marc and Shirine Salerno
LA BASTIDONNE

Marc Salerno has farmed his family land in the Luberon organically since 1972. He has produced wines for home consumption since the 1990s, and has used zero winemaking additives since 1999. Yet it is only since 2011 that he and his partner, Shirine, have commercialized their thought-provoking, iconoclastic wines. Even the terms "farming" and "winemaking" are arguably too strong for what the Salernos do: they place their wild-farmed grapes in glass demijohns, where the marc ferments in the open air, then bury them beneath the earth for aging in the spring. The dense, intense results have captured the imagination of vignerons and wine lovers in search of ever-greater simplicity in wine production.

Julien Besson (*right*) and his companion, Charlotte Villemade, at Domaine de la Cavalière

Vineyards at Bertrand Gautherot's Vouette et Sorbée estate in Buxières-sur-Arce in the Aube

16. Champagne

Natural wine lists contain Champagne, and natural vignerons throughout the world enjoy Champagne. But Champagne can (almost) never be natural in the way other natural wines are natural. The celebrated Champagne method itself consists of at least two additives—sugar and yeast.

This contradiction did not stop the emergence, in the 1990s and 2000s, of a certain loosely defined natural style of Champagne vinification, increasingly favored by the region's growing number of organic and biodynamic estates.

To a greater extent even than in the Loire Valley, Champagne vignerons seeking a more natural style drew inspiration from biodynamic circles. Yet Champagne vignerons, unlike those in the Loire and elsewhere, have had little incentive to cultivate natural wine clientele, because in recent years even the most minor procedural distinction from the work of the big négociant houses has been enough to stoke wild demand on export markets.

As a result, Champagne today is home to a host of excellent biodynamic estates who practice natural-style Champagne vinification, but who interact in only the most limited way with the wider natural wine community. Such estates haven't been included here because they exert little influence within the natural wine world.

Profiled instead are the veteran Champagne vignerons who have chosen, for diverse reasons of their own, to spend time cultivating natural wine markets: appearing at natural wine tastings, organizing events with natural wine retailers, producing collaborative cuvées with natural wine retailers, and so on. Theirs are the estates that can be most trusted to align their practices with the wider ethos of natural wine.

Their ranks include larger longtime organic and biodynamic estates, like Champagne André Beaufort and Champagne Fleury. They also include Vouette & Sorbée's Bertrand Gautherot, an acolyte of the influential Champagne vigneron Anselme Selosse. Vincent Laval and David Léclapart were almost alone in their generation in the Marne to embrace both organic agriculture and natural-style vinification on the scale of a micro-vigneron. Finally there was Montgueux upstart Emmanuel Lassaigne, who joined the ranks of early natural Champagne producers not after any ecological epiphany, but after acquiring a taste for natural wine at Troyes wine bistro Aux Crieurs de Vin.

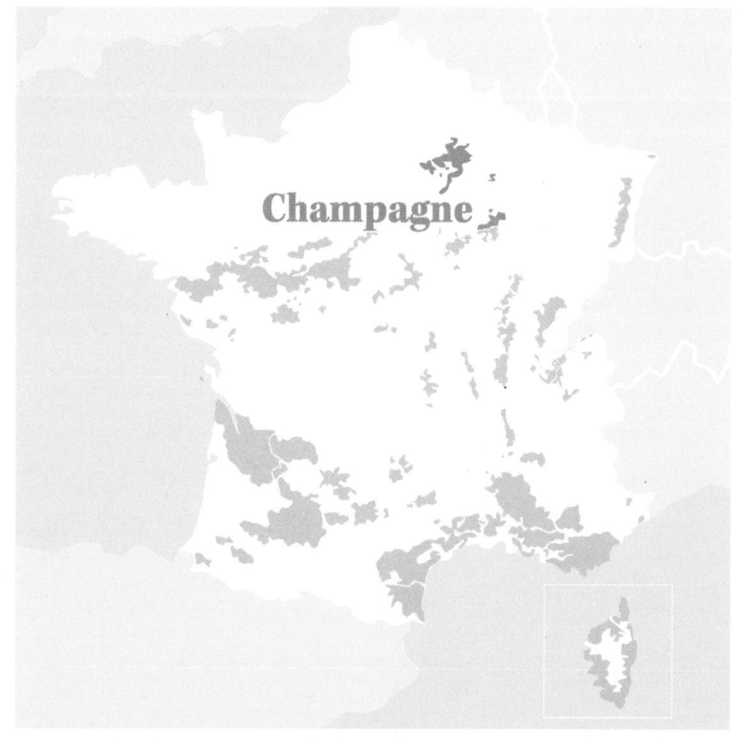

Total vineyard surface:
33,800ha (83,522 acres)[1]

Rate of organic agriculture:
3.5 percent[1]

Key grapes:
Chardonnay, pinot noir, meunier

---

1 CIVC

Wines to Know

Many are bottled without *liqueur de dosage*, so natural-style Champagnes tend to be drier and possess more acidity than conventional Champagnes. They also tend to be even more expensive, the result of more labor in the vineyards, lower yields, and more labor in the cellar. Natural-style Champagnes are less consistent from year to year than heavily blended Champagnes from the big houses. This is a feature, not a bug. It's because natural-style Champagnes tend to include less reserve wine; often they are composed of the wine of just one vintage, like the wines of Vincent Laval or David Léclapart. These wines show what is often called a "vinous" character: they are winelike, richer and more intense than conventional Champagnes made with higher yields.

Champagne Georges Laval | Champagne Premier Cru–Cumières

Laval's Cumières, an undosed vintage blend of all his parcels within the village, comprises about 70 percent of his tiny production in a given year. Typically composed of half chardonnay, with a quarter each of pinot noir and meunier, the wine is persistent, generous, dazzling in its purity and the clear, starry-sky-like detail of its chalky white fruit.

Champagne Jacques Lassaigne | Champagne Extra Brut–Les Vignes de Montgueux

An undosed blend of wines from nine different unplowed, chalk-based chardonnay vineyards in the eponymous village, Emmanuel Lassaigne's Les Vignes de Montgueux is the most finely calibrated and pure Champagne in its price point, a by-the-glass fixture at natural wine destinations worldwide. In recent years, it has become a blend of three consecutive vintages, discernible by a lot number on the label. Incisive and crisp, it offers a summary of Lassaigne's house style: lacy bubble structure, emphatic chalk notes, and the bracing refreshment of a spring rain.

Vouette & Sorbée | Champagne Brut Nature–Fidèle

"Faithful" is what biodynamic Aube vigneron Bertrand Gautherot titled his flagship wine, a name that often elicits awkward jokes from couples ordering it at restaurants. He envisioned a Champagne that would be faithful to its origins, one that would reflect every gesture and practice in the vineyard, as in the cellar. Fidèle is a rich and vinous blanc de noirs, exuberant in personality yet bone-dry. About 7 percent of each bottling derives from a réserve perpetuelle; the rest derives from a single given vintage.

CHAMPAGNE ANDRÉ BEAUFORT

Constant and Réol Beaufort

POLISY

Champagne André Beaufort's Jacques Beaufort was among the first to forswear synthetic chemical vineyard treatments in Champagne, following an allergic episode shortly after taking the reins of the estate in 1969. Beaufort began treating his vines with essential oils in the 1980s, even managing to eliminate sulfur use for many years on his parcel in Ambonnay.

The estate is split between 1.6 hectares (4 acres) of vines in Ambonnay and 6.2 hectares (15.3 acres) in Polisy, in the Aube. Today it is also divided among Jacques Beaufort's many sons. At time of writing, Constant and Réol Beaufort are the only two brothers remaining at the estate.

"In terms of organics, I'm one of the ancients that keeps up the traditions. Nowadays there are products permitted in organics that I would never use."

—Jacques Beaufort

Pinot noir represents three-quarters of the estate's vineyards; chardonnay the rest. The estate ages base wines for twelve months before secondary fermentation, which, in a rarity for the region, is initiated using native yeast starters, prepared by blending the year's base wine with concentrated must and fresh juice from the following vintage. In his time, Jacques Beaufort often produced Champagnes without sulfite addition, but in recent years the estate has resumed minimal sulfite addition at bottling.

THE WINE TO TRY

Champagne Grand Cru–Ambonnay Brut

Even ripe vintages of the Grand Cru Ambonnay, typically composed of 80 percent pinot noir and 20 percent chardonnay, can be slender and refined, showing notes of tarragon and a marked maritime air.

Jacques Beaufort in his tasting room in Ambonnay

Since: The estate is in its third generation. Its renown was built by Jacques Beaufort, who retired in 2007.

Winemakers: Constant and Réol Beaufort

Vineyard surface: 7.8ha (19.3 acres)

Viticulture: Practicing organic since 1971. Certified organic since 1997.

Négociant work: Yes, but chiefly to purchase grapes belonging separately to the various brothers.

Grapes: Pinot noir, chardonnay

Appellations: Champagne

Filtration/Fining: None

Sulfites: Doses in the range of 20–30mg at pressing and disgorgement

CHAMPAGNE FLEURY

Jean-Sébastien, Benoît, and Morgane Fleury

COURTERON

Jean-Pierre Fleury was the first in Champagne to begin practicing biodynamics back in 1989. He followed this farsighted farming effort with the more ambiguous decision to embrace sales to big-box European retail outlets in 1992. The estate's filtered, nonvintage brut cuvée still comprises 80 percent of production.

"I like to say we're on Chablis terroir. In truth, I feel more Bourguignonne or Chablisienne than Champenoise."

—Morgane Fleury

Since joining the estate in 2004, Fleury's eldest son, Jean-Sébastien, has made progress improving the estate's winemaking. He introduced aging in 60-hectoliter oak foudres for 45 percent of the estate's wines, and in 2009, he produced their first unsulfited cuvée, Sonate. Today all bottles are hand-disgorged. Secondary fermentation is initiated with a yeast isolated from the Fleurys' own vineyards.

Jean-Sébastien's brother Benoît joined the estate in 2010; today he oversees vineyard work and is introducing elements of agroforestry. The estate's primarily Kimmeridgian marl vineyards are planted with 85 percent pinot noir and 10 percent chardonnay, with the rest made up of more obscure Champagne varieties.

The Fleury brothers' disarming sister, Paris natural wine bar owner Morgane Fleury, made its wines a fixture in natural wine bars in Paris during the 2000s and 2010s.

Morgane Fleury conducts a tasting in the family's vineyards

THE WINE TO TRY

Champagne Extra Brut–Sonate

A graceful, red-fruited blend of 60 percent pinot noir and 40 percent chardonnay, Sonate illustrates the formidable potential of this historic estate.

Since: Estate founded in 1895. Estate bottling since 1929 under Robert Fleury. Jean-Pierre Fleury took over in 1970.

Winemaker: Jean-Sébastien Fleury since 2004

Vineyard surface: 15ha (37 acres). The equivalent of another 15ha is purchased.

Viticulture: Practicing organic since 1970. Certified biodynamic since 1992.

Négociant work: Yes, since 1996, to purchase a neighbor's biodynamic grapes

Grapes: Pinot noir, chardonnay, pinot blanc, and more

Appellations: Champagne

Filtration/Fining: The basic nonvintage brut sees filtration.

Sulfites: The cuvée Sonate is unsulfited. All other wines see sulfite addition at pressing and after disgorgement.

CHAMPAGNE JACQUES LASSAIGNE

Emmanuel Lassaigne

MONTGUEUX

Bored with the packaging industry, Emmanuel Lassaigne returned to help his parents save their struggling Champagne estate in 1999. He never formally studied wine, but after discovering natural wine thanks to Nicolas Vauthier and Jean-Michel Wilmes of nearby Aux Crieurs de Vin, he emerged in the 2000s as a talented and ambitious vigneron and an inquisitive vinification technician. In addition to his aerial and brisk Les Vignes de Montgueux and other regularly produced cuvées, Lassaigne experiments with microquantities of everything from skin-macerated Champagne (Le Flacon d'Incertitude) to Champagne aged in barrels obtained from friends like Jean-François Ganevat (Le Clos Sainte-Sophie) to, in 2019, a free-run rosé of pinot noir—"for barbecues," as he puts it.

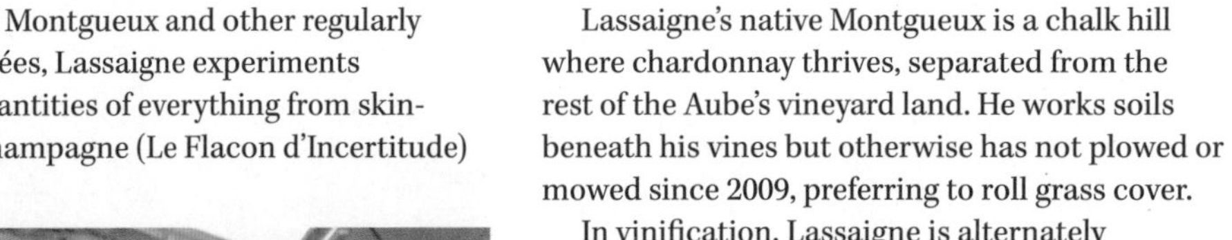

"We were lucky in Troyes to have one of the first natural wine bistros. I was among the first clients around the time I started to make wine."

Lassaigne's native Montgueux is a chalk hill where chardonnay thrives, separated from the rest of the Aube's vineyard land. He works soils beneath his vines but otherwise has not plowed or mowed since 2009, preferring to roll grass cover.

In vinification, Lassaigne is alternately pragmatic and painstakingly artisanal. He adds a light sulfite dose at pressing to ward off mouse, and never again throughout the production process. He does not fine or filter at any stage. Musts are fermented in a mixture of barrel and tank. Secondary fermentation takes place at low temperatures (10° to 12°C/50° to 53.6°F) with concentrated must and the same yeast strain used by his colleagues at Champagne Fleury. Lassaigne takes pride in manually disgorging each bottle. In recent years, he has used no *liqueur de dosage*.

THE WINE TO TRY

See Wines to Know: Champagne, *page 333*.

Since: Founded by Jacques Lassaigne in 1964. Emmanuel Lassaigne returned to the estate in 1999.

Winemaker: Emmanuel Lassaigne since 2002

Vineyard surface: 3.75ha (9.2 acres)

Viticulture: Practicing organic since 2002

Négociant work: Yes, from neighbors and relatives in Montgueux

Grapes: Estate grapes are all chardonnay since 2011. Pinot noir and pinot meunier are purchased for certain cuvées.

Appellations: Champagne

Filtration/Fining: None

Sulfites: Light sulfite addition at pressing

VOUETTE & SORBÉE

Bertrand Gautherot

BUXIÈRES-SUR-ARCE

During a tasting of his vins clairs, Bertrand Gautherot will tell you not to smell the wine, and not to swish it around much in the mouth, either. He'll suggest instead that you focus on *feeling* the physicality of the terroir, noticing the way the mineral salts in the wine hit your palate before

"From the beginning, I decided not to mix grapes, not to mix the years, not to buy sugar, not to buy yeasts. Because I wanted to understand what I did. All the actions that I do to the wine, I'll find them in the wine."

you spit. This is how he distinguishes between his two key terroirs, the Kimmerdgian marl of Vouette and the Portlandian marl of Sorbée. (He also maintains another parcel called Biaunes, the source of his wiry chardonnay cuvée Blanc d'Argile.)

Driven by a passion for vineyard work, Gautherot left a career in luxury cosmetics packaging to take on some of his father's vines in 1993. A pinot noir from Alsace vigneron Jean-Pierre Frick inspired him to reappraise the Champagnes he knew, which he'd never overly enjoyed. He transitioned to biodynamics, and with encouragement from his friend Anselme Selosse, he began producing wine in 2001. After a disappointing experience that year, he swore off all dosage, and soon earned a place at the vanguard of natural Champagne by experimenting with practices like initiating secondary fermentation with frozen must (in the rare Sobre cuvée). Sulfites are added only at the press, and wines are never fined or filtered. Only his primary wine, Fidèle, sees a small addition of reserve wine; all other cuvées are vintage dated. Pinot noir ages in Burgundy barrels, while his chardonnay tends to age in larger demi-muids. (He also experiments with aging in amphora.)

Gautherot's is an intellectual approach to both biodynamic farming and natural vinification, informed by his impressive breadth of knowledge and his own paysan origins. It yields some of the most vinous and affecting Champagnes on the market, particularly his concentrated and profound Saignée de Sorbée.

THE WINE TO TRY

See Wines to Know: Champagne, *page 333.*

Since: Gautherot took on his father's vines in 1993. He began estate bottling in 2001.

Winemaker: Bertrand Gautherot

Vineyard surface: 5ha (12.3 acres)

Viticulture: Certified biodynamic since 1998

Négociant work: Yes, as a separate project (titled Champagne Clandestin, with Benoît Doussot)

Grapes: Pinot noir, chardonnay, pinot blanc

Appellations: Champagne

Filtration/Fining: None

Sulfites: 20–30mg/L at pressing

CHAMPAGNE RUPPERT-LEROY

Bénédicte and Emmanuel Leroy

ESSOYES

Bénédicte and Emmanuel Leroy are a couple of former gym teachers who throughout the 2010s have revealed themselves to be the most driven, reflective, and radical Champagne vignerons of their generation. They decided to adopt organic agriculture immediately upon Bénédicte's taking over from her retiring parents in 2010. Notably, she studied enology in Beaune, not Champagne—and this after having lived in the Jura for a spell. In establishing their estate, the Leroys brought an open-mindedness and a herculean can-do spirit almost unheard of in the region. (Emmanuel, before joining the estate full-time in 2014, studied carpentry and constructed the couple's striking prairie-roofed timber house.) Biodynamic conversion followed for the couple soon after. Plowing and treatments are minimal, with the latter limited to low doses of copper and a whey-sulfur dilution; since 2014 the couple have braided vines in summer, rather than trim them.

The Leroy's most radical departure from Champagne orthodoxy came in 2013, when they eliminated all sulfite addition from every stage of the winemaking process. They employ a pneumatic press, followed by native yeast fermentation and aging principally in barrel. Secondary fermentation is initiated with addition of concentrated must and exogenous yeast; wines never see dosage. The core of Ruppert-Leroy's production are three scintillating single-vintage parcel cuvées (Martin Fontaine, Fosse-Grely, and Les Cognaux) and the autumnal 11, 12, 13 . . . , produced from a *réserve perpétuelle*. But the couple also push stylistic boundaries with lesser-known cuvées, like their divine, tannic, almost-red Rosé.

Emmanuel (*left*) and Bénédicte Leroy in front of the prairie-roofed home they constructed in Essoyes

THE WINE TO TRY

Champagne–Martin Fontaine

From half a hectare (1.2 acres) of chardonnay on steep, thin-soiled limestone in the village of Noé-les-Mallets, Martin Fontaine often shows a captivating, noble reduction on the nose, only to reveal a racy, pure, and surprisingly red-berried fruit.

Since: 2010

Winemakers: Bénédicte and Emmanuel Leroy

Vineyard surface: 3.8ha (9.4 acres)

Viticulture: Organic since 2013, biodynamic since 2014

Négociant work: Technically the estate is a négociant structure. They purchase their own fruit from a holding company jointly owned by themselves along with several other family members.

Grapes: Chardonnay, pinot noir

Appellations: Champagne

Filtration/Fining: None

Sulfites: None added since 2013

Jean-Michel Wilmès of Aux Crieurs de Vin on Natural Wine in Champagne

A native of the textile town of Troyes in southern Champagne, Jean-Michel Wilmès cofounded the seminal natural cave à manger Aux Crieurs de Vin (see page 342) in 1998 with his erstwhile business partner Nicolas Vauthier (see page 197). Wilmès and Vauthier discovered natural wine in the 1990s, thanks partly to an eye-opening visit with Marcel Lapierre. At Aux Crieurs de Vin, they would become responsible for bringing the anarchic, free-thinking ethos of natural wine to rural Champagne, inspiring local vignerons like Emmanual Lassaigne and Bertrand Gautherot along the way.

Jean-Michel Wilmès (*left*) and his new associate Nicolas Ubanowicz

"A lot of vignerons in Aube and the Marne play with the natural thing. But Champagne can't really be natural, because it's yeasted. And the majority of vignerons in Champagne, even the ones we like, add sugar and sulfites. But it's in digesting the wine that you understand if it's natural."

"For me, dosage is impossible. I can't drink dosed Champagnes. But you can find naturally made Champagnes with dosage."

"What's important is the feeling when you drink a natural wine. You shouldn't feel the sulfur. It should be as close as possible to the juice. You'll feel that freedom sometimes in the vins clairs. But when you drink the finished Champagnes, so many things have changed."

"In Champagne, there are vignerons who make *vins libres* [wines that taste free in a spiritual sense], and the wines sell well, and they have momentum. That could make the appellation evolve. Like Lapierre and Overnoy, who changed their regions. It can happen in Champagne."

"You can't do anything you want in Champagne. You can't make Vin de France. It's Champagne, and it passes the appellation tasting committee, otherwise it goes to the distillery."

CHAMPAGNE GEORGES LAVAL

Vincent Laval

CUMIÈRES

In the south-facing, premier cru village of Cumières, vigneron Georges Laval never adopted the synthetic herbicides and pesticides that conquered the Marne in the 1970s, and eventually accepted official certification of his farming in the mid-1980s. His son Vincent Laval, since joining the estate in 1991 and taking over vinification in 1996, engaged in a series of land exchanges with neighbors that allowed him to build manageable blocks of vines in this region known for highly fragmented vineyard ownership. Vincent's terroir-focused work in the chalk-soil lieux-dit of Les Hautes Chèvres, Les Longues Violes, and Les Chênes has since cemented the estate's reputation at the pinnacle of natural-style Champagne. Vineyard work is meticulous, employing three full-time employees for just 3.4 hectares (8.4 acres) of vines.

High-sited Les Hautes Chèvres, the tiny mid-slope Les Longues Violes, and the lower-lying Les Chênes, along with the Cumières village blend, are all produced as vintage wines without dosage. (Only Les Garennes, from vines in Chambrecy, is produced with reserve wine and sees light dosage.)

"The way we work, vintages are super marked. The light, the rain, the sun exposure, the temperatures weren't the same. So it gives different wines."

Laval harvests very ripe for Champagne, between 11 and 11.5 percent potential alcohol. Wines are fermented in Burgundy and Bordeaux barrels and are bottled for secondary fermentation after relatively short aging periods of six to eight months. Bottle aging on lees ranges from one and a half to four years. Wines are hand-disgorged, without sulfite addition at disgorgement. Laval's diligence at all stages of production is matched only by his evident love for his métier and his heritage, which extends to collecting vintage wine presses and maintaining the traditional basket-weaving practiced by his grandfather.

THE WINE TO TRY

See Wines to Know: Champagne, *page 333.*

Since: Estate bottling began in 1971. Vincent Laval joined the estate in 1991.

Winemaker: Vincent Laval since 1996

Vineyard surface: 3.4ha (8.4 acres)

Viticulture: Practicing organic since 1971. Certified since 1985.

Négociant work: None

Grapes: Chardonnay, pinot meunier, pinot noir

Appellations: Champagne Premier Cru

Filtration/Fining: None

Sulfites: 20mg/L at pressing

CHAMPAGNE DAVID LÉCLAPART

TRÉPAIL

Following an early interest in organic foods, Marne biodynamics trailblazer David Léclapart discovered the influential École d'agrobiologie de Beaujeu, where he went on to study with famed professors including Max Léglise, Pierre Rabhi, and Claude Bourguignon. It was a foundational experience that led him to install biodynamic agriculture when he took the reins of his small family estate in Trépail following the death of his father in 1996. Léclapart's first releases were in 1998, an alliterative trio of undosed, vintage chardonnays (L'Amateur, L'Artiste, and L'Apôtre) that soon rose to international acclaim.

Like his friend Vincent Laval, Léclapart seeks maximum ripeness at harvest, typically bringing in grapes between 11 and 11.5 percent potential alcohol. Léclapart's twenty-two parcels are primarily chardonnay, with four tiny plots of pinot noir.

Wines are typically bottled for secondary fermentation after about ten months. The enameled steel–fermented L'Amateur sees the shortest aging period on secondary lees, at two years, while the barrel-fermented, single-vineyard L'Apôtre sees the longest, at four years. Léclapart prefers organic cane sugar for secondary fermentation, finding it a purer product than concentrated grape must. With the exception of two successful efforts, in 2005 and 2006, to employ native yeast starters for secondary fermentation, Léclapart has otherwise used commercial yeasts. For the 2020 vintage, he inaugurated a new chapter at the estate, employing a cultured yeast selected from throughout his own vineyard parcels.

"I don't want people to seek my Champagne for the David Léclapart taste. I want to translate the taste of nature, here in Trépail, according to the vintage, which is always different."

THE WINE TO TRY

Champagne Premier Cru–L'Amateur

L'Amateur is fermented in enameled steel tanks before two years' aging on secondary lees. In warm vintages, it combines the baroque mineral structure and stirring purity common to all Léclapart's Champagnes with a particularly inviting freshness and immediacy.

Since: Estate bottling began in 1998.

Winemaker: David Léclapart

Vineyard surface: 2.97ha (7.34 acres)

Viticulture: Biodynamic

Négociant work: None

Grapes: Chardonnay, pinot noir

Appellations: Champagne Premier Cru

Filtration/Fining: None

Sulfites: 20mg/L at pressing

CHAMPAGNE

Natural Wine Dining

Obscured by the touristic luxury getaways of Reims and Épernay is the Champagne region's rich, no-frills historical cuisine. Its most iconic dishes might carry an "ick" factor for non-French diners. One is *tête de veau*, or calf's head. Sometimes it consists of flesh picked from the head, rolled with herbs, and poached or roasted; in the more up-front versions, it's simply chunks of poached meat in which one can discern the outlines of cheek and muzzle. A vinegary *sauce ravigote* cuts the fat, as does an undosed blanc de blancs. Meanwhile, the fragrant pork-intestine sausage known as andouillette attains perfection at Charcuterie Thierry in the town of Troyes. For a truly memorable experience, try it prepared beneath a melting snow-white landslide of local Chaource cow's-milk cheese.

Aux Crieurs de Vin

TROYES

Founded in 1998 by ex-chain-wine-shop clerks Jean-Michel Wilmes and Nicolas Vauthier, Troyes cave à manger Aux Crieurs de Vin is a landmark for natural wine at large, as well as for its local Aube Champagne scene. For many years the kitchen was overseen by Wilmes's mother, a virtuoso of well-sourced bistro comfort food. More recently, young Troyes chef Théophile Duc has ably taken up the task. The owners also maintain a tiny wine shop in the market hall in Troyes, where you can enjoy brilliant natural Chablis alongside trays of oysters, whelks, and shrimp from the nearby fishmonger.

L'Épicerie au Bon Manger

REIMS

Former Parisians Aline and Eric Serva opened this charming cave à manger and gourmet grocer in 2008, bringing the cutting edge of natural wine to the conservative confines of Reims. The unlikely wine selection features the crème de la crème of cult biodynamic Champagne alongside unsulfited fruit bombs like the wines of Auvergne natural vignerons Catherine Dumora and Manu Duveau. The short menu emphasizes high-quality épicerie foodstuffs, from sublime foie gras to spiced smoked salmon, plated and heated (where necessary) with care.

Le Garde Champêtre

GYÉ-SUR-SEINE

Opened in 2018, Le Garde Champêtre is a farm-to-table restaurant and bed-and-breakfast housed in a former train station, handsomely renovated by an ownership team that includes veteran Paris restaurateur and wine retailer Juan Sanchez (of Semilla, La Dernière Goutte, Fish La Boissonnerie, etc.) and Côte des Bar vignerons Cédric and Emilie Bouchard. With its choice location midway between Paris and Burgundy—and a stone's throw from many of the Aube's most talented natural and biodynamic vignerons—Le Garde Champêtre has made a wine lover's destination out of this long-overlooked corner of Champagne.

Le Garde Champêtre in Gyé-sur-Seine

CHAMPAGNE

Legends in the Making

The agricultural and winemaking reform that began in the Aube has long since spread to the Marne, as more and more savvy Champagne vignerons convert to organics and biodynamics and embrace more winelike, natural-style Champagne vinification, the better to distinguish their product from the oceans of commodified Champagne produced at the large houses.

Here are three younger Champagne estates that have gained a notable following in the natural wine world.

Bulles de Comptoir
Charles Dufour

LANDREVILLE

Young Aube vigneron Charles Dufour's unorthodox approach to Champagne is evident in his informal business name, which translates to "bubbles for the bar." He took on 6 hectares (14.8 acres) of family vines in 2008 and converted to organic agriculture by 2010. Until 2016, he produced parcel cuvées, but has since stopped in favor of enriching his evolving Bulles de Comptoir range, which consists of a different label and wine name each year, always an undosed, unfiltered, single-vintage wine assembled with wine from a *réserve perpétuelle* maintained since 2010.

Champagne Val Frison
Valérie Frison

VILLE-SUR-ARCE

Having established her wine-growing estate in 1997 on 1 hectare (2.5 acres) of family vines, Valérie Frison converted to organics in 2003 and first began bottling wine with erstwhile partner Thierry de Marne in 2007. After de Marne's departure in 2012, Frison rebaptized the estate, and with some initial natural winemaking advice from Burgundy's Dominique Derain, she soon made a name for her bold, single-vintage Champagne. Lightly sulfited only at press, unfined, unfiltered, and undosed Champagnes like Traverse and Portlandia are rich, full-fruited beauties.

Charles Dufour outside his cellar, with a glass of his pinot noir

Champagne Amaury Beaufort
Bar-sur-Seine

As of 2018, Amaury Beaufort left Champagne André Beaufort, where he had worked alongside his brothers since the early 2000s, taking 0.88 hectare (2.17 acres) of family vines (80 percent pinot noir, 20 percent chardonnay) in a parcel called Le Jardinot in Polisy. From Le Jardinot, adjacent to vines still farmed by his brothers, he produces about six thousand bottles of undosed, vintage Champagne, remarkable for its very long (around 21-month) demi-muid aging before secondary fermentation. Beaufort also practices his father's protocol for initiating secondary fermentation with native yeast starters produced with concentrated must. In addition to his refined Champagne from Le Jardinot, Beaufort produces an entry-level Champagne from purchased pinot noir, and a slim range of Chablis, also from purchased fruit.

Pergola-trained vines at Azienda Agricola Emidio Pepe in Abruzzo, Italy

17. Europe and the Caucasus

Natural wine is about more than just France, of course. Throughout the world, wherever wine growing formed part of the lifestyle of a paysan community, there is a history of what we can call "natural wine."

This is winemaking as it was practiced before the contemporary globalized wine market created financial incentives that encourage winemakers to produce ever more wine, ever more cheaply, and ever more adapted to mass tastes.

In recent years, as natural wine drinkers have rediscovered the value of this paysan heritage, they have found a cultural bonanza in the winemaking traditions of much of Europe and parts of the Caucasus. From Portugal to Armenia and everywhere in between, a natural wine revolution is either happening, or lying in wait.

As within France, however, certain regions have proven more fertile terrain for natural winemaking than others. This is mostly the fault of history, not geography.

In general, natural wine has flourished where land ownership is fragmented. It has also thrived in places where vineyard land remains accessible and affordable (as in Catalonia, versus, say, Napa), and where social support from the state, community, or family unit encourages the individual risk tolerance required for natural farming and vinification (as in France and Italy, but not in the United States).

The success of early natural wine communities in France, Spain, and Italy has since initiated what we might call the post-paysan era of natural winemaking. In the 2000s, many figures separated from paysan winemaking communities due to cultural factors like socioeconomic status and/or foreign identity began to see success in natural winemaking.

The post-paysan era has seen outsiders establish influential estates in previously overlooked regions, particularly those with impressive viticultural heritages. An archetypal example is the Belgian former fine-wine dealer Frank Cornelissen, who moved to Sicily's Mount Etna in 2001, drawn by its striking terroir and the availability of pre-phylloxera vineyards.

In post-Soviet states in particular, neo-vignerons have embraced natural winemaking in the post-paysan era as a celebration of a once repressed regional or national heritage. The most obvious example is Georgia, where winemaking on a family scale persisted despite Soviet occupation. Paysan winemaking abounds in Georgia. But many of the country's influential natural vignerons led fruitful careers in other sectors before turning to winemaking (American John Wurdeman was a painter; John Okruashvili, a telecommunications executive). A similar dynamic has played out in Slovenia, where construction magnate Zmago Petrič has spared no expense creating his impressive Vipava Valley natural wine estate, Guerila.

Eduard Tscheppe (*right*) and Stephanie Tscheppe-Eselböck of Gut Oggau beside an experimental vineyard planted to crossings between hybrid grape varieties and traditional Austrian *Vitis vinifera* varieties

In recent years, prosperous wine and hospitality families throughout Europe have also made transformational pivots to natural winemaking. Movia's Aleč Kristančič was far ahead of the curve, producing undisgorged pétillants naturels back in the early 2000s at his historically well-connected estate in Slovenia's Brda region. In 2007 in Austria, Eduard Tscheppe and Stephanie Tscheppe-Eselböck founded the influential Burgenland natural wine estate Gut Oggau, drawing from their respective experiences working at Eduard's family's conventional estate in Styria and at Stephanie's family's magnificent two-Michelin-star restaurant, Taubenkobel.

Mount Etna vigneron Frank Cornelissen in his tasting room

Natural wine's popularity has risen in tandem with increasing awareness, among wine drinkers and vignerons alike, of the luxury of small-scale organic farming and natural vinification. Banks deem both practices risk-intensive, compared to their alternatives, and accordingly both have essentially disappeared from the mass market. Since the 2000s, demand for natural wine has instead been driven by a sophisticated, international network of high-end natural wine retailers.

Today, natural vignerons of all backgrounds are confronted with two, increasingly distinct markets: one lucrative and international, the other regional and limited by the social obligations and economic constraints of the immediate agricultural community. Post-paysan natural winemaking prioritizes the former over the latter.

Regardless of the socioeconomic profiles of the natural wine communities outside of France, one thing, at least, is certain: they represent the future of natural wine. Here's an overview of the most dynamic and influential natural wine communities in Europe and the Caucasus in the twenty-first century.

Construction on Frank Cornelissen's new home and cellar on Mount Etna, August 2020

Natural Wine in Spain

Laureano Serres in his terraced vineyards in Terra Alta

Led by the cultural dynamism of Catalonia and the sophisticated dining scene of its capital, Barcelona, Spanish natural wine is spring-loaded for a renaissance. The scene gathered momentum throughout the 2000s and coalesced into its present state when Catalan vignerons Laureano Serres and Joan Ramón Escoda founded the first Spanish natural wine association, Productores de Vinos Naturales (PVN), in 2009.

Among those who have led Spain's natural winemakers into the international limelight is Girona's three-Michelin-star restaurant El Celler de Can Roca, a longtime supporter of natural wines from both sides of the Catalonian border.

Today's Spanish natural wines offer a stark departure from oaky modern Rioja or filtered albariño. Natural reds can be wiry and lean or keenly volatile, while many whites see a period of skin maceration. The Catalan natural wine scene is, if anything, even more radically opposed to sulfite addition and filtration than its counterparts in France.

Here are four estates to use as North Stars for further exploring Spanish natural wine.

Celler Laureano Serres Montagut
Laureano Serres

EL PINELL DE BRAI, CATALONIA

Armed with a wild-man enthusiasm and an indefatigable sense of experimentation, Laureano Serres plunged into winemaking in his native El Pinell de Brai in 1996 after abandoning a career in computer programming. Since beginning to bottle his own wines in 1999, usually under the "Mendall" label, Serres has become a key Catalonian link to France's natural winemaking community, forging close ties with elders like Thierry Puzelat as well as young guns like La Sorga's Antony Tortul (with whom Serres and his friend Joan Ramón Escoda created the Brutal Wine phenomenon; see page 278). Serres and Escoda also founded the PVN (Asociación de Productores de Vinos Naturales) and the H2O Vegetal wine salon. Serres's organically farmed, unfiltered wines, from 6 hectares (14.8 acres) of north- and northwest-facing clay-limestone parcels on thin soils in the Terra Alta, have had zero sulfites added since 2004. He uses a range of vessels and vinification styles, working with native grapes and a few Bordeaux varieties. His wines are confrontational, frequently volatile, spritzy, oxidative, and often truly inspired.

Celler Escoda-Sanahuja
Joan Ramón Escoda and Mari Carme Sanahuja

CONCA DE BARBERÀ, CATALONIA

With a silver bowl cut and a grin as wide as his belt, Joan Ramón Escoda is a towering figure in Catalonian natural winemaking. He and his companion, Mari Carme Sanahuja, farm a 15-hectare (37-acre) mixed-agricultural estate, founded in 1997, planted with a host of French varieties along with native grapes like sumoll and garnatxa (grenache). Red and whites macerate mostly in clay amphorae of varying sizes; élevage occurs in a mix of amphora, steel tanks, and a handful of oak barrels. Escoda is the founder of the PVN and the H2O Vegetal wine salon alongside Laureano Serres, and co-owns Barcelona natural wine hot spot Bar Brutal (see page 406), making his influence inescapable in Spanish natural wine. Most recently, he and Sanahuja have opened Tossal Gros, a gastronomic farm-to-table restaurant at their estate.

Joan Ramón Escoda in his cellar

Barranco Oscuro

Manuel and Lorenzo Valenzuela

CÁDIAR, ANDALUSIA

Barranco Oscuro's Manuel Valenzuela founded his mixed-agricultural estate on the high, clay-schist hills of the Alpujarras in 1979, farming organically from the start. At an average of 1,300 meters (4,265 feet), his vines are among the highest in Europe, and benefit from the drying Mediterranean breeze. After selling grapes to a local cooperative for many years, Valenzuela began making and bottling wine in the mid-1990s, soon after obtaining Spain's first organic certification. Today Manuel and his son Lorenzo have abandoned organic certification, finding it insufficiently rigorous. They farm 12 hectares (29.6 acres) of international varieties planted in collaboration with a local university, as well as native ones. Of particular interest is the rare native white vigiriega grape, which sees snappy, saline expression in the cuvée V for Valenzuela.

Partida Creus

BONASTRE

In 2001, Italian architects Massimo Marchiori and Antonella Gerosa moved from Barcelona to the low mountains of Bonastre Massif, where they set about acquiring tiny, all-but-abandoned parcels and working to conserve local Catalan grape varieties on the verge of extinction. Today they farm 6 hectares (14.8 acres). The couple began commercializing their wines in 2007, and soon became known for their iconic (and gnomic) two-letter labels, as well as a gloriously light touch in whole-cluster red vinification. Sulfite addition was eliminated in 2013. Following the precise grape blends within each cuvée is a fool's errand for all but serious aficionados of Catalan ampelography (cartoixa vermell, trepat, garrut, anyone?). This is an estate that knows the value of mystery. Of unmistakable interest is the couple's white AA, or Anonimo Ancestrale, a delicate, low-alcohol pétillant naturel packing a life-of-the-party personality.

FURTHER TASTING

Oriol Artigas, Nuria Renom, Erik Rosdahl, Vinyas d'Empremta, Ismael Gozalo, Jordi Llorens

Natural Wine in Italy

Since arising as an ideology in the early 2000s, Italian natural wine has faced a typically Italian challenge: there are too many leaders. In contrast to France, where most involved agree that Marcel Lapierre was the movement's first coalition-builder, Italian natural *vignaioli* have always been divided among several competing camps—or refused to be part of any camp whatsoever.

In 2001, the farsighted wine-and-spirits distributor Luca Gargano created his Triple "A" association. The three As stand for *agricoltori, artigiani e artisti* (farmers, artisans, and artists), the threefold identity he sought in winemakers as he rebuilt his import portfolio. Triple "A" is often considered an Italian counterpart to Nicolas Joly's La Renaissance des Appellations association. But Gargano merely endorses and does not insist on organic or biodynamic certification. In practice his association includes a connoisseur's selection of excellent natural estates, along with certain conventional estates whose wines happen to sell well in Italy.

In 2003, four leading natural vignaioli—Umbria's Giampiero Bea, Friuli's Stanko Radikon, Tuscany's Fabrizio Niccolaini, and the Veneto's Angiolino Maule—founded the natural wine association ViniVeri ("True Wines"), which organized Italy's first natural wine salon. Frustrated with what he saw as insufficient enforcement of ViniVeri's ideals, Angiolino Maule split off in 2006 to form VinNatur, an association devoted to iterating and enforcing a charter of certified organic natural wine production for its members.

Subsequent years have brought even more Italian natural wine associations, but Triple "A," ViniVeri, and VinNatur remain the most important today. The estates that follow include their most prominent members, and a few nonmembers; all rank among the most influential estates in the continuing saga of Italian natural wine.

Azienda Agricola Paolo Bea

Paolo, Giuseppe, and Giampiero Bea

MONTEFALCO, UMBRIA

Architect turned vignaiolo Giampiero Bea returned to his family's mixed-agriculture estate in the mid-1980s. Having never studied winemaking, he adopted the methods his father, Paolo, had been using until then, farming without chemicals and vinifying without artifice. His faith in these practices, paired with a gift for communication, made Giampiero Bea one of a select few Italian vignerons of his era willing to oppose the prevailing fashions of modern enology, including international grape varieties, new oak, filtration, and yeasting. In 2003, Bea helped found ViniVeri. Today he and his brother Giuseppe farm 5 hectares (12.3 acres) of vines, principally the dark, polyphenol-rich native sagrantino, which in their hands attains a rare detail and finesse. Grapes are destemmed and undergo long fermentations with regular punchdowns in stainless steel, before aging for up to five years in large oak *botti.* Sulfites are added only at bottling, arriving at total sulfite levels of between 40 and 50mg/L.

Cascina degli Ulivi

Stefano Bellotti, Ilaria Bellotti

NOVI LIGURE, PIEDMONT

The radical, freethinking Gavi vignaiolo Stefano Bellotti was for many years the public face of biodynamics and natural wine in Italy. An instinctive organic farmer, Bellotti rehabilitated his family farm starting in 1977, and discovered biodynamics in 1984. He subsequently became president of the Italian chapter of Nicolas Joly's La Renaissance des Appellations. Bellotti was ahead of his time within biodynamics circles, practicing unfiltered, unyeasted, unsulfited natural vinification for most of his production long before it came into fashion. A friend and collaborator of the distributor Luca Gargano, Bellotti was also central to the founding of Gargano's Triple "A" natural wine association in 2001. Before his death in 2018, Bellotti received accolades not only for voluptuous, long-aged corteses like Filagnotti and his crisp Semplicemente bianco and rosso

Stefano Bellotti's daughter Ilaria Bellotti at Cascina degli Ulivi

bottlings, but for the utopic working model of his biodynamic farm, Cascina degli Ulivi, where livestock, spelt, fruits, and vegetables intermingle with grapevines. Today the 22-hectare (54.3-acre) estate is run in the same spirit by his widow, Zita; his daughter, Ilaria; and her companion, former sommmelier Filippo Mammone.

Le Coste
Clémentine Bouveron and Gianmarco Antonuzzi

GRADOLI, LAZIO

French-Italian couple Clémentine Bouveron and Gianmarco Antonuzzi worked with a veritable Who's Who of natural wine legends in France, including Dominique Hauvette, Philippe Pacalet, Bruno Schueller, Dard et Ribo, and Didier Barral, before creating their polycultural, practicing-biodynamic estate on volcanic terroirs in the northeast corner of Lazio in 2004. Today it extends over 14 hectares (34.6 acres), of which 6.5 hectares (16 acres) are planted with a teeming diversity of varieties, both foreign (cabernet and merlot) and ultra-local (aleatico, procanico, ansonica, verdello, and roscetto, among others). The couple emphasizes experimentation in vinification and rarely repeat the same recipe from year to year. This, combined with their high prices, can be off-putting: depending on the cuvée and the year, the wines can be transcendent, exotic masterpieces, or opaque, volatile disappointments. More consistent are the couple's inexpensive and popular Litrozzo cuvées, which are by-the-glass fixtures at natural wine bars around the world and beloved for their unsulfited gluggability.

Azienda Agricola Frank Cornelissen
Frank Cornelissen

PASSOPISCIARO, SICILY

Frank Cornelissen, Belgian fine-wine dealer turned Etna vigneron, debuted in the early 2000s as a charismatic, well-connected Anglophone armed with an unimpeachable familiarity with the world's great wines. At that time he presented murky, earth-moving nerello mascalese from what was then an exciting new frontier of Italian wine. Cornelissen's unfiltered and unsulfited early wines found placement on Italian wine lists with no natural focus, giving many outside France their first taste of natural wine. His iconic cuvée Magma, from the pre-phylloxera Contrada Barbabecchi at 910 meters (2,985 feet) of altitude, has always been undeniably complex, with ashen tannins and dense, highly spiced black fruit. After years of adoring press coverage in most markets, a backlash was inevitable. But by the time parts of the natural wine world soured on Cornelissen, he had soured on the natural wine world, too. In 2016, he began filtering his wines, and two years later decided to systematically add small doses of sulfites to all wines. Never a naïf when it comes to business, Cornelissen now devotes his impressive intellect toward the quixotic goal of routinizing the production of his natural wine, without caring whether it is still called "natural" at the end of the day.

Josko Gravner

OSLAVIA, FRIULI-VENEZIA GIULIA

Josko Gravner adopted Georgian qvevri for fermentation in the early 2000s, sparking an explosion of international interest in the method. Alongside peers like Stanko Radikon, Gravner is also largely responsible for the present-day popularity of orange wine, having successfully introduced it to Western markets via the high-end

Italian restaurants of the early 2000s. Gravner took over his father's estate straddling the Italian-Slovenian border in 1975, and spent the rest of the century earning accolades for conventional winemaking. His interest in natural wine ideals in the late 1990s came out of his disgust for the winemaking he encountered on a visit to California. He began prioritizing old methods: organic farming (uncertified), native yeast fermentation, wood vessels, and, soon enough, Georgian qvevri. Gravner farms 18 hectares (44.5 acres) along with his daughter Mateja and grandson Gregor. They practice macerations of around seven months in buried qvevri, and very long aging in oak foudres before release. Unlike Radikon, Gravner has never abandoned sulfite addition. More significantly, he and his family evince a grandiose disdain for fellow winemakers that is at odds with the communal spirit of natural wine.

Above right: Giuseppe Maga at his tasting room

Below: Wine importer and vignaiolo Eric Narioo in his vines on Mount Etna

Azienda Agricola Barbacarlo di Lino Maga

Lino and Giuseppe Maga

BRONI, LOMBARDY

Until his death in 2022 at age 90, Oltrepò Pavese vignaiolo Lino Maga was a singular figure in this unheralded stretch of northern Italy, bearing few traces of influence other than his own family history. (The estate has existed in his family since 1886.) Maga farmed two vineyards of croatina, uve rare, vespolina, and ughetta grapes: Barbacarlo and Montebuona. Both parcel cuvées are nervy, ashen, hugely characterful reds, as long-lived as they are capricious and spritzy in youth. Today the estate is run in the same spirit by Maga's loquacious, chain-smoking son Giuseppe. The wines have always attained curiously high total sulfite levels (often in the 70mg/L range), despite the Magas' insistence that no sulfites are ever added.

La Biancara

Angiolino Maule

GAMBELLARA, VENETO

Sincere and inquisitive, Veneto *pizzaiolo* turned vignaiolo Angiolino Maule is the chief organizing force in natural winemaking in Italy. He founded La Biancara estate in the late 1980s, and from the start resisted the cynical farming and winemaking he witnessed in neighboring Soave, insisting on native yeast fermentation, additive-free vinification, and nonfiltration, as he reduced sulfite doses to their present very low level (he adds none at all for most cuvées). In 2006, Maule founded the VinNatur association after deciding that his cohorts in the ViniVeri association (which he also cofounded) would never accept certification of their stated ideals. Yet at heart, Maule is neither a radical nor a utopist. He is a precise technician, both in the vines (which see minimal irrigation) and in the cellar. He has earned acclaim within both high-end Italian wine circles and natural wine circles: a mark of his mastery is his flagship Pico, a lightly skin-macerated, foudre-aged garganega ripe with notes of pear and lemon curd.

Angiolino Maule in his cellar

Vino di Anna

Anna Martens and Eric Narioo

SOLICCHIATA, SICILY

Founded in just 2010, Vino di Anna is a collaboration between Australian winemaker Anna Martens, formerly a winemaker at Passopisciaro on Mount Etna, and her husband, Eric Narioo, whose UK import business Les Caves de Pyrene has helped define the French natural wine community since 1988. Their working methods at Vino di Anna meet the highest standards of thoughtfulness and purity in both agriculture and winemaking. The high, pre-phylloxera terraced vineyards they've recovered are treated without Bordeaux mixture; they forgo mowing in favor of mulching and manual soil work. Some wines are vinified in a restored traditional Sicilian *palmento*, a vinification room consisting of graduated levels for foot-treading grapes and drawing off juice. Many wines age in a magnificent qvevri cellar, carved into the hillside beneath their house. Their efforts are best realized in the cuvée G, an unsulfited, partially skin-macerated selection-level grecanico, concentrated, saline, and bristling with wild herbal aromas.

Massa Vecchia

Fabrizio, Vasco, Antonino, and Tosca Niccolaini

SOLICCHIATA, SICILY

Since 1985, when he took over vines his father planted in 1972, Fabrizio Niccolaini's commitment to small-scale, chemical-free, polyculture farming and low-sulfite vinification at Massa Vecchia has held an unflattering mirror up to the modern "Super Tuscan" scene that exploded around it in the Maremma. Organic certified since 1993, the estate is renowned for its minuscule production of patiently extracted, ferrous, low-yield, very-low-sulfite sangiovese-based reds and, more recently, its rich, long-macerated whites. Niccolaini was among the founders of ViniVeri in 2003. His estate's legend endured throughout an unusual arrangement in which his partner Patrizia Bartolini's daughter Francesca Sfrondrini (along with partners Daniel and Ines Wattenhofer, Thomas Frischknecht, and Rocco Delli Colli) ran the estate, with Niccolaini providing counsel, from 2009 until she retired in 2019. As of 2019, Niccolaini has resumed running the now 5-hectare (12.3-acre) estate along with his children, Vasco, Antonino, and Tosca.

Azienda Agricola Emidio Pepe

Emidio, Sofia, Daniela, and Chiara Pepe

TORANO NUOVO, ABRUZZO

Born in 1932, Emidio Pepe took over his family estate in 1964, becoming the repository for a family winemaking tradition dating back to 1889. His key insight was to insist, unwaveringly throughout his career, on the nobility and age-worthiness of Montepulciano d'Abruzzo, a wine most professionals continued to consider gas

station plonk until well into the 2000s. He would be proved correct by the seeming immortality of his wines.

Nothing is quite normal at this 16-hectare (39.5-acre) estate on the pergola-trained, clay-limestone hills of Torano Nuovo. Its montepulciano undergoes hand-destemming before fermentation and aging in glass-lined concrete tanks. Malolactic fermentation, curiously, tends to occur in bottle, prompting Pepe's unusual practice of disgorgement and decanting of lees into a second bottle. Wines (including the trebbiano and the lightly macerated pecorino) are generally sulfited to bring them up to a certain level in analysis, unless that level has been attained by naturally occurring sulfites.

Radikon
Stanko, Suzana, Saša, and Ivana Radikon

OSLAVIA, FRIULI-VENEZIA GIULIA

Alongside his friend and neighbor Josko Gravner, Stanko Radikon revived orange winemaking for the Western world, with long-macerated ribolla gialla–based wines that hit the palates of Italian wine buyers like depth charges. From the start, his aims diverged from those of Gravner, who sought influence from abroad. When Radikon began experimenting with skin maceration in 1995, he wanted to recapture a historical wine style, the way his grandfather vinified in the 1930s. Accordingly, he didn't use qvevri; instead, he achieved a similar wine style through destemmed three-month macerations of white grapes in wooden vats with frequent punchdowns. He was among the first in Italy to publicly forswear sulfite addition as far back as 2003. (He also reconceived the ideal bottle size, bottling his own wines in specially designed bottles of 500ml and 1L.) His family, led by his son Saša, now runs the estate, which today comprises 17 hectares (42 acres) on both sides of the Italian-Slovenian border. About half of the production goes into a new "S" range of shorter-maceration wines in standard 750ml bottles.

FURTHER TASTING

Nadia Verrua, Cantina Giardino, Fabio Gea, Barbacàn, Camillo Donati, Podere Pradarolo, Lamoresca, Colombaia, Denis Montanar, La Stoppa, Vina Čotar, Bera Vittorio e Figli, Dario Princic, Damijan Podversic

Top: Emidio Pepe and his family at a party for the estate's fiftieth anniversary in 2014

Right: Saša Radikon in his tasting room

Natural Wine in the Czech Republic

Three decades after the Velvet Revolution did away with the productivist Soviet influence on viticulture, the rediscovery of high-quality native wine culture commands a certain reverence in the Czech Republic, where a band of winemakers has been promoting natural wine principles since the late 2000s. Prague natural wine impresario Bogdan Trojak, a restaurateur, winemaker, and wine distributor, founded the Autentisté association in 2008 in collaboration with his erstwhile friend, the vigneron Richard Stavek. Trojak's Prague wine bar Veltlin was the city's first natural wine bar, and he followed it in 2014 with the nation's first natural wine festival, Prague Drinks Wine. Onetime Veltlin employee Jan Čulik has brought the natural wine movement to Tabor, south of Prague, with his wine bar Thir and its associated natural wine fair, Bottled Alive.

Today natural vignerons in the Czech Republic number in the dozens, based overwhelmingly in Moravia. (Just 4 percent of Czech winemaking occurs north of Prague in Bohemia.)

Bohemian whites tend to be pitched at the decibel of birdsong, while many fleshier Moravian whites see light skin maceration. Bright, rustic reds often hover at the lower limits of ripeness and extraction.

Dobrá Vinice
Petr Nejedlík

ZNOJMO, MORAVIA

Founded in 1994 by managing director Petr Nejedlík and Jaroslav Osička, Dobrá Vinice is sited mostly within the Podyjí National Park, covering 15 hectares (37 acres) on quartz, flint, granite, and gneiss soils. Since it was planted from scratch, the first vintage was in 2000. Nejedlík farms organically (without certification), emphasizing yield limits of under 35 hectoliters per hectare.

Today Dobrá Vinice is considered the Czech Republic's forerunner of quality natural winemaking. Of a truly world-class stature are his exuberantly pure, undisgorged pétillants naturels, Crème de Parc National and Crème de Pinot Noir.

Jaroslav Osička

VELKÉ ŽERNOSEKY, BOHEMIA

A former viticultural school professor who helped found Dobrá Vinice, Jaroslav Osička has farmed his own 3 hectares (7.4 acres) of vines (including chardonnay, pinot gris, gewürztraminer, and pinot noir) in Velké Bílovice organically since the 1980s. Low yields, light skin macerations, barrel-aging, and lees stirring are employed in service of a notably Chablisien winemaking aesthetic, chiseled and austere. Wines are unfiltered and sulfited at 20mg/L at bottling.

Richard Stávek

NĚMČIČKY, MORAVIA

A former food-and-wine journalist, Richard Stávek began producing wine for himself and his friends in the mid-1990s. Later he went on to help found the Autentisté group with his erstwhile friend Bogdan Trojak. His 4.7 hectares (11.6 acres) of vines include a host of extremely obscure varieties, including sevar and rubinet. His idiosyncratic wine range, meanwhile, includes cuvées that are not entirely grape wine, like Medový Muškátek, made with a dollop of his farm's own honey production.

FURTHER TASTING

Milan Nestarec, Porta Bohemica, Bogdan Trojak, Martin Vajčner

Richard Stávek in his cellar

Natural Wine in Austria

Austria's natural wine community first arose in Styria in the early 2000s, and has flourished throughout the 2010s in concert with the Karakterre wine tasting, an influential salon featuring natural wines from Eastern Europe organized by Croatian natural wine importer Marko Kovač. Since 2018, Karakterre has benefited from funding from the Austrian Wine Marketing Board, a rare instance of governmental support for natural wine.

Today, thanks to the dynamism of vignerons like Styria's Maria and Sepp Muster and Burgenland's Eduard Tscheppe and Stephanie Tscheppe-Eselböck, Austrian natural wine is the toast of natural wine bars from Tokyo to Brooklyn. For outsiders, there is something truly novel about Austrian natural wine culture, which has less in common with the culture of its biggest trading partner, Germany, than with those of the neighboring post-Soviet states of Slovenia and Hungary.

Natural wine fans have thrilled to the nation's skin-macerated whites and red wines. The latter, in particular, appeal to drinkers seeking ethereal reds without the overt fruitiness common to carbonic macerations in warmer climes.

Weingut Muster
Maria and Sepp Muster

LEUTSCHACH, STYRIA

Marie and Sepp Muster's 10 hectares (25 acres) of vineyards sit on clay-silt soils just north of the Slovenian border in the Südsteiermark. Neighbors and contemporaries of Karl and Eva Schnabel, the Musters took over their family estate in 2000 and obtained Demeter certification in 2003. While reds are by now produced entirely without sulfites (or filtration, of course), most whites see a light sulfite dose after racking. Reds and whites both see destemming and fermentation in large wooden vats, then extended aging in foudre.

Weingut Karl Schnabel
Karl and Eva Schnabel

MAIERHOF, STYRIA

Biodynamic certified since 2003 and entirely zero-zero in vinification since 2007, Weingut Karl Schnabel comprises 5 hectares (12.3 acres) of vineyards on siliceous rock on the slopes of the Sausal mountains in the Südsteiermark. The Schnabels practice cow pasturage in their vines and maintain a herd of cattle, one of whose faces beams out from their wine labels. Their pristine and supple reds from pinot noir, blaufränkisch, and Zweigelt see destemming, fermentation in open vats, and aging in used oak barrels.

Strohmeier
Christine and Franz Strohmeier

STAINZ, STYRIA

Though he declines organic or biodynamic certification, Franz Strohmeier has practiced a radical and highly attentive sort of gardening on his 8 hectares (19.7 acres) of vines north of the village of Stainz since 2003. He experiments with what he calls minimal pruning, allowing his high-trained vines to branch out extravagantly, in search of a purity akin to that found in truly wild grapes. Yields rarely exceed 20 hectoliters per hectare. In his low-alcohol, crystalline white wines—particularly those from the unsulfited Trauben, Liebe und Zeit range—one senses the

Sepp Muster in his cellar

Franz Strohmeier amid his minimally pruned vines

delicacy of his work, the unhurriedness with which his grapes attain their modest maturity. Strohmeier's soft touch in the vines continues into the cellar; most wines see minimal maceration. They always emerge with a stirring sense of direction and a whipcrack of acidity.

Gut Oggau

Eduard Tscheppe and Stephanie Tscheppe-Eselböck

OGGAU, BURGENLAND

Magnetic young couple Eduard Tscheppe and Stephanie Tscheppe-Eselböck took over this neglected seventeenth-century estate in 2007 and soon turned Gut Oggau into a tremendous branding success. Their whimsical concept was to create fictional characters for each of their wines, comprising one large fictional family whose faces adorn the wines' labels in handsome sketches by the artist Jung von Matt. The Tscheppes quickly found renown for their ambitiously priced, unfiltered, unsulfited wines, which typically see partially destemmed fermentations before élevage in oak barrels. Exotic, concentrated whites like Timotheus have long been more successful than the estate's reds, but recent vintages have evinced a laudable new direction in the latter, emphasizing lift and brilliant acidity over extraction. In 2018, the couple launched a new range of wines called Masquerade, which comprises wines produced from recently acquired parcels still in mid-conversion to biodynamic agriculture.

Christian Tschida

ILLMITZ, BURGENLAND

Since taking over from his father in 2003, fourth-generation Neusiedlersee vigneron Christian Tschida has come to captivate the natural wine world with his delicate, perfumed reds and rosés and his soaring whites, produced from 14 hectares (34.6 acres) of vineyards by Lake Neusiedler. He stopped adding sulfites after the 2013 harvest. His reds and rosés, from blaufränkisch, pinot noir, cabernet sauvignon, cabernet franc, syrah, and zweigelt, are known for very low extraction, which derives from whole-cluster maceration and an exceptionally soft press. Most wines age in Stockinger foudres. Tschida is a fanatical craftsman with a fearsome attention to detail in both the cellar and his vineyards; in the latter he takes particular care with made-to-measure compost treatments.

FURTHER TASTING

Meinklang, Weingut Werlitsch, Claus Preisinger, Andert, Alexander and Maria Koppitsch, Andreas Tscheppe

Natural Wine in the Republic of Georgia

Keti Berishvili of Gogo Wine driving a tractor home during harvest

Even natural wine diehards, accustomed to the colorful *paysannerie* of Europe's wine communities, usually find themselves overwhelmed by an initial encounter with Georgia's vast, ancient wine culture.

Here at wine's origins in the Caucasus, historical factors have preserved wine production at the scale of family consumption, a situation without precedent elsewhere in the world. Husbands and grandfathers in Georgia know how to make wine for the same reason grandmothers in rural Italy know how to make pasta: it has remained part of their traditional family hospitality; something shared, but not necessarily sold. In many villages, no dwelling is without a trellis of grapevines pitched above an outdoor dining area, providing, along with shade in the summer, just enough grapes to fill a qvevri, the buried clay amphora common throughout Georgia (see page 104). Without the financial pressures involved in monocultural grape farming or wine production on a commercial scale, many of the country's home winemakers work naturally, as they have always done.

Wine unites Georgians' attachment to both their Christian religion (which distinguishes them from neighbors in Turkey and Azerbaijan) and their oft-invaded lands, so throughout history, it has been a defining symbol of Georgian identity. It was only the country's Soviet-era repression that kept its rich native wine culture a secret from Western markets for so long. That all began to change in the 2000s. Italian vignaioli like Josko Gravner and Giusto Occhipinti were among the first to visit and draw inspiration from traditional Georgian qvevri vinification. Importers like Chris Terrell and Sune Rosforth and sommeliers and wine writers like Lisa Granik and Isabelle Legeron followed later that decade. Yet it wasn't until New York wine author Alice Feiring introduced the American expat and Georgian natural wine impresario John Wurdeman to influential Loire vigneron Thierry Puzelat in 2011 that Georgia's nascent network of vocational natural winemakers encountered the broader natural wine market that would soon embrace them wholeheartedly.

For vignerons and natural wine fans throughout the world, the discovery of Georgian wine culture has been like uncovering a time capsule: one that contains a recipe for traditional natural winemaking divorced from—not just conducted in defiance of—the market pressures that contributed to winemaking's industrialization almost everywhere else. If in France, natural winemakers have learned to adapt their winemaking to do without the sulfites and additives that stabilized their wines in the bottle, Georgia's natural winemakers are instead learning to adapt unsulfited and additive-free qvevri winemaking to the trauma of the bottling process. (Traditionally, qvevri wines were served within the home, from pitchers, and rarely bottled for sale.)

Today Georgia's natural wine culture—the commercially available part, at least—centers around a seemingly limitless number of restaurant and tourism projects launched by John Wurdeman and a long list of Georgian collaborators. Wurdeman is also a cofounder, with his friend and fellow vigneron John Okruashvili, of Zero Compromise, a Georgian natural wine fair, and the Natural Wine Association, which was established in 2017 to unite Georgian natural vignerons.

Prepare for a steep learning curve of grape varieties. But prepare also for your horizons in wine appreciation to be altered beyond recognition. It's hard to conceive that one country with a population under four million can contain such immense regional diversity, in everything from grape varieties to wine growing to winemaking. There are the famous orange wines, of course, that seem to nourish rather than intoxicate. But Georgian natural wine also offers delicate direct-press whites from grapes like tsolikouri and chinuri. Most (but by no means all) of the nation's reds tend to be rather dark and intense, particularly those from the teeth-staining saperavi grape.

Pheasant's Tears

Gela Patalishvili and John Wurdeman

SIGHNAGHI, KAKHETI

Georgia's most highly visible, scaled-up producer of natural qvevri wine, Pheasant's Tears is a collaboration between eighth-generation winemaker Gela Patalishvili and the American painter John Wurdeman. The pair met in 2005 while Wurdeman was painting a landscape beside Patalishvili's vines and began to collaborate in 2007, with the help of Scandinavian investors. Today Pheasant's Tears farms over 15 hectares (37 acres), and also produces wine from grape purchases from Imereti and Kartli as well as its local Kakheti region. All wines are produced in a hangar-size qvevri winery built adjacent to their vineyard site near Sighnaghi. (The estate also comprises a restaurant and tasting room in the village of Sighnaghi.) Patalishvili oversees actual production, while Wurdeman travels tirelessly throughout the world promoting Pheasant's Tears and other projects. Wines see anywhere from three weeks to six months of maceration, and are bottled unfined, unfiltered, and usually unsulfited.

John Wurdeman hosting a lunch for US importer Chris Terrell at the Pheasant's Tears restaurant in Sighnaghi

Nikoladzeebis Marani

Ramaz Nikoladze

NAKHSHIRGELE, IMERETI

Enthusiastically experimental yet ever precise, Ramaz Nikoladze is a winemaker's winemaker. Since 2007, Nikoladze has farmed 1.5 hectares (3.7 acres) of family vines without treatments or plowing, supplementing the small production with fruit from his wife's uncle's vines, as well as grape purchases from neighboring growers. The traditional winemaking style of his native Imereti (a cooler, more humid region) emphasizes far less skin contact on whites than in Kakheti, and Nikoladze has proven adept at experimenting with French practices like barrel aging and the production of pétillants naturels from the local tsolikouri grape. His direct-press, qvevri-fermented tsolikouri (made with just a bucket of skins) is a delicate, luminous beauty, evoking stone fruit and white tea.

Giorgi Natenadze

AKHALTSIKHE, MESKHETI

Giorgi Natenadze's natural wine production is intimately linked to the history of his native Meskheti region, which was occupied by the Ottomans for four centuries. The occupation obliterated Meskheti's local wine culture—almost. Natenadze began rambling in the woods at a young age, foraging for wild grapes to vinify, and began marketing the resulting wines in 2009. The vines of this extremely remote region on the Turkish border grow up trees at altitudes that begin at 1,000 meters (3,280 feet) above sea level; since the vines are wild, there are no treatments whatsoever. One particular White Horse Breast vine is over four hundred years old and still producing fruit. His Meskhuri red is slender, crisp, and beautifully herbaceous.

Imereti vigneron Archil Guniava (*left*) leads a tasting from his qvevri.

Iago's Wine and Mandili

Iago Bitarishvili and Marina Kurtanidze

CHARDAHKI, KARTLI

Iago Bitarishvili farms a small surface of chinuri vines obtained by his family after land reforms in 1992. His were among the first certified organic vines in Georgia. Like many winemakers in Georgia, Bitarishvili has qvevri buried both inside the small cellar beside his conspicuously well-appointed tasting room and outdoors in the open air. He makes skin-macerated and direct-press versions of chinuri; both are splendid, evoking ripe peach and fresh almond. His wife, Marina Kurtanidze, also produces a terrific mtsvane from grapes purchased in Kakheti.

Artanuli Gvino/Gogo Wine

Kakha and Ketevan Berishvili

ARTANA, KAKHETI

Former concert violinist Kakha Berishvili was among the first Georgian neo-vignerons to embrace traditional qvevri winemaking in the mid-2000s, working in his native village of Artana. His daughter Ketevan left a career in banking to join him at the estate in 2015, and launched her own wine production under the Gogo Wine label in 2017. Today Kakha produces an intense, inky saperavi, while beside him in the qvevri cellar Ketevan produces orange wines from rkatsiteli and mtsvane, fermented as a mixture of direct-press and destemmed fruit. Her Cuvée Bébé, in particular, is a fine-grained, apricot-toned highlight of the work of a savvy new generation of Georgian natural winemakers.

FURTHER TASTING

Tamuna Bidzinashvili, Okro's Wine, Didgori Winemaking, Giorgi Revazashvili, Ének Peterson, Shota Lagazidze, Archil Guniava

Natural Wine in Germany

Biodynamic thought originated in Germany, and is so well established in German society that even humdrum supermarket chains stock Demeter-certified bean sprouts and biodynamic wines. Natural wine, however, is almost nowhere to be found. As in nearby Switzerland, vignerons and consumers have remained attached to an outmoded, sterile notion of perfection in enology, so the aesthetics of natural wine have only begun to gain traction in Germany in the past decade.

The first glimmers of natural winemaking in Germany occurred only in the late 2000s, among several Mosel vignerons who were encouraged by Scandinavian sommeliers and wine buyers. Though he doesn't farm organically, Alf vigneron Dr. Ulli Stein began experimenting with nonfiltration and sulfite-free vinification as far back as 2007. Germany's true pioneers in wholly natural winemaking are found a half hour's drive farther south in Kinheim. Rudolf and Rita Trossen have farmed biodynamically since 1978, and began producing a range of unsulfited, unfiltered wines in 2011 to great acclaim. This inspired organic and biodynamic colleagues like Thorsten Melsheimer and Jan Matthias Klein to begin producing wines in a similar style. Since the 2010s, a host of mostly young German vignerons from the Pfalz to Franconia have embraced natural winemaking.

With the exception of the more considered and terroir-driven work of Rita and Rudolf Trossen and a few others, German natural wine today seems tailor-made to the tastes of contemporary natural wine bars. Light, crushable white wines abound, often released cloudy with lees after very short aging periods.

Rudolf Trossen in his Madonna vineyard

Weingut Rita and Rudolf Trossen

KINHEIM, MOSEL

Rita and Rudolf Trossen took over Rudolf's small family estate at a young age; their great act of rebellion was installing biodynamic viticulture amid the conservative chemical agriculture of the middle Mosel. Rudolf Trossen credits an eccentric Belgian wine collector named William Vermeylen, who once bicycled by the Trossen estate by chance and happened to stop, with turning him on to the wines of Pierre Overnoy and others in the French natural wine movement. Encouraged by this collector, along with a handful of Danish sommeliers, Trossen began bottling his Purus range of natural wines in 2011. No fan of skin maceration, Trossen produces almost exclusively direct-press parcel cuvées of riesling, along with reds from dornfelder and pinot noir. Wines ferment in steel tanks before continuing in large oak barrels. Trossen bottles them under crown cap when he feels they have completed their biological cycle, often with residual sugar still remaining. Low yields from pre-phylloxera vines on sloped blue schist soils that have long been farmed biodynamically help make these among the most moving and luminous rieslings on earth.

FURTHER TASTING

Jakob Tennstedt, Philip Lardot, Jan Matthias Klein, Thorsten Melsheimer, Andi Mann, Stefan Vetter, Brand Brothers, Julien Renard, Christine Pieroth

PART III

Enjoying Natural Wine

Previous pages: Saint-Georges-sur-Allier, Auvergne. In the aftermath of a near-miss hail storm, vigneron Jean Maupertuis polishes off a bottle of his neighbor Patrick Bouju's pétillant naturel.

This page: Harvest in 2015 for La Cuvée des Copines, a collaborative wine gleaned from abandoned vines, produced by a group of young women from the Beaujolais

Enjoying natural wine can be easy—even instinctive. But how we taste something, and whether we like it, depends upon our prior experiences.

There is, unfortunately, no overestimating the extent to which our palates have been conditioned by the conventions of what we might as well call "unnatural wine." Many basic features of natural wine—from native yeast fermentation to nonfiltration and low sulfite levels—have been absent for generations in many wine markets, particularly outside Europe. As a result, for many people, tasting natural wine can be an unsettling and exciting new experience, like visiting an outdoor market in a foreign country heaped with unknown foodstuffs.

Natural wine calls for slight modifications in how we shop for, serve, and taste wine. This section offers suggestions on how to provide the right context for natural wine. Some are based on observations of how natural wines behave compared to their conventional counterparts when you open the bottle. Others are purely matters of style, to encourage wine lovers to enjoy natural wine in the spirit in which it was made.

Anjou vigneron Sébastien Dervieux, aka Babass, enjoys a glass of chenin at the end of a day's harvest.

18.
How to Taste Natural Wine

How Tasting Has Changed

We're often taught to envision wine tasting as a moment of truth.

The work of the vigneron (as both farmer and winemaker) is examined by the taster, who becomes the de facto judge of both winemaking and agricultural methods.

The validity of this examination depends on the experience of the taster. Unfortunately, for most of modern history, tasters haven't had much access to critical information about agriculture and winemaking. You could glean such information if you happened to live near winemaking communities and were inclined to visit vignerons. Otherwise, all you had were tasting notes. Learning wine from tasting notes is like learning filmmaking from reading film reviews.

Throughout the twentieth century, an entire edifice of international wine education was constructed on the shifting sands of taster subjectivity, simply because no other approach seemed possible. We are encouraged to describe what the sensory experience of consuming a wine is like and whether it pleases us. But rarely do we dare to hazard more than a vague guess as to *why*.

Traditionally, only the most passionate and keen wine tasters could be expected to care about farming and winemaking techniques and their effects on a wine's organoleptic profile. Today we call such tasters natural wine fans.

It's no coincidence that the notion of naturalness in wine has gone mainstream in concert with the rise of the internet. In 1985, if you wanted to know what the process of racking a wine looked like, you probably had to enroll in winemaking school. Today, a YouTube search suffices. As wine tasters, we now have access to an unprecedented range of information about a subject we formerly knew only from what was in our glasses.

This new dynamic suggests a wholly different conception of wine tasting. The moment of tasting is no longer simply the end of a process with the taster as final arbiter. It is more of a beginning: a jumping-off point for a taster to learn more about a given wine and the culture surrounding it. It's not a chance to judge vignerons; it's a chance to learn from them. If we can accept the premise that the circumstances of a wine's production—its terroir, its farming, its vinification—determine its profile, it's no great leap to conclude that the ultimate sources of knowledge about the wines of a given region are the vignerons of that region. Knowing how they work, and why, lends form and purpose to tasting notes. It's still important to know whether you like something or not. And it's useful to give expression to our sensations in tasting notes. But there's no need to stop your inquiry there, before it's even begun.

Tasters at a village wine fair in Chinon in the Loire valley, 1957

Unlearning Wine

Champagne vigneron Bertrand Gautherot explaining the illustrations on his cellar walls. Gautherot believes many tasters are often "dazzled by aesthetics," causing them to miss the physical sensory impact of a wine's mineral salts.

Wine tasting experience is just muscle memory. You taste widely and try your best to remember what you've tasted, assembling a database of sensory impressions. Sooner or later your palate knows the profile of, say, gamay from Fleurie, the same way your fingers know the feel of cashmere or leather.

Gaining a knowledge of natural wine involves learning how such wines behave in a glass: what they look like, how they smell, how they taste, and how these attributes evolve over time. But it also involves *unlearning.*

Our impressions of wines always refer to wines we've already experienced. So coming to grips with natural wine often involves unlearning many expectations specific to conventional wines, from color and limpidity to aroma and flavor to the way a wine changes, or fails to, after the bottle is opened.

Just about all aspects of contemporary conventional wine are manipulated to avoid risk in production, stabilize it for transport, and ensure mass appeal in the marketplace. Wines that have been overly stabilized become biologically inert—or simply put, dead.

Tasting natural wine is a different experience, with a different set of aesthetics. It involves learning to appreciate wine while it is still vibrant and alive.

The Colors of Natural Wine

This is the part sommeliers call, rather heavily, the "visual examination." You do it before sniffing or sipping. It consists of looking at the wine in the glass, ideally in a well-lit place against a white background.

It's a moment of mild hilarity in most wine tastings, because it's unavoidably comical when people engage in simple gestures like *regarding something* in an exaggerated, performative way, like mimes. Someone invariably comments on a wine's "legs," a colloquial term for the semitransparent rivulets that cling to the wall of a wineglass after the wine has been swirled. (The French call them *larmes*, or "tears.") Legs aren't meaningless: they give an indication of the glycerolic content of a wine, which hints at how richly sunny its fruit will feel in the mouth. But they aren't worth excessive attention.

Look instead at the way light plays within the wine itself. Heavily filtered wines have a flat aspect: nothing shimmers. Sometimes they verge on tropical neons. They are optically empty.

Well-made natural wines tend to be limpid, too, but they evince an altogether different aspect: there is a vivacity, a vitality to their color, a shimmer deriving, presumably, from subtle variations of microparticulate matter suspended within the liquid itself.

Meanwhile, nothing prevents a young, slightly cloudy natural wine from being altogether enjoyable in certain contexts. It'll be leesy and straightforward, but sometimes very satisfying. (Cloudiness in older wines is more foreboding.)

Sometimes it can be tempting to dismiss the entire "visual examination" stage as an anachronism intended to enforce filtration by maligning any trace of cloudiness. But it's worth paying attention to what wine looks like. With an uncanny regularity, truly great wines announce themselves right out of the bottle with a memorable color display.

Pierre Beauger's Un Point C'est Tout! gamay

Color Enhancement

For much of the history of wine commerce, consumers have been wary of wines that appear too pale. Such a wine, the wisdom goes, may have been overcropped, harvested unripe, or, worst of all, literally diluted. This prejudice among consumers grows more outdated every year, but it is responsible for a long history of color enhancement in winemaking.

You can see an ancestral version of color enhancement if you gaze at certain old-vine parcels of gamay in the Beaujolais in autumn. Among the yellow leaves of the traditional gamay noir à jus blanc variety, you'll see vines whose leaves turn crimson instead. They're usually what are called *teinturier* varieties, dark-juiced variants of gamay planted among the traditional grapes to deepen the color of the finished wine.

Most efforts to alter the color of wines, though, are less natural.

Blending in darker wines from southern regions like the Languedoc and Spain has been common throughout history, but today this technique has been surpassed by modern enology. Contemporary winemaking prescribes solutions ranging from aggressive temperature control (for a "cold soak" before fermentation to allow colors to deepen) to micro-oxygenation (to stabilize color) to adding enzymes selected to extract color, like Lallzyme EX. By far the most notorious color-enhancing additive is Mega Purple, one of a range of grape juice concentrates marketed by Canandaigua Concentrate and widely employed in mass-market budget winemaking. None of these color-changing processes and additives can be applied to a wine without changing its organoleptic profile, too.

In recent years, global warming has rendered consumer concerns about thin, pale wines increasingly obsolete. Modern enology is increasingly tasked with ameliorating the opposite problem: excessively rich, colorful, high-alcohol musts.

Natural wines see no enological color enhancement: nothing is added, nothing is taken away.

Gamay teinturier vines visible among the other gamay in a parcel of Guy Breton's Régnié

In the Wineglass: Fining and Filtration

Spend a few years drinking exclusively unfiltered natural wines, and you'll probably gain a superpower: the ability to discern manipulated wines on sight, the moment they're poured. You begin to realize there are some colors of wine, certain ways it catches the light, that simply aren't found in nature.

"All filtered wines are diminished. I prefer a wine that is intact. You feel something's been removed from a filtered wine. Something's missing."

—Morgon vigneron Jean Foillard

Sometimes filtration is obvious to the eye, as in the contrast between these two Anjou chenins. On the left is a 2018 Brutal!!! from Les Vignes de Babass. On the right is an organic, filtered Anjou chenin from 2020.

It's not just about clarity, of course. Some filtered wines can become cloudy or leave large deposits in a bottle, and some unfiltered wines can achieve a remarkable limpidity. It's about gaining a feel for how wine looks in a natural state.

That being said, it's often easier to discern filtered wines by tasting them. The organoleptic effects of filtration differ according to many factors, including the tightness of the filtration and, of course, all the characteristics of a wine before it was filtered. Generally speaking, richer reds tolerate filtration better than lighter reds, for the simple reason that they possess more textural material to begin with, so removing some via filtration still leaves a wine with a sense of body and heft. Even a light diatomaceous earth filtration on reds from gamay and pinot noir, by contrast, leaves them feeling brittle and skeletal on the palate.

Sometimes filtration is less obvious to the eye, as in the contrast, or lack thereof, between these two Rhône syrahs. On the left is a 2018 Cuvée Larmande from Domaine du Mazel. On the right is a biodynamic filtered Vaucluse syrah from 2018.

Filtration in white and rosé wines can be harder to detect because it is so much more widespread. Almost all whites and rosés are filtered, some many times. Yet well-made, unfiltered whites and rosés reveal a substantially greater texture and weight on the palate and a more measured rhythm. (Some very accomplished natural winemakers, like Stéphane Tissot in the Jura, nonetheless prefer to filter white wine, finding that it "purifies" the wine.)

Aromas and Flavors

A classical approach to wine tasting examines a wine's aromas separately from its flavors. But the qualities that distinguish a natural wine's aromas from those of conventional wines apply equally to its flavors. Natural wine aromas and flavors differ from their conventional counterparts in two ways: diversity and temporality.

Natural wines encompass a wider and more dynamic variety of aromas and flavors, because the complex work of native yeast fermentation has been neither supplanted by powerful commercial yeasts nor stunted by excessive sulfite addition. Conventional enology systematically excludes just about everything besides fruit and floral aromas and flavors, considering earthy, vegetal, or animalic components a threat to an artificial ideal of cleanliness. Natural wines, by contrast, offer the fruits and the flowers alongside the rest of the teeming aromatic and gustatory smorgasbord of life: rainfall, civet, porcini mushrooms, leather jackets, sweet onion, andouillette, and so on. Some of these aromas and flavors might strike you as more appetizing than others. But it's important to recall that none arrive alone, or without the supporting context of other aromas and flavors. Natural wines are rarely one-note.

The aromas and flavors of natural wines also tend to evolve at a different rate than those of conventional wines. Natural wines are alive in the same sense as active-culture yogurt or kombucha. When the environment changes, they respond! Low-sulfite or zero-sulfite-added wines can sometimes be fragile or highly changeable upon air contact. More important, nonfiltration tends to yield wines in which lees are unevenly distributed throughout the bottle. As a result, the aromas and flavors of natural wines are often more ephemeral, appearing and receding, only to reemerge with some new inflection. They reward careful attention not just on the first sniff or the first sip, but through the duration of a glass or, better yet, a bottle. Conventional wines, conditioned for stability with filtration, fining, and sulfite addition, tend to offer a more monotonous, if more consistent, experience.

Christine Strohmeier of Styria's Weingut Franz Strohmeier picking blueberries in her garden

Texture

What we call "flavors" are combinations of tactile sensations and olfactory sensations (aromas). To understand the notion of texture in wine, try to focus exclusively on its tactile quality. The texture of a wine defines the timing of delivery of its aromas, or its rhythm.

Texture encompasses the elemental flavor categories (acidic, salty, bitter, sweet), which are in essence tactile sensations perceived on the tongue. A wine's chemistry will affect its texture. But texture also depends on three components that are often drastically adjusted in conventional winemaking: tannins, lees, and CO_2.

Tannins

Tannins clench the tongue. But not all tannins are created equal: ripe tannins are preferable to green tannins, and natural tannins to artificial ones. Natural wines, it goes without saying, contain only natural tannins, and occasionally tannins imparted by aging in oak. Tannins in conventional wines are liable to be artificial, or transformed in some way, kneaded to a pulp by excessive micro-oxygenation (see page 87). Ripe natural tannins in a well-made wine move with the grace of a canoe oared through still waters.

Lees

Lees are the solid particulate matter that collects at the bottom of a tank or barrel as a wine ages. Unfiltered natural wines—even limpid whites—contain trace amounts of fine lees in suspension. These lend a wine its grain, its palpability. Too many fine lees in suspension is undesirable: it lends the wine an opaque, chalky, reductive profile. But eliminating fine lees through fining and filtration is often worse, leaving wines stripped, fragile, and wan.

Carbon dioxide

Carbon dioxide is a natural by-product of fermentation. Wines emerge from alcoholic fermentation and malolactic fermentation saturated in it. It's a good thing, in that it protects wines against oxidation. Natural winemakers often like to leave a little more CO_2 embedded in their wine at time of bottling, because it allows them to use lower doses of sulfites, or none at all. (Conventional winemakers tend to eliminate more CO_2 by degassing; see page 108.) This means that natural wines are often peppier and livelier on the palate than conventional wines. There is sometimes even a full-on spritz. Sometimes it is undesirable, since it can keep wines in reductive states that can mask their fruit expression. At other times, a lightly saline zip of CO_2 provides perfect context for a wine's flavors (see page 378).

Macerating juice drawn from a buried qvevri at Domaine des Miquettes

The Feeling

Chambre Noire owner Oliver Lomeli and his employee prepare a healthy lunch at his pop-up space in Paris's 11th arrondissement.

Often overlooked in the emphasis on description and identification in wine tasting is wine's capacity to nourish. Wine is not a perfume: it is a foodstuff that we incorporate into our bodies. A harmonious, unadulterated natural wine should leave you feeling sated and aglow. At worst, conventional wines are like fast food, leaving a sensation of greasiness and fatigue. (As with processed food, so with processed wine.)

It can be hard to determine how you feel, physically, after just a sip or two. Usually it takes at least a glass. This is perhaps why this stage of wine assessment is so often neglected. By the time you realize a wine is making you feel bad, you're usually halfway into the bottle and the purchase can't in good conscience be undone. It's worth listening closely to your personal physical response to wines, however. Sooner or later, you get a feel for what goes down harmoniously and what doesn't. Natural wine teaches us to extend the notion of our palate to encompass the whole human organism.

To be fair, chemical farming, winemaking additives, and transformative processing aren't the only things that can render a wine indigestible. Shoddily made or precipitously consumed natural wines often possess discomfiting flaws. Some of the most routine culprits are oxidation, excessive or poorly integrated volatile acidity, volatile sulfur compounds, and brettanomyces. (See page 376 for more on wine flaws.)

NATURAL /LEX·I·CON/

DIGESTE

We don't often discuss our digestion in polite English. The French, however, have a delightfully discreet adjective for foods and beverages that are digestively harmonious: they call them *digeste*. This is a term worth adopting into English, as it's among the highest compliments one can pay a wine.

Mussels and white beans at Paris restaurant Le Cadoret, with Auvergne vigneron Aurelien Lefort's Ergastoline, a négociant wine produced from roussanne from Savoie

Flaws and Problematic Phases

"Flaws" is a heavy word. It's often misapplied, with a definitive finality, to criticize traits in a wine that are just passing phases. This is particularly true within the context of natural wines. If conventional wines are not subject to these problematic phases, it is because conventional enology has "solved" them, at the expense of the wines' integrity.

Depending on their severity, and the tolerances of individual wine tasters, natural flaws can range from innocuous to irredeemable. But none are as black-and-white as environmental flaws like cork taint or heat damage.

Volatile acidity

Volatile acidity—VA for short—is a measure of the gaseous acids in a wine. In France it is expressed in grams per liter of sulfuric acid. All wines have a little volatile acidity. Higher levels manifest themselves as vinegary notes from acetic acid. (Ethyl acetates are often produced in parallel.) High levels of VA are often the result of sluggish fermentations, when lactic acid bacteria consume sugars.

But volatility is not always a bad thing—far from it, in fact.

In its most noble manifestations, volatile acidity can integrate into the architecture of a wine, lending it aromatic lift as well as an impressive capacity to age. It's the secret to many of the most beautiful and long-lived southern European reds. But it becomes problematic in wines that lack the stuffing to counterbalance their volatility. They become harsh and grating.

This is why wine laws throughout the world regulate acceptable levels of VA. The legal limits, however, have declined in concert with the rise of enology in winemaking, so today many naturally made wines find themselves literally outside the law. (In France in 1921, wines were permitted up to 2.5g/L of volatile acidity. Today the legal limit in France for red wines is 0.98g/L.) This is a case-by-case issue: some highly volatile wines simply need more time in the bottle, while others are lost causes. Sometimes it's a phase, sometimes a flaw.

"Often you have wines that are fruity and fat, and they feel fabricated. Because they lack a bit of volatility. They're good, but only because they're not bad."

—Beaujolais vigneron Philippe Jambon

Cloudiness

This isn't a flaw so much as an observation of the state of a wine. That said, cloudiness can bespeak certain other flaws. Wines undergoing malolactic fermentation are cloudy, but you will almost never taste a wine mid-malo from a bottle. Cloudiness can result from short aging periods or clumsy racking, in which case it is caused by suspended lees. Mild cloudiness is nothing to fuss about. It only becomes irritating when the reductive nature of the lees blocks the sensation of fruit in a wine.

Reduction

Reduction is not, in itself, a flaw. Reduction, like oxidation, is a broad physical-chemical phenomenon that affects all wines to varying degrees. In certain forms, it can lend enjoyable grace notes (gunflint, new car) to wine, even in youth. It only becomes unpleasant if it manifests itself in burnt matchstick or rubbery aromas that persist and erase sensations of fruit. This can be caused by excessive sulfitage (which is never the case in natural wines), or by a predominance of reductive lees in suspension (which is, sadly, often the case). Ultimately, reduction is a phase, one more excusable in wines intended to age.

Volatile sulfur compounds

Much of the unpleasantness attributed, imprecisely, to reduction is in fact the result of volatile sulfur compounds, including hydrogen sulfide (rotten eggs), dimethyl sulfide (cabbage), and mercaptans (oniony notes). Volatile sulfur compounds can sometimes diminish with aeration. At other times, they're tenaciously embedded in a wine's suspended lees, in which case decanting can make things worse. This is a true flaw, often the result of an unfortunate or poorly managed fermentation. Volatile sulfur compounds are the reason certain disastrous wines can seem to be at once reduced *and* oxidized.

Ropiness or oiliness

This is less disgusting than it sounds. It's when certain species of bacteria cause harmless glucan polymers to form within a young white (or, more rarely, rosé) wine, rendering it somewhat thicker than wine. Ropiness doesn't affect a wine's flavor profile, but it does make it lugubrious and unpleasant to drink.

In principle, the polymer chains can be broken "mechanically"; that is, by shaking vigorously in a decanter or by putting the wine in a blender. In practice, though, these actions do as much harm as good, because they simultaneously exhaust a wine's entrapped CO_2, which can also give wine an impression of thickness.

You'll rarely find ropiness outside the realm of unfiltered, unsulfited white wines. It's among the most frustrating of all the problematic phases, because the condition usually passes after three or six months' aging in a cellar. A ropey wine was always opened too soon.

Brettanomyces

Brettanomyces—"brett" for short—is a genus of wild yeasts that degrade phenols in a wine into phenolic components, which can create an animalic range of aromas along with significant volatile acidity. Some tasters train themselves to be maniacally sensitive to brettanomyces; tasting with such people is like a military bunk inspection. More enlightened palates tend to allow that brett can contribute to complexity. If in bad cases it smells like horse stables, sweat, and old Band-Aids, in milder doses it can evoke fresh leather, musk, and a warm embrace. Sometimes brett can integrate with bottle aging; in other cases, it can conquer a wine entirely, with calamitous consequences. At best it's a phase, at worst it's a flaw.

Acetates

This is the colloquial term for ethyl acetate, the esterification of acetic acid, which is responsible for annoying aromas of nail polish and Scotch tape in wines. Because acetic acid is the key factor in a wine's volatile acidity, acetates are usually linked to high levels of volatile acidity. Yet certain wines show one symptom without the other. Acetates are generally a flaw, but usually an ephemeral one, for their defining feature is volatility: they tend to dissipate with aeration. In rare instances, those solvent aromas somehow provide lift and pleasant counterpoint to certain grassy or floral notes, evoking the Comme des Garçons perfume Odeur 53.

Oxidation

In a chemical sense, oxidation is a loss of hydrogen atoms in a wine's constituent molecules, resulting from contact with oxygen in the air. Serving wine—opening the bottle and pouring its contents into wineglasses—rapidly oxidizes wine that has been housed in a reductive state. This is why wine can be so changeable with aeration.

Think of the nutty and appley aromas and flavors of oxidation like a color palette: autumnal reds, browns, yellows, auburns, and burnt siennas. In the right configuration, they can be as astounding and moving as the display of forests in the fall. It's when the browns and yellows predominate that we speak of oxidation as a flaw. The vista grows flat and obscure; we smell only simple aromas of cider, bruised apple, and dry citrus peel. In wines fatigued by oxidation, acidity plummets. The wine becomes brief, and fruit disappears.

Popular wisdom suggests that natural wines are more prone to oxidation because they see less (or no) sulfite addition. Some do conform to the stereotype. Yet the presence of fine lees (due to nonfiltration) and embedded CO_2 can also help to preserve wines against oxidation.

"If we don't want oxidation and volatility anymore, it's because in the 1960s, enology arrived, and enology made us add products to wines, and everyone still does the same thing. But we know very well that wines used to go in all sorts of directions."

—Alsace vigneron Bruno Schueller

Fizz

A little fizz is not a flaw at all. Carbon dioxide is a normal by-product of fermentation that helps preserve wine against oxidation; by the end of their fermentation cycle, most wines are suffused with it. Yet there remains a widespread misperception among wine drinkers that a still wine that shows a bit of fizz is necessarily "refermenting" and should not be drunk. In reality, few still wines continue fermenting in the bottle in any significant way. That spritz is just embedded CO_2, which often has salutary effects on the palate. It heightens sensations of acidity and salinity, which can provide a pleasant counterpoint to a wine's fruit.

Fizz only becomes problematic if it disrupts the harmonious expansion of a wine's flavors on the palate, or if it renders reductive aromas more tenacious. Happily, it's one of the rare issues that can be addressed in the moment by simply aerating the wine.

"Today, stories of mouse appear because there's more liberty in the wines. And we find ourselves in the same difficulties people confronted at the start of the nineteenth century, or the end of the eighteenth century."

—Anjou vigneron Richard Leroy

Mousiness

Mousiness, or just "mouse," is a catchall term for a range of retro-olfactory aromas that can appear in unfiltered, low-sulfite wines after malolactic fermentation and after bottling. The name comes from the unpleasant odor of mouse droppings in cupboards, but the phenomenon also manifests itself in aromas of peanuts, boiled green beans, canned black olives, and more.

Where does mousiness come from? Enologists have implicated lactic acid bacteria, whose activity can sometimes produce volatile compounds in the family of tetrahydropyrimidines that are responsible for some of the odors. There is also evidence to suggest that sensitivity to mouse varies widely among individual tasters. But the science remains hazy.

"Me? I don't care about mouse at all. I know in six months it'll be gone."

—Jura vigneron Jean-François Ganevat

Empirically, natural winemakers have observed that mousiness can occur in tank or in bottle, often worsens in the second half of an open bottle, and often appears shortly after bottling, then disappears weeks or a month later. This last phenomenon occurs often enough that winemakers have given it an affectionate name: *la petite souris de la mise*, or "the little mouse of bottling." Mouse can arise during transport, with some wine buyers attesting that wines they tasted in bottle at an estate proved mousy upon receipt months later. The good news is that mousiness is rarely a permanent condition; it usually disappears after six months to two years of bottle aging, unless nervous winemakers affix it in place with an ill-timed sulfite addition. (Just as the appearance of mousiness is the effect of a biological process, so too is its disappearance. Sulfites can inhibit both.)

At the end of the day, encountering mouse in wine is like encountering a mouse in real life—it's not worth causing a scene about. Of graver concern is the incipient weaponization of the catchall term "mousiness" by critics and wine professionals to disparage natural wine, simply because they are unfamiliar with its normal behavior, which includes the subtle evolution of a living biome within the bottle.

The Limits of Tasting

Natural wines are not inert creations—they are biologically alive, so they differ from conventional wines in their relationship to time. Some change from one sip to the next, like the oxidative Sancerre wines of Sébastien Riffault. Others, like the wildly abstract Alsatian wines of Bruno Schueller, can show immaculate stability when stored in a half-empty jeroboam in a corner of the cellar for four years. Almost all will pass through difficult phases in the bottle at some point or another.

A good natural wine taster knows they are getting only a glimpse of the wine at a moment in time. The wine tasting itself is a social construct; unlike, say, a three-course meal or a long drinking session, a tasting offers only a momentary perspective on a given natural wine. It's often less than worthwhile to rush to judgment on the basis of a sip.

By no coincidence, natural wine culture prioritizes the act of *drinking* over the act of tasting. You spend more time with a wine when you're drinking. You digest it, and you gauge your response. Your body is your palate.

Of course, you don't have to finish an entire bottle to understand every natural wine (that would be an unwise strategy, in fact). It just helps, when considering natural wine, to bear in mind the limits of wine tasting.

"Natural wine is a living thing. So you have to accept that sometimes it tastes good, sometimes less good. And some wines are more fragile than others."

—Roussillon vigneron Jean-François Nicq

Open bottles at the cellar of Guy Breton

Sommelier Sebastiano Simeoni serving wine at his former workplace, natural wine bar Deviant in Paris's 10th arrondissement

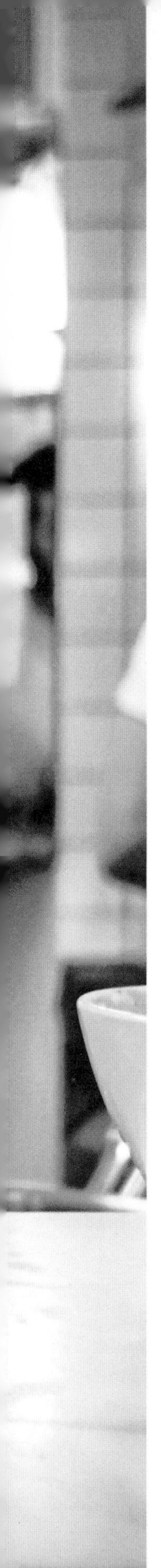

19. How to Serve Natural Wine

Unlearning Wine Service

Perhaps the greatest principle for wine service—and hospitality in general—comes from a 1528 book by Italian diplomat Baldassare Castiglione, *The Book of the Courtier*. He called it "sprezzatura," and defined it as a way of behaving "so as to conceal all art and make whatever one does or says appear to be without effort and almost without any thought about it." He advised readers to "avoid affectation in every way possible as though it were some rough and dangerous reef."

Natural wine service calls for its own sort of naturalness in style and bearing. It takes inspiration from the fancy-free paysan farming communities from which the wines derive, as well as from the rational care involved in appreciating the great wines of yesteryear.

In practical terms, avoiding affectation in wine service means avoiding gratuitous displays of luxury. Enormous fragile glassware, overwrought decanters, and expensive Coravin devices are not necessary for wine appreciation.

Avoiding this frippery can be a challenge in the context of today's international wine market. Natural wines brought into the world as simple, inexpensive thirst-quenchers can become sought-after wallet-busters on the export market. In the restaurants and wine bars where such wines are served, the general inclination is to provide clients with a luxurious setting to encourage heavy spending. Yet nothing could be further from the spirit of natural wine, which has its roots in neighborhood bistros, not three-Michelin-star hotel-restaurants.

Natural wine can indeed be expensive, depending on where you encounter it. But to serve it as a luxury product usually betrays the spirit in which it was made.

Bar service at the former location of Paris wine bar Chambre Noire

Liter Bottles

Liter bottles of wine have long been associated with downmarket Edelzwicker and California chablis. But the format has proved a perfect fit for natural wine estates aiming to highlight their wines' accessibility. Nowadays liter bottles from winemakers like Louis-Antoine Luyt, Beaujolais émigrés Raphaël Beysang and Émelie Hurtubise, and Lazio's Le Coste are the toast of natural wine bars from Paris to San Francisco.

Initial credit for redeeming liter bottles within natural wine circles must go to Friuli's Stanko Radikon and Edi Kante, who embraced the format in the 2000s. They reasoned that a mere 750 milliliters of their orange wines would not last the duration of a meal for two.

Liter bottles by southern Beaujolais vignerons Raphaël Beysang and Émelie Hurtubise

Beyond the Bottle

The wine bottle is iconic. Its dimensions are instantly recognizable, and all the subtly transformative air contact that modern wines undergo in their bottling, storage, and service is based on a volume of 750 milliliters.

But the wine bottle is not eternal.

The modern wine bottle was invented by English diplomat Sir Kenelm Digby, who in 1632 refined a method at his glassworks to create stronger bottles that offered protection against light. Yet bottles remained expensive and were reserved for the aristocracy until the end of the eighteenth century. Even then, estate bottling was rare. Wine was instead transported in barrels, then bottled and brought to market by négociants, or consumed directly from the barrel at bistros. (Barrels themselves were an invention of the Gauls; Romans and other ancient civilizations transported wine in amphorae.)

In Georgia, meanwhile, bottling wine is an even more recent development. Traditionally, wine was served directly from qvevri in pitchers to guests in the home.

All this is to say that natural wine existed long before the wine bottle, and it will exist long after. In today's era of growing environmental consciousness, estates, retailers, and consumers alike are questioning the wisdom of shipping wine around the world in a container that accounts for an average of 40 percent of its weight.

Languedoc vigneron Olivier Cohen has experimented with putting his wines in kegs in recent years.

Shaking Things Up

If you frequent natural wine spots, you may encounter servers and sommeliers eager not only to decant the wine you order, but to shake the living heck out of the decanter, as if it were a can of spray paint. The practice is meant to violently aerate a wine and remove entrapped CO_2. It appears to have been born of a marriage of convenience between brash young natural winemakers, impatient to bottle their wines, and servers responding to the irrational consumer fear of spritzy wines. It is folly.

In a better world, all wines would be aged long enough to release entrapped CO_2 in a controlled manner, and all winemakers would bottle their wines with enough finesse to leave precisely the right level of CO_2 in the bottled wine. In this better world, wine drinkers would also be aware that entrapped CO_2 is usually a good thing, as it prevents oxidation and lends pleasant acidity and crunch to a wine. Natural wine purveyors in this better world would not sell wines until their entrapped CO_2 had sufficiently integrated with bottle aging.

In the world as it is, though, certain winemakers and sommeliers are apparently so terrified of delivering a slightly fizzy wine to a client that they're willing to sacrifice the wine itself. Shaking does indeed remove CO_2, but in unfiltered wines, it also throws all particulate matter into suspension. This tends to lend an unfocused, matte texture to the wine, while also foregrounding any volatile sulfur compounds that may be housed in its reductive lees. It is the wine-service equivalent of locking yourself out of the house, then breaking down the door instead of awaiting a locksmith.

Think Before Decanting

Most decanters are lovely objects, like little glass monuments on the tabletop. It is understandable that wine drinkers would be eager to use them if decanters are on hand. Decanting natural wine, however, is different from decanting conventional wines, and requires more forethought.

Most agree that older wines should be decanted with caution, because they tend to be fragile and their fleeting, complex aromas can be lost with precipitous air exposure. This is as true of natural wines as it is of conventional wines, if not more so.

Conventional wisdom says there's no harm in decanting young wines, and this is perhaps true of young conventional wines. Young natural wines require more attention. Because they see less (or zero) added sulfites, they can be more prone to change with air exposure. More significantly, natural wines tend to be unfiltered, so they require greater care when decanting. Clumsy decanting will unsettle a wine in much the same way shaking it will, sending lees into suspension and encouraging the expression of volatile sulfur compounds, all while administering a thwack of oxidation that can rapidly degrade more fragile natural wines.

Decanting in a cellar

Modest Glassware

Pineau d'aunis in INAO glasses in the château of Sarthe vigneron Philippe Chidaine

First-time visitors to natural wine bistros in France are often surprised at the modest size of their wineglasses, which can be puny compared to their counterparts in conventional luxury restaurants the world over. Variations on the standard 7-ounce INAO glass are common at restaurants ranging from Paris's Aux Deux Amis and Le Baratin, to the many wine bars and restaurants they've inspired, like Oakland's Ordinaire and New York's Frenchette.

Why so tiny?

Modest glassware is practical: cheaper, easier to wash, easier to store. But it's also a political statement: how we serve wine is reflective of wine's role in a given society. In the viticultural regions of Europe and the Caucasus, wine has roots in peasant agriculture. Famous, aristocratic wines have always coexisted with wine's societal role as a workingperson's beverage. So in many of these areas, it's considered ludicrous to serve wine with fancy, ostentatious glassware anywhere outside the context of a luxury restaurant.

In the New World, the UK, and Asia, where wine has no roots in peasant culture, it has come to be associated almost exclusively with the notion of fine dining. This has created an expectation that all dining establishments—even informal bistros and wine bars—ought to provide ritzy "restaurant" glassware for wine. Glassware manufacturers and enterprising wine experts alike have rushed to fill this "need" by proposing a myriad of all-purpose glassware, region-specific glassware, and grape-specific glassware, all at premium prices.

In refusing to join this nonsensical international restaurant arms race of luxury glassware, natural wine bistros remind us that before natural wine was auctioned online, it was served at lunch tables on farms. When tasting in their cellars, the pragmatic and thrifty vignerons who actually make the wine typically employ simple, sturdy glassware. Why should we need anything more complicated to appreciate it?

The Biodynamic Calendar

Many biodynamic vignerons believe a wine's organoleptic expression can vary according to the lunar cycle and astrological sign of a given calendar day. In the biodynamic calendar, days of the year are divided into root days, flower days, leaf days, and fruit days, and each is thought to influence a wine's expression in its own way. Today there is a small industry of competing lunar calendar books and apps, all presenting slight variations from one another, by authors like Matthias Thun and Michel Gros.

In theory, wines should taste at their fullest on fruit days. Flower days highlight aromatic characteristics, while leaf days foreground mineral and earth characteristics. Root days are notorious among believers for causing wines to taste subdued.

Is any of this worth your attention? Only if you are the sort of supremely organized person who finds the task of arranging opportunities to taste wine amid the usual melee of work and family engagements insufficiently challenging.

Consider Temperature

Temperature is critical in wine appreciation. Wines that are too cold seem muted and opaque, while wines that are too warm seem blurry and slightly nauseating. The ideal service temperature of any given bottle should be assessed on an individual basis.

The natural wine scene has popularized light reds like poulsard, cinsault, and gamay. Many of these can be enjoyed chilled, but not ice-cold.

Orange wines often show best at the temperature of a light red, which is to say, slightly warmer than a typical white wine.

Serve sparkling wines colder than other wines. It helps preserve the bubbles, because CO_2 dissipates faster at higher temperatures. (This is also why an insufficiently chilled bottle of Champagne tends to erupt when opened.)

Do yourself a favor and ignore any charts you might encounter that claim to tell the precise ideal serving temperatures for a wine based on its appellation. Appellation no longer transmits much meaningful information about the wine inside a bottle—especially when it comes to natural wines, which rarely conform to outmoded stylistic norms. For example, you can find many light, crystalline natural reds that require chilling in the southern Rhône, a region known for huge reds in the popular imagination.

Tavel bottles chilling in the irrigation channels in the old village center

Bring Reinforcements

Natural wines are like people—their behavior can sometimes surprise you. Even good friends sometimes let you down or behave like strangers. Usually it's through no fault of their own. You just caught them at a bad time. Maybe they're late to work, or on the verge of a breakup.

So it is with natural wine. Once you accept that natural wine is living its life in the bottle, it becomes clear why natural wines can, and should, present different profiles on different days. A wine that shows beautifully in November might taste like strawberry yogurt in February, while a wine that was hard and ungainly the year after its release might be harmonious and suave two years later. Try as you might to affect the profile of an open bottle by decanting, sometimes the wine just wasn't in a good place when you opened it.

The best thing to do in such circumstances is admit defeat and open a different wine. By all means, recork the wine that wasn't showing well and try it again in a day or two. But in the meantime, don't get caught with nothing else on hand. It's always a good idea to prepare more wine than you need for a given occasion.

Finished bottles outside the cellar of Georgian vigneron Shota Lagazi

Zero Waste

It's never pleasant to pour wine down the drain—so don't! Unfinished bottles can always be repurposed for cooking. Surplus unsulfited, unfiltered natural wine is also well suited to making vinegar, which, as winemakers like to point out, is the destiny of all grape juice without human intervention. (Conventional wine is less appropriate for making vinegar, because sulfites inhibit the activity of acetic acid bacteria.)

In the most basic sense, all you have to do is affix cheesecloth around the mouth of the unfinished bottle (to keep out flies while allowing airflow) and leave it in a warm place for a few weeks. A vinegar "mother"—an agglomeration of cellulose and acetic acid bacteria—should form on its own, and can be saved and used to kick-start subsequent vinegars. Note that it's best to do this in a well-aerated space away from your living area, as the aromas of acetic bacterial activity are unwelcome in most social situations.

Vinegar production at Le Petit Gimios in the Minervois

When to Age Natural Wine

The myth that natural wine can't withstand aging has by now been so thoroughly debunked that we should no longer even need to address it. Let's call it what it is: a tribal critique based on an archaic understanding of wine, generally used by older wine drinkers to disparage the tastes of a new generation. Those who say natural wine can't age just haven't tasted enough natural wine.

In reality, whether a wine can improve with age has little to do with its naturalness. The intention of the wine's creator is what counts. The vast majority of wines, even expensive wines, are produced today for immediate consumption. They might improve over five years in the bottle, but they're rarely built for the long haul. The creation of truly age-worthy wines begins in the vineyard with exceptionally low yields, and continues in the cellar with exacting, often quite extractive vinification. Often such wines are impacted and undrinkable upon release. Foreboding aspects like volatile acidity, intense reduction, vicious tannicity and even brettanomyces can, with the passage of time, harmonize into something beautiful.

Unfortunately, winemakers are also known to invoke age-worthiness as an alibi for clumsy, overextractive vinification. Sometimes wines that offer no pleasure or fruit in youth never attain those traits with age. Aging wine is not a morality play; there is no automatic reward for patience.

As you can see, there are no rules, which makes collecting and aging wine a terrible investment. The ultimate value of a long-aged wine depends heavily on its storage conditions, along with other intangibles, like the integrity of the corks used by the estate in a given year, and the inscrutable caprices of the living wine itself.

Do you have adequate cellar conditions (see opposite) at your disposal? Are you at a more or less stable point in your life? Do you have money to burn? If so, by all means, age some wine. It is a sentimental investment, often rewarding, but rarely in an economic sense.

There remains one persuasive argument in favor of collecting and aging natural wine. The fact is, good natural wine is in high demand today, which creates a situation in which very few wine estates, restaurants, or retailers bother with the cost or risk of retaining stock of back vintages. It can be damnably hard to source well-aged natural wines. Once a vintage is sold out, it's often gone for good—unless you yourself have squirrelled away a six-pack or two.

Syrah in Hervé Souhaut's cellar in Ardèche

How to Store Natural Wine

Darkness. Stillness. Humidity. Chill. Inertia. These are the prerequisites for the ideal wine cellar. Historically, most cellars have been underground, because it's a natural way to achieve all these conditions.

Of course, underground wine storage isn't possible in all parts of the world, let alone in all lifestyles in all parts of the world.

Today many wine collectors must rely on air-conditioning to maintain cool temperatures at their storage sites. This is less than ideal for many reasons: it has a tendency to dry out the air (and thus the corks), it creates vibrations, and it wastes energy. The same drawbacks apply to the wine storage fridges popular with restaurants and urban professionals alike. Fridges are great for keeping wine ready to serve, but they're terrible for conserving wine long term.

The ideal wine storage conditions might be out of reach for most of us. But here are a few guidelines to keep in mind for safe short- to medium-term natural wine storage in the suboptimal conditions of a normal house or apartment.

The carefree organization of Doubs winemaker Georges Comte's personal cellar

1. Avoid heat.

Keep wine away from radiators, windows, and any walls that absorb a lot of heat from the sun. Heat tends to encourage unruly bacterial activity in unsulfited, unfiltered wines.

2. Avoid light—especially sunlight.

Keep wine away from sources of light. Sunlight is especially dangerous, because UV rays kill bacteria, including beneficial bacteria.

3. Avoid vibrations.

Keep wine away from mechanical vibrations from washers, dryers, air conditioners, dehumidifiers, and the like. Vibrations—even on the electromagnetic scale—can cause a wine's lees to remain in suspension and perturb the biological processes implicated in aging natural wine.

4. Avoid dryness.

Long-term exposure to fridge fans, for example, tends to dry out corks and cause labels to become unstuck.

5. Avoid sudden shifts in any of these things.

A bottle of natural wine is a biosphere. Sudden shifts in environmental circumstances can discombobulate the community living within it. Even changes in atmospheric pressure, like the arrival of a summer storm, can temporarily transform wine.

Mosel vigneron Thorsten Melsheimer's riesling vines at the top of the Mullay-Hofberg vineyard

20. How to Find Natural Wine

The Rarity of Natural Wine

Natural wine is swell. The bad news is, it accounts for a vanishingly small proportion of wine produced worldwide. Even organic viticulture, which is considered a prerequisite of natural winemaking, but which is not at all the same thing, accounts for just 5 percent of vineyard surface worldwide. There are no official statistics for wine produced naturally. Perhaps a hundredth of that 5 percent?

If you want to explore natural wine further, it can help to adjust your expectations. Unless you happen to live in a wine region or a developed metropolitan wine market, natural wine might be hard to find in daily life. It's not like kosher or gluten-free food; there aren't sections devoted to it in every major supermarket. (And where there are, approach with caution. See page 399.) You often have to seek out dedicated natural wine retailers.

Depending on your budget and where in the world you happen to be, sometimes you might find yourself settling for natural-*enough* wine. Don't get discouraged, though. Settling for natural-enough is what every natural wine importer does when assembling a portfolio of estates, and what most retailers and sommeliers do when assembling the wine selections for their shops and restaurants. The important thing is to make the effort to support natural methods in wine production, and to know why they matter.

Le Verre Galant, Jean-Jacques Maleysson's renowned natural wine cave à manger in the Massif Central mining town of Saint-Étienne

How to Read a Natural Wine Label

1. Name of estate/vigneron

Sometimes the name of the estate is the same as that of the vigneron, sometimes not.

2. Status of producer

A phrase intended to tell consumers about the commercialization conditions of a wine. Is it grown at the estate? Or is it a négociant wine?

3. Location of the estate

For wines produced without an appellation (e.g., Vin de France), this may refer only to the site where bottling occurred.

4. Natural claim

Some natural winemakers include label text to let consumers know that their wines exceed the purity standards set by organic and biodynamic certification.

5. Organic (and sometimes biodynamic) certification

Many natural wines are also certified organic and/or biodynamic, but many are not (for reasons why, see page 55).

6. Alcohol content

Be aware that winemakers have a half degree of legal wiggle room on this figure, and many bend the facts even further.

7. Lot number

In wines where vintage information is not displayed, it can usually be found in the lot number.

8. Vintage

Until 2009, it was illegal for wines released as Vin de France to display a vintage or grape variety. Nowadays both are possible, but many natural vignerons prefer to continue omitting this information.

9. Cuvée name

Sometimes the cuvée name is made up (the French enjoy tenuous puns). At other times, it refers to a specific parcel or lieu-dit (place-name).

10. Appellation

Some natural vignerons decline to use their local appellations. It's a way of protesting the stylistic conformity enforced by appellation tasting panels hostile to natural winemaking.

Name of importer and distributor (not pictured)

These are usually featured on a wine's rear label. It's worth remembering the names of importers. The best ones strive to uphold a standard of quality and naturalness across their portfolios.

Natural Wine Shops

In recent years, shops dedicated to natural wine have opened in just about all the world's developed metropolitan wine markets, and in wine regions that benefit from significant tourism. Radical selections of diverse natural wines can be found from Brooklyn's Foret Wines and Manhattan's Discovery Wines, to Chicago's Diversey Wine, to Oakland's Ordinaire Wine and Los Angeles's Domaine LA and Psychic Wines. (For more, see Resources, beginning on page 421.)

But natural wine can also be found in established quality wine shops without a particular natural focus. Famous examples include Kermit Lynch Wine Merchant in Berkeley and Chambers Street Wines in Manhattan. There are, after all, many natural estates among the "greats" that have enjoyed worldwide acclaim in the mainstream wine media. To know which is which, it helps to get to know the reputations of individual wine estates. Failing that, ask the wine shop staff what they have in the realm of natural wine.

Just be aware that the question can inspire resentment in career wine professionals who until recently have not been asked to consider the naturalness of the wine they sell. It's best to read the room before asking, in the same way you might hesitate before asking for vegan options at a steak house.

Top left: Reims restaurant and natural Champagne destination Au Bon Manger, opened in 2008 by Aline and Eric Serva

Top right: Milbrae, California's original natural wine shop, Vineyard Gate, run since 1998 by Alex Bernardo

Bottom left: Olivier Roblin's Paris Left Bank natural wine shop Les Caves du Panthéon

Bottom right: Paris wine shop and wine bar Ma Cave Fleury, run since 2009 by Morgane Fleury of Champagne Fleury

Importers

Wine importation long predates the question of naturalness in wine, so the portfolios of most well-established quality wine importers in the United States include natural estates alongside conventional estates. Famous examples include Kermit Lynch, whose portfolio includes natural wine luminaries like Domaine Gramenon and Marcel Lapierre alongside many conventional estates; and Weygandt-Metzler, whose extremely conventional portfolio nonetheless includes natural vignerons like Richard Leroy. This just reflects the reality of doing business in most markets. It's pointless to import all the greatest artisanal natural winemakers if an importer can't keep their business afloat with more reliable, high-volume sales from conventional estates.

Importers who work exclusively with natural estates are a more recent phenomenon, dating to the early 2000s. Today these are the importers most inclined to communicate clearly about the naturalness of wine. Their websites are often treasure troves of useful information about the viticulture and winemaking at their estates (to be fair, so is the Kermit Lynch website). Following natural wine importers on social media is a great way to stay up to speed about natural wine wherever you are in the world.

For a list of natural wine importers operating where you live, see Resources, beginning on page 421.

Left: Eric Narioo of UK-based importer Les Caves de Pyrene in his vineyards on Mount Etna in Sicily

Below: US importer Chris Terrell embracing Georgia winemaker John Wurdeman

Restaurants and Wine Bars

Wherever you find a developed wine market, you'll find restaurants and wine bars devoted to natural wine. This is particularly true in areas that have experienced striking gentrification in the last thirty years; places like Brooklyn, East London, and Oakland. In east Paris, the ethos of natural wine has so conquered the dining scene that almost every ambitious restaurant opening is accompanied by a natural wine program.

In many ways, establishments with knowledgeable servers and sommeliers are the ideal place to experience natural wines, because it allows for feedback and discussion in the moment. For vignerons, placing wines on restaurant wine lists offers the potential for great visibility (since just about everyone photographs their meal for social media these days), as well as some assurance that the wines will be served correctly and in positive contexts.

Today natural wine is catching on throughout the hospitality industry, from dive bars to Michelin three stars. This is no small achievement given the normal capriciousness of natural wines, and the amount of guest reeducation needed to serve them. Much credit is due to the work of a relatively small, international coterie of enlightened chefs and sommeliers.

For more on the most influential natural wine restaurants and wine bars in various world capitals, see pages 402–415 and the Resources, beginning on page 421.

Left: London's 40 Maltby Street is a destination for radical natural wines from owner Raef Hodgson's import company, Gergovie Wines.

Below: British chef Harry Lester's Le Saint Eutrope in Auvernat capital of Clermont-Ferrand draws natural wine diehards from around the world.

Bottom: David Loyola's popular Paris natural wine bistro Aux Deux Amis offers a raucous vibe and a retro décor.

Above: Manhattan's Frenchette boasts a plethora of exclusive collaborative cuvées, thanks to sommelier Jorge Riera's long roots in the natural wine scene.

Michel Tolmer and "Épaulé Jeté"

Above: Michel Tolmer at a Portes Ouvertes event hosted at Domaine Breton

Right: Tolmer's iconic "Epaulé Jeté" poster, outside the restroom of Arles natural wine bistro Le Gibolin

The work of Parisian illustrator Michel Tolmer occupies a singular place in the iconography of natural wine. His iconic "Épaulé Jeté" image has become the closest thing the natural wine world has to a flag, displayed on the walls of natural wine bistros and bars from Tokyo to Tbilisi. Initially produced in 2002 as a series of three hundred signed, numbered prints, the image was a commission for Bourgeuil vignerons Catherine and Pierre Breton, who offered it as a gift to favored clients. Demand for the image far outpaced the initial run, however, and Tolmer soon gave the Bretons permission to reproduce it as a poster for wider release.

Born in Paris in 1960, Tolmer obtained a degree at the city's École Professionelle Supérieure d'Arts Graphiques before embarking on a career as a freelance illustrator in the early 1980s. A hobby interest in wine led him to frequent the bistros and wine shops of the early proponents of unsulfited wines: François Morel's Les Envierges, Luc Desrousseaux's L'Echanson, and Bernard Pontonnier's Café de la Nouvelle Mairie. In 1987, Tolmer received his first commission for the creation of a wine label, from Cheverny vigneron Charles Guerbois. A label for Catherine and Pierre Breton soon followed, and wine label design—for estates ranging from Champagne Jacques Selosse to Château Moulin Pey-Labrie—gradually became Tolmer's chief business.

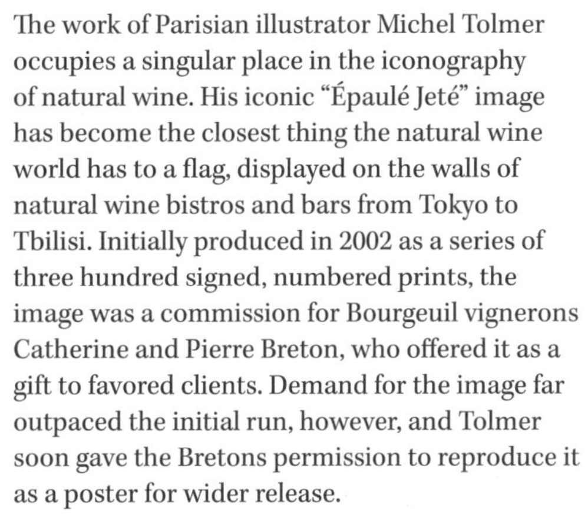

Alongside his artwork for renowned bistros including Paris's Bistrot Paul Bert and Jaja, his promotional artwork for wine importers like Les Caves de Pyrene and Japan's Racines, and his illustrations in many books about wine and cuisine, Tolmer has published three books of cartoons revolving around the fictional wine tasters Mimi, Fifi, and Glouglou.

"I started by drinking the worst. When we had a break between classes at school, we went to the bistrot next door and drank truly infamous Côtes du Rhône. Then one day someone brought a Médoc and I found it extraordinary. Today I'd find that, too, very mediocre. But I had a curiosity. Each time I made contact with wine, I was inspired."

Speculation

The minuscule production of certain famous natural wine estates has led to speculation on secondary markets. This is nothing new when it comes to wine collecting, but the phenomenon has accelerated thanks to the hype-stoking power of the internet. Today it's common to see a wine bottle purchased from a vigneron for €20 fetch upwards of €350 on a wine list overseas.

It's a galling scenario for the vignerons, most of whom want their wines to remain affordable. Drastically raising their prices is tantamount to declaring they produce wine only for the wealthy, an unpopular sentiment in rural communities. But retailers have reasons for raising prices. Cult wines on shop shelves at standard markups are prone to get snapped up by cherry-picking wine dealers and resold at higher prices. Sommeliers, too, know that if they list sought-after wines at normal margins on their wine lists, those wines will be the first to disappear.

As a result, natural wines from the likes of Maison Overnoy-Houillon, Clos Rougeard, Domaine des Miroirs, and Pierre Beauger have effectively disappeared from shop shelves and wine lists. If you want to try these wines today, you usually have to befriend a well-connected sommelier, importer, or wine retailer.

Empty Domaine des Miroirs bottles after a tasting

The Wine Media

Among the greatest obstacles preventing a wider understanding of natural wine is the mass-market wine media: wine magazines, newspapers, and brand-name wine critics. The editorial conventions of these institutions—everything from vintage reports to buying guides to wine awards—are particularly ill suited to coverage of natural wine.

How so? Editors of mass-market wine publications, like their advertisers, seek to reach the broadest possible audience by covering the wines most available to the greatest number of readers. This results in a systemic bias toward large-scale conventional wine producers, who fabricate wine on a scale incompatible with the artisanal quality of natural winemaking. Such producers engage public relations firms, purchase ad space, and provide mass-market wine journalists with perks ranging from wine samples to free travel.

Natural wine, by contrast, is a phenomenon that has arisen through the efforts of small-scale family estates. It is almost unheard of for such estates to engage public relations firms, purchase ad space, or sponsor journalists. Few have the financial means to do so. Just as important, few see any need to, since overproduction is rarely an issue for estates practicing low-yield organic agriculture and natural vinification.

With few exceptions (notably Eric Asimov's excellent writing for the *New York Times*), the mass-market wine media can be trusted to cover only mass-market wine. To find the best writing about natural wine, turn to newsletters, magazines, and websites that specialize in it. (For a list, see Further Reading, page 421.)

Fake Natural Wines

How do you fake a natural wine?

It's easy. From supermarket brands to large négociants to individual wine estates, natural-fakers exploit consumer inexperience by suggesting an equivalence between "no added sulfites" and "natural wine." Contemporary enology leans heavily on sulfite addition because sulfites are cheap and effective, but they're far from the only way to denature a wine.

Here are a few additives regularly used by natural-fakers. Often used in combination with sterile filtration, they help ensure that a fake-natural unsulfited wine is just as biologically stable (i.e., dead), as a heavily sulfited wine.

Dimethyl dicarbonate (DMDC)

Dimethyl dicarbonate is used to stabilize beverages by inactivating yeasts.

Lysozyme

This is an enzyme derived from egg albumen that inhibits microbial activity (see page 115).

Sorbic acid and potassium sorbate

These are often used as yeast inhibitors in the production of sweet wines.

Chitosan

Derived from crustacean shells, chitosan is used to inhibit yeast activity and to aid in wine clarification.

Natural Wine Disinformation

As awareness of natural wine has risen, unscrupulous players in the wine industry have responded with specious claims about what "naturalness" in wine means. Here are some common examples of natural wine disinformation.

Biohacking

Certain mailing lists market natural wine by associating it with popular diet trends. At least one such mailing list promotes a definition of natural wine based upon an analysis of residual sugar and alcoholic degree, an arbitrary set of criteria that betrays a blithe disregard for the reality of natural winemaking practice.

Vegan wines

A "vegan" wine has not been fined with antiquated, animal-based fining agents (which include egg whites and fish bladder), and has not been vinified with lysozyme derived from egg whites. The designation has nothing to do with natural wine. Many wines boasting a vegan-friendly status are not even organic.

"Clean" wines

A bevy of wine marketing scams have begun employing the word "clean" as a natural-adjacent euphemism aimed at gullible consumers. They adopt the rhetoric of natural wine, criticizing the wine industry's lack of transparency, while offering minimal transparency about the origins of their own mass-produced products, which have nothing whatsoever to do with natural wine.

"No added sulfites"

Be wary of wine bottles bearing cuvée names touting normal natural production practices, particularly with regards to sulfite addition. There are, after all, many other ways, besides sulfites, to denature a wine. (See above.)

How to Visit Natural Wine Estates

Late Loir-et-Cher vigneron Olivier Lemasson (*left*) conducting a barrel-tasting

Why visit wineries? Visiting estates provides a fascinating perspective on the working methods—and the personality—of a given vigneron. Depending on the estate, visits can also be opportunities to purchase wine.

But the customs surrounding winery visits vary greatly between nations, between regions, and between individual estates. In the US and in parts of Champagne, Bordeaux, and Jerez, for example, it is customary for wine estates to maintain actual tourist facilities, and visitors are often charged a fee for visits. While this can be a valuable source of revenue for wine estates, the commodification and the routinization of human interaction is antithetical to the spirit of natural wine.

If you arrange to visit a natural vigneron in the Beaujolais or the Languedoc or the Loire, for example, in most cases it will be the vigneron or a family member who receives you, often enough at a kitchen table or in a makeshift tasting room. (This is also because labor law in France and certain other parts of Europe encourages a reticence among business owners to hire permanent staff.) It can be a real pleasure discussing wine with a vigneron whose wines have moved you. But bear in mind a vigneron's time is valuable. Here are a few tips for arranging estate visits and for making them pleasant experiences for all involved.

1. Make an appointment.

This might seem obvious. But unconscientious wine tourists the world over persist in trying to simply drop by, as if a vigneron were just waiting around for visitors. Email first, and if you receive no response, try calling just before or after lunch hours a few days in advance.

2. Do some research.

It really helps to have a basic knowledge of a vigneron's background and wine production when you visit an estate. (Imagine how many times vignerons must be obliged to trot out the same family history before visitors.)

3. Be clear about the experience you're looking for.

Some visitors want to taste a few wines. Others want to take photos of vineyards and tour the cellar. Still others just want to buy wines and leave. Tell the vigneron what brings you to their doorstep that day.

4. But go with the flow.

Sometimes you catch a vigneron in an expansive or generous mood. What begins as a brief tasting can, in the company of certain vignerons, at certain times of year, become a daylong odyssey of vineyard tours and old vintages.

5. Know that buying wine is not always an option.

Estates whose wine production is tightly allocated often do not offer wine purchases to visitors. If you see a price list on display, it's probably okay to purchase wine. If you don't, they might not be for sale—but it's worth asking nicely.

TASTING DESTINATIONS
NATURAL WINE SALONS

To the extent that the natural wine scene is organized at all, it is organized around tasting salons. By bringing like-minded natural vignerons together, they illustrate the different communities within natural wine, each with its own criteria for naturalness in wine. Some, like Bojalien and H2O Vegetal, unite radically natural vignerons working with zero added sulfites and no filtration. Others, like Isabelle Legeron's profitable RAW WINE fairs, take a big-tent approach, permitting entry to more conventional estates that are just beginning to work toward more natural vinification.

Together, the world's natural wine salons constitute an international tasting circuit that attracts the most enthusiastic natural wine importers, wine buyers, and wine journalists. If you have the time, the travel budget, and a whole lot of patience for hilariously inconsistent event administration, why not join the fun?

New natural wine salons are created every year. What follows is a by-no-means-complete list. Use it as a guide for further explorations.

Karakterre: Vienna
VinNatur: Genoa
Le Nez dans le Vert: The Jura and Paris
ViniVeri: Verona
H2O Vegetal: El Pinell de Brai, Tarragona
El Saló dels Vins Naturals: Barcelona
Vella Terra: Barcelona
RAW WINE Fair: London and beyond
Real Wine Fair: London
Sous les Pavés la Vigne: Paris
Salon des Débouchées: Lyon
La Dive Bouteille: Saumur
Les Pénitentes: Angers
Bojalien: Saint-Étienne-des-Ouillères
Bien Boire en Beaujolais: The Beaujolais
Les Greniers Saint-Jean: Angers
Les Vins Anonymes: Angers
ViniCircus: Diverse locations in Brittany
Brumaire: Oakland, California
Prague Drinks Wine: Prague
Zero Compromise: Tbilisi
La Remise: Arles
Le Vin de Mes Amis: Montpellier and Paris
Brut(es): Mulhouse
Les Affranchis: Montpellier and Paris
Indigènes: Perpignan
Les Vignerons de l'Irréel: Montpellier
Canons, Le Salon: Nantes
Bottled Alive: Tabor, Czech Republic

The Vella Terra salon in Barcelona

Paris

La Cave des Papilles in the 14th arrondissement

Paris is where the first community of natural wine appreciation took root among a handful of influential proponents of *vin sans soufre* in the mid-1980s (see page 32). Today it's still the most dynamic and influential market for natural wine.

Historically, many natural wine destinations could be found in the affluent Left Bank. But times have changed. Today Paris's natural wine scene is concentrated in the eastern half of the Right Bank, particularly the densely populated 10th and 11th arrondissements. Here are some of the most influential Paris natural wine establishments of recent years.

Chambre Noire

Mexican émigré Oliver Lomeli (with the aid of his Franco-Japanese associate Remi Kaneko) has rejuvenated Paris's natural wine scene since 2015 with his ever-evolving, unpredictable, multisited series of radical wine shops and wine bars, all titled Chambre Noire. What began as a thronged bar on the 11th arrondissement's rue de la Folie-Méricourt moved to a corner space on rue Saint-Maur in 2020. Lomeli opened a wine boutique in another location on rue de la Folie-Méricourt the same year, following it with a collaborative pop-up with Pierre Jancou (see page 43) held in Lomeli's apartment down the road. The constant, throughout changes of site and personnel, is Lomeli's talent for creating a stylish, carefree, inviting vibe around his deep cellar of wines from Europe's zero-zero natural upstarts.

La Cave des Papilles

La Cave des Papilles was founded as a cave à manger on rue Gay-Lussac in 1996 by Gérard Katz, an acolyte of Jean-Pierre Robinot (see page 155). In 2001, Katz transitioned the business to its present location, where, aided by his lieutenant Florian Aubertin, he built it into the most well-respected natural wine retailer in the city. The shop boasts peerless allocations from the most cult natural vignerons of France; it's also beloved for its biannual block parties featuring live bands and shellfish stands. Katz recently sold his share of the business to influential sommelier Ewan Lemoigne, formerly of Saturne and Clown Bar.

Les Caves du Panthéon

Wine merchant Olivier Roblin started working at this historic Left Bank wine shop (est. 1944) in 2001, and purchased the business in 2009. Roblin oversees a vast natural wine selection spanning everything from veterans like Hervé Souhaut to newcomers like Yann Bertrand. He also maintains an off-site cellar in Touraine where he performs the true work of a *caviste*: aging natural wine until it's ready to drink.

Le Chateaubriand

A storied address in the annals of Paris natural wine, Le Chateaubriand passed through the hands of natural wine pioneer Olivier Camus (see page 35) before Basque chef Iñaki Aizpitarte took it over in 2006 and transformed it into one of the city's most ambitious natural wine tasting menu destinations. It's a place where grain-encrusted duck hearts share table space with unsulfited Languedoc elixirs from La Sorga. Aizpitarte also oversees its sister wine bar, Le Dauphin, and collaborates with the adjacent wine shop, Le Cave, opened by his longtime sommelier Sébastien Chatillon (see page 243).

Crus et Découvertes

In his comically cramped and disorganized wine shop on rue Paul Bert, the preternaturally low-key retailer Michael Lemasle has since 2003 championed the radical unsulfited vanguard of natural wine, from Anjou's Olivier Cousin to Ardèche négociant Daniel Sage.

Le Repaire de Cartouche

The towering, garrulous Normand chef Rodolphe Paquin opened this timeless bistro in 1998 with the help of a loan from Domaine Gramenon legend Philippe Laurent. Over two decades later it remains a destination for hearty traditional cuisine that reaches its zenith during game season, and for Paquin's vast cellar of natural wine treasures, including back vintages from the likes of Dard et Ribo, Château Sainte-Anne, and Domaine Gramenon.

Septime

Former L'Arpège chef Bertrand Grébaut and his partner, sommelier Théo Pourriat, opened Septime in the 11th arrondissement in 2011 and quickly earned plaudits as some of the most talented and ambitious natural wine restaurateurs of their generation. Their restaurant group today encompasses the Michelin-starred mothership Septime, the thronged shellfish bar Clamato, and the tiny cave à manger Septime La Cave kitty-corner across the street. More recently the duo opened a countryside auberge getaway called D'Une Île in Parc Naturel du Perche, and a pastry shop called Tappisserie.

Le Verre Volé

Cyril Bordarier's cozy, defiantly informal cave à manger Le Verre Volé was founded in 2000 on what was then a downtrodden patch of the Canal Saint-Martin. As gentrification transformed the surrounding neighborhood, Bordarier and his lieutenant Thomas Vicente transformed their ad hoc kitchen into a natural wine dining destination with a series of talented foreign chefs. Today the iconic purple Le Verre Volé brand also encompasses a wine shop on rue Oberkampf and a gourmet grocer on rue de la Folie-Méricourt.

NATURAL HISTORY

LE BARATIN

Le Baratin is the last holdout of Paris's first-generation natural wine bistros, run since 1987 by Argentine chef Raquel Carena. Carena was joined in 2002 by her companion, Philippe Pinoteau. Carena's sublime traditional cuisine and her effortless good cheer have done wonders to transmit the gospel of natural wine to subsequent generations of Parisian chefs and restaurateurs, as have Pinoteau's caustic wit and immaculate taste in natural wine. Expect lots of wine from the couple's renowned friends Jean and Agnès Foillard and Eric and Marie Pfifferling.

Raquel Carena (*left*) and Philippe Pinoteau behind the bar at Le Baratin

Copenhagen

Copenhagen's twenty-first-century emergence as a key natural wine market owes a lot to importer Sune Rosforth, who began importing the wines of Clos Rougeard (see page 169) and other Loire estates in the mid-1990s, while still in his early twenties. His company, Rosforth & Rosforth, laid the groundwork for the Copenhagen dining scene's wholesale embrace of natural wine in recent years.

The phenomenon was hastened by original Noma sommelier Pontus Elofsson, who began concentrating on natural wine in 2006. Noma's subsequent ascendance to the luxury food media stratosphere brought about a transformation in attitudes toward natural wine in Denmark and beyond. In the 2010s, former Noma cook Christian Puglisi also helped spread the gospel with his renowned restaurants Relæ and Manfreds, before shuttering both in the wake of COVID-19.

Today both Rosforth & Rosforth and their peers Jeppe Gustavsen and Philip Laustsen of Lieu-Dit imports have branched out to numerous restaurant projects. The Copenhagen natural wine scene's twin hearts remain Noma, where sommelier Mads Kleppe has continued and refined Elofsson's work, and Rosforth & Rosforth's funny, ramshackle wine shop and hang-out space beneath the Knippelsbro bridge.

Den Vandrette

Kitted out like a cross between a tiki bar and the hold of a ship, Rosforth & Rosforth sister restaurant and wine shop Den Vandrette pairs a glamorous, youthful ambience with joyful, informal hospitality and small plates of staggering craft. Texan chef Dave Allen Harrison did a stint at Paris's Au Passage before perceptibly raising his game upon moving to Copenhagen. Order everything.

Gaarden & Gaden

Restaurant scene veterans love Thomas Spelling's Nørrebro natural wine canteen

Den Vandrette's semi-subterranean space on the waterfront

NOMA

Copenhagen's Noma was not the first Michelin-starred restaurant to support natural wine (that honor went to Restaurant Alain Chapel, back in the 1980s). But Réné Redzepi's celebrated New Nordic destination was the one that validated natural wine for a contemporary generation of wealthy, well-traveled gastronomes—not to mention for its own numerous and highly entrepreneurial chef and sommelier alumni. Early Noma wine director Pontus Elofsson's embrace of natural wine around 2006 helped initiate a sea change in how such wines were seen, suggesting an inherent kinship between Redzepi's sensational, confrontational forager cuisine and the vinous art brut filling the restaurant's stemware.

Elofsson's successor Mads Kleppe, who has run the wine program since 2010, has cemented the restaurant's kingmaker influence in the natural wine world, cultivating close relationships with favored vigneron friends in Burgenland, Catalonia, and the Roussillon. His superhuman hospitality and synesthetic approach to natural wine description are as memorable as Redzepi's baroque menu.

Noma's dining room in the Christiana neighborhood, where the restaurant relocated in 2018

for its astute wine cellar, its unpretentious menu of hearty, well-executed international cuisine (shepherd's pie!), and its welcoming, spontaneous vibe. Spelling formerly managed Den Vandrette and did a stint with Rhône vigneron Marcel Richaud before opening the bar in 2015.

Omegn & Venner

Tucked beside a side entrance to Copenhagen's tony Torvehallerne food hall is this subdued and intimate gourmet grocer and all-day dining bar, offering chef and co-owner Denny Vangsted's simple, immaculately executed contemporary Danish comfort food: oysters, lobster omelets, linguine carbonara, and more. Given that the restaurant comprises just ten barstools and is not open for dinner, co-owner Sarah Backer-Vangsted's well-thumbed natural wine list is impressively long, featuring cult favorites from the likes of Jean-François Chéné and Daniel Sage. The couple also run a nearby bakery in the food hall in collaboration with Berlin bakery Albatross (called, naturally, Albatross & Venner).

Pompette

Well-traveled sommelier Martin Ho and friends opened their dimly lit Nørrebro wine bar in 2018, focusing on the vanguard of unsulfited wines from France and Italy. Youthful and snack-oriented, Pompette feels like a mash-up of global influences, ranging from Paris's Chambre Noire to Manhattan's Ten Bells and Tokyo's Ahiru Store.

Barabba

Venetian sommelier and Noma's 108 alum Riccardo Marcon reigns over this delightfully punk pasta destination, armed with a generous and idiosyncratic wine list packed with radicals from France and Italy. Chef Marco Cappelletti's liquid-Parmesan-filled ravioli are unforgettable.

Rosforth & Rosforth

A true hybrid space, Rosforth & Rosforth's location beneath Knippelsbro bridge is at once an office, a wine storage site, a wine distribution hub for visiting Swedes, and a wine bar. It comes to life as the latter in the summertime, when the sprawl of natural wine lovers reaches right out to the riverbank and beyond.

Barcelona

As a major economic center with a *very* independent cultural identity and massive annual tourism, Barcelona is a ripe market for natural wine, which exploded in popularity in the city during the 2010s. Barcelona's natural wine community also benefits from the city's proximity to wine regions to the north and south along the Mediterranean coast, including Tarragona, Penedès, and Conca de Barberà, as well as to the thriving natural wine scene of the Roussillon in France.

L'Ànima del Vi

French expat Benoît Valée opened L'Ànima del Vi as Barcelona's first natural wine shop back in 2006, after experiences working for Didier Barral and Jeff Coutelou. With a location change in 2010, he transformed it into a poky cave à manger in the Parisian style. Today it's a hot spot for apéro and light meals over cult natural wines, offering a homespun, studenty atmosphere on a central side street in El Born.

Bar Brutal

Founded by Max and Stefano Colombo, Joan Valencia, and Conca de Barberà winemaker Joan Ramón Escoda (see pages 348–349) in 2013 in the space of the historic Can Cisa wine shop, Bar Brutal quickly became the beating heart of the Barcelona natural wine world. Its labyrinthine interior houses an ongoing bacchanal of sterling Catalan cuisine and cutting-edge natural wines.

Bar del Pla

Named for owner Jaume Pla, Bar del Pla is a popular Barcelona institution known, since 2008, for its unpretentious, classic décor, its tourist-friendly traditional Calatan cuisine, and its impressive natural wine list, smartly overseen by sommelier Sergi Ruiz.

Garage Bar

Garage Bar is a terraced wine bar and wine shop in the Sant Antoni neighborhood run by Ale Delfino and Stefano Fraternali, organizers of the city's Vella Terra natural wine salon. It offers a sprawling natural wine selection alongside rich Catalan comfort food in an informal, industrial-chic setting.

Monocrom

Opened in 2016 by brothers Janina and Xavi Rutia on the quiet Plaça de Cardona, Monocrom is a thoughtful and tasteful terraced natural wine bistro, which delights a mostly native clientele with its broad, inventive menu and connoisseur's selection of Spanish and French natural wines.

Bar del Pla, a respected natural wine destination in the guise of a vintage tapas bar in Barcelona's El Born neighborhood

Rome

Litro in Rome's Monteverde district specializes in natural wine—and mezcal.

Italy's capital is the world's second most visited city after Paris. Yet dedicated natural wine establishments were slow to take root in Rome. The city's recent natural wine revival owes much to the work of native lawyer, wine journalist, and wine distributor Tiziana Gallo, who began writing about natural wine in the early 2000s alongside her mentor Sandro Sangiorgi. In 2005, she began distributing natural wines, and founded the long-running natural wine fair Vignaoli Naturale à Roma in 2009. Here, as elsewhere, the 2010s saw a tidal shift in the city's wine establishments, with many historic wine addresses like Roscioli and Enoteca Bulzoni broadening their selections to include a partial emphasis on natural wine. Dedicated natural wine purveyors also began appearing early that decade. Here are the best of the latter today.

Les Vignerons

Rome's first natural wine shop, Antonio Marino and Marisa Gabbianelli's Les Vignerons opened in 2010 in Tor Pignattara district before relocating to the Trastevere area in 2016. It remains the retail reference for lovers of natural wine and craft beer in Rome.

Litro

Founded in 2013 by Maurizio Bistocchi, Andrea Baroni, and Alessio Ceccotti and located off the tourist circuit in Monteverde, Litro offers a quirky and sophisticated mélange of radical natural wine, coffee, and mezcal-happy cocktails.

Enoteca Mostò

Just a few blocks north of the MAXXI museum, Ciro Borrielli and Pasquale Livieri's cozy, low-lit cave à manger offers cheese, charcuterie, and lauded tartares and carpaccios alongside the crema of Italian natural wine.

SO_2 Distribuzione ed Enoteca

Natural wine distributors Alfonso Scarpato and Katia Frontino opened their event space, tasting room, and wine bar in Pigneto in 2018. Situated right next to their warehouse, the enoteca boasts a vast selection of Italian wines, plus draft beer, charcuterie, and cheeses.

SantoPalato

Abbruzese chef Sarah Cicolini's modern trattoria in the San Giovanni has wowed the most discerning natural wine aficionados since opening in 2017. Expect immaculate, fresh interpretations of classics like carbonara and amatriciana alongside wines from radical estates such as Cantina Giardina and Le Coste.

London

London's innovative natural wine bar P.Franco hides in plain sight behind the facade of a Chinese grocery store.

Wine culture in the United Kingdom has long had to contend with deep-seated class issues surrounding wine consumption. Yet the increasing success of natural wine in the capital has helped overturn the stereotypes, showing Londoners that wine isn't just for special occasions at snooty restaurants. Eric Narioo's Les Caves de Pyrene import company, founded back in 1988 in Guildford, was the first to break natural wine into the UK market; it remains among the most dynamic wine sales companies worldwide, with branches in Italy and Australia. In 2012, Les Caves de Pyrene's marketing director, Doug Wregg, founded the city's Real Wine Fair, the same year as the enterprising sommelier and wine writer Isabelle Legeron founded her own roving RAW WINE fair series. The natural wine movement gathered steam throughout the rest of the 2010s, and has achieved a success in the more progressive parts of East London that mirrors its earlier conquest of east Paris.

40 Maltby Street

Chef Harry Lester (see page 322) and Raef Hodgson's Gergovie Wines has earned acclaim for its Auvergne-heavy porfolio of radical natural wines, as well as for its flagship wine bar, 40 Maltby Street, a model of discretion and well-priced good taste amid the arches of Maltby Street Market. The bar's unusual hours, far-flung location, and outstanding cuisine make it a true destination.

Brawn

Opened in 2010 by Les Caves de Pyrene on the chic Columbia Road in Hackney, Brawn was subsequently purchased by original chef Ed Wilson, who radicalized the wine list even further. A decade on, the restaurant is an East London natural wine institution, beloved for its inventive small plates and the sense of community among its regulars.

P.Franco

P.Franco was opened as a wine shop in 2014 by the duo behind the Noble Fine Liquor wine shops, Liam Kelleher and James Noble. It blossomed in 2015 with the arrival of manager Phil Bracey, who initiated an inspired guest chef program that brought the vibe of Parisian caves à manger like Vivant to London. The response was overwhelming. Londoners, it turned out, were simply waiting for someone to suggest they stand throughout a stunning small-plates meal and use empty shelf space as a makeshift bar.

Sager + Wilde

Swiss-born Londoner Michael Sager cut his teeth in the city's fine cocktail scene before opening his eponymous Hackney Road wine bar in 2013 with then wife Charlotte Wilde. The couple opened another location in Paradise Row before splitting. While not exclusively focused on natural wines, the Sager + Wilde bars are known for their impressive selection of back vintages from high-end natural heavyweights like Jean-Michel Stephan.

Montreal

The wine industry in Quebec's Francophone capital has long enjoyed a strong cultural exchange with its counterpart in Paris, and vice versa. So it's no surprise that Montreal has followed Paris in becoming a destination for natural wine.

The city's enlightened importers, sommeliers, and restaurateurs have demonstrated that a state monopoly on alcohol sales is no obstacle to the development of a thriving natural wine scene. Today the Société des alcools du Québec (SAQ) even promotes natural wine on its website. Natural wine–focused restaurants and bars abound throughout the city, with more than a few deriving from the lineage of the renowned Joe Beef restaurant group. Its sommelier-turned-wine-importer, Vanya Filipovic, has been among the most influential supporters of natural wine in Montreal since the mid-2000s, alongside importers like Aurelia Filion of Oenopole and sommeliers like Emily Campeau. Many cite the influence of sommelier Jeannot Gingras, who began championing what would become known as natural wines back in the 1990s.

Candide

Egalitarian communal seating exemplifies the progressive approach of this refined farm-to-table restaurant in a former church refectory, where sommelier (and former chef) Emily Campeau's list features divine natural wines from Franz Strohmeier, Andreas Tscheppe, and more.

Larry's

Part café, part bistro, part wine bar, Larry's is a breakfast-till-dinner, no-reservations establishment, offering everything from craft coffee, eggs, and pastries to oysters, beef tartare, and, of course, natural wine. Owners Sefi Amir, Marc Cohen, Annika Krausz, and Ethan Wills also run nearby restaurant Lawrence, and its butcher shop, Boucherie Lawrence.

Le Vin Papillon

A Joe Beef sibling restaurant born in 2013, Le Vin Papillon is a destination for Vanya Filipovic's natural wine selection and its warm, neighborhood atmosphere.

Le Petit Alep

Sisters Tania and Chahla Frangié's restaurant in the Villeray-Petite-Patrie neighborhood offers something unique: heartwarming Syrian and Armenian cuisine paired with a broad list of vins naturels. The latter was once the work of their childhood friend Jeannot Gingras; today it is overseen by sommelier Alain Paillassard.

Mon Lapin

Opened in 2018 by longtime Joe Beef sommelier Vanya Filipovic and her husband, chef Marc-Olivier Frappier, Mon Lapin (aka Vin Mon Lapin) is a stylish, terraced bistro in Petite Italie foregrounding locally sourced ingredients and natural wines from Filipovic's own Importations Dame-Jeanne. The restaurant's name is an homage to Jura vigneron Jean-François Ganevat's habit of calling everyone *mon lapin*, or "my rabbit."

Mon Lapin, in Petite Italie

Los Angeles

The City of Angels has only embraced natural wine with gusto since the mid-2010s, but the transformation has been astounding. Throughout the 2000s, nary a natural wine could be found; no one in the city even spoke in those terms. A decade later, even the supermarkets in LA advertise (questionable) natural wine. Things began to change in 2009, when former film producer Jill Bernheimer opened Domaine LA, the city's first natural wine shop. Natural wine then advanced in tandem with the galloping gentrification of Los Feliz, Echo Park, and Downtown, reaching a new generation of open-minded wine drinkers in these rapidly transforming neighborhoods.

Psychic Wines

Former Ordinaire employee Quinn Kimsey-White opened Psychic Wines on a quiet street of Echo Park in 2018, bringing a thoughtful, boutique approach to LA wine retail. Within its spartan, woodsy retail space, you can find the radical unsulfited avant-garde wines of California, France, and beyond.

Domaine LA

Los Angeles's pioneering natural wine retailer evolved out of owner Jill Bernheimer's previous online wine sales business. Soon after opening in 2009, Domaine LA began making a mark in its Melrose neighborhood for its comprehensive selection of natural wines that draws equally from California and Europe.

Night + Market Song

Chef Kris Yenbamroong's daring reworkings of Thai bar food got most of the attention when his restaurant opened in 2014. But he's since made an equal impact with his statement natural wine lists. (Yenbamroong goes the distance for natural wine. He once cooked a meal chez Jean Foillard in the Beaujolais.)

Bar Bandini

Roomy, low-lit Bar Bandini opened on Echo Park's Sunset Boulevard in 2015, featuring an extra-long bar where a range of gluggable natural wines gush forth from KeyKegs. Owners Michael Lippman and Jason Piggott have created a sexier nightlife atmosphere than you'll find in more academic natural wine destinations.

Voodoo Vin

Owners Natalie and Michael Hekmat spent years traveling the wine regions of France and beyond before opening this airy, terraced natural wine shop and wine bistrot in Virgil Village in 2020. Its tall wine retail shelving nods to Parisian inspirations like Vivant Cave and Le Verre Volé, while chef Travis Hayden (formerly of Santa Monica farm-to-table restaurant Rustic Canyon) brings a can-do spirit to the in-house preparation of everything from pasta to charcuterie.

Lou Wine Shop

Lou Amdur ran an eponymous restaurant from 2005 to 2012, where he championed obscure, misfit wines in a boldly low-budget setting. When he opened his wine shop in Los Feliz in 2014, he decided to double down on the natural wine aesthetics he'd gradually come to support, choosing "Natural + Unusual" as his shop's motto.

Echo Park's Psychic Wines

San Francisco Bay Area

Aran Healy's Ruby Wine in Potrero Hill

Thanks to its booming and youthful tech economy, its proximity to California wine communities, and its history of environmental activism, the San Francisco Bay Area surpassed New York in the 2010s as home to America's most energetic and dynamic natural wine scene. Much credit is due to the organizers of Brumaire, an exclusive annual natural wine festival that unites French natural vignerons with the young California négociants they've inspired in a weekend-long bacchanal.

Ordinaire

Ordinaire founder Bradford Taylor wanted to re-create the beautiful simplicity in natural wine service that he knew from visits to Paris's Le Verre Volé. So in an inspired move, he hired ex–Le Verre Volé server Quentin Jeanroy to manage his Oakland shop and wine bar. Ordinaire offers a masterful selection of natural wines balanced between local Californian upstarts and the most radical natural vignerons of Europe. It is also known for its brilliant dinner events, many of which involve formerly Paris-based chef Kosuke Tada. Bradford Taylor also co-owns Chicago's Diversey Wine, and was one of the founders of Brumaire.

The Punchdown

Former California cellarhand D.C. Looney and his companion Lisa Costa opened the first iteration of this natural wine bar in 2009, before losing their lease in 2013 and finally reopening in a roomier bar space in 2016. Beneath the antlerlike auspices of uprooted grapevines hung on the walls, Looney and Costa offer up to thirty glass pours, ranging from California natural wine institutions like Coturri Winery to the low-yield unsulfited northern Beaujolais gamay of Philippe Jambon.

Ruby Wine

Former California winemaker Aran Healy's tiny Ruby Wine resembles less a bottle shop than a sort of tiki-lit speakeasy for natural wine initiates. Charming and surprisingly radical, Healy stands out for his ironclad commitment to small-scale organic, unfiltered, unsulfited winemaking amid a generation of natural wine enthusiasts scheming on how to scale up.

Terroir

Terroir, San Francisco's first natural wine bar, opened its doors in 2007 as the project of Guilhaume Gerard, Dagan Ministero, and Luc Ertoran. Spacious and informal, it doubles as a bottle shop, and has become a touchstone for a generation of Bay Area natural wine enthusiasts. Gerard has since left to import natural wine as Selection Massale.

Vineyard Gate

Founded in 1998 in Millbrae, Alex Bernardo's Vineyard Gate was the first US wine shop to focus on natural wines. A reflective taster and a noted sake expert, Bernardo is the monkish uncle of natural wine in California, known for his longtime support of such radical natural vignerons as Romain des Grottes and Patrick Desplats.

New York City

Williamsburg's The Four Horsemen

America's most cosmopolitan and developed wine market, New York enjoys the first pick of wine pallets arriving from Europe. Yet the rise of natural wine in the city was a slow burn. Longtime wine buyers like David Lillie of Chambers Street Wines and the late Joe Dressner of Louis/Dressner Imports arrived at natural wine independently, tracing a line from the great winemaking of yesteryear. Much of the city's natural wine scene germinated at Balthazar, which opened in 1997 and employed many future natural wine players.

The mid-2000s was a pivotal time for New York natural wine culture. In 2003, Jenny Lefcourt and François Ecot began importing exclusively natural wine in the city. In 2004, Balthazar alum Arnaud Erhart opened the short-lived but influential 360 restaurant in Red Hook, which helped turn *New York Times* wine critic Eric Asimov on to natural wine. Since the mid-2000s, a generation of enlightened wine buyers, including Jorge Riera, Lee Campbell, Justin Chearno, Bill Fitch, Severine Perru, and Pascaline Lepeltier—along with the journalist and author Alice Feiring—have helped natural wine achieve a critical mass among the city's most forward-thinking wine establishments.

Chambers Street Wines

David Lillie and Jamie Wolff founded Chambers Street Wines in Manhattan in 2001, taking a bold stand in favor of artisanal European wines. The shop would become the premier hub for serious wine appreciation in New York, always illustrating, in its astute selection, the evolutionary links between timeless, quality-focused winemaking and what we now call natural wine.

The Four Horsemen

Opened in 2015 by a quartet of partners including longtime natural wine buyer (and Turing Machine guitarist) Justin Chearno and LCD Soundsystem frontman James Murphy, the Four Horsemen soon distinguished itself among the most distinctive and forward-thinking natural wine restaurants in the US. Its interior is a mash-up of Le Verre Volé and an Aesop skincare store; its wine list as thick and idealistic as Don Quixote; its cuisine a cultural melting pot of everything from French bar food to masterful crudo.

Frenchette

Two decades after working together at Balthazar, chefs Riad Nasr and Lee Hanson and sommelier Jorge Riera joined forces again at Frenchette, an iconoclastic power-brasserie on West Broadway offering stylish twists on hearty French cuisine and a bold wine list that reflects Riera's long experience in natural wine.

Ops

On a nondescript corner in Bushwick, Ops partners Mike Fadem, Gavin Compton, and Marie Tribouilloy have demonstrated since 2016 how a smart natural wine selection can ennoble even the most basic restaurant concept. It helps that Fadem's pizzas are up there with the city's best. In 2018, Tribouilloy also opened her own natural wine boutique, Foret Wines, specializing in the unsulfited French elixirs she helped popularize at Ops.

June

Chef-restaurateur Tom Kearney has run this well-appointed Cobble Hill natural wine destination since 2015 with encouragement from his friends at Jenny & François Selections. June's low-key charms include a secluded interior courtyard, an ambitious cocktail program, and far more vegetarian options than your average natural wine spot.

Contra

Chef Jeremiah Stone, who previously worked for Paris chef Giovanni Passerini at the natural wine bistro RiNo, brought New York–size ambition and budget to the discreet flair of the modern Parisian tasting menu when he opened Contra in 2014. A succession of well-informed natural wine sommeliers including Linda Milagros Violago, Jorge Riera, and Sam Anderson has helped put Contra—and its sibling restaurants, Wildair and Una Pizza Napoletana—at the forefront of high-end natural wine dining in Manhattan.

The Ten Bells

Opened by French natural wine aficionado Fifi Essome in 2008, the Ten Bells has long been the downtown social hub of New York's natural wine scene, a place where sommeliers and wine nerds gather to argue and imbibe without the distractions of a well-crafted meal. Dark enough to require night-vision goggles, the Ten Bells marries dive-bar informality with bona fide natural wine sophistication and, especially by New York standards, exceptionally kind prices.

NATURAL HISTORY

THE BALTHAZAR CONNECTION

Opened on Spring Street in 1997 by restaurateur Keith McNally, Balthazar was not—and is not—a natural wine spot. Yet it casts a long shadow in the natural wine world all the same, thanks to its renowned alumni.

Film director and author Jonathan Nossiter designed the opening wine list in 1997. He'd go on to exert huge influence with his wine films *Mondovino* and *Natural Resistance*, along with wine books like *Cultural Insurrection* and *Liquid Memory*.

Balthazar manager Arnaud Erhart went on to open the influential Red Hook restaurant 360, the first natural wine restaurant in the US, where he worked alongside former Balthazar server Jorge Riera.

Riera later did stints at natural wine restaurants the Ten Bells, Wildair, and Contra. Today he runs the wine program at the acclaimed restaurant Frenchette. *Food & Wine* magazine named him a Sommelier of the Year in 2019.

Balthazar server Laurent Saillard (see page 175) would go on to open Fort Greene natural wine restaurant Ici in 2004, and became a Loire Valley vigneron in 2013.

Balthazar server Yann Le Pollotec also returned to France after years working in Norway, and now brews beer following natural wine principles at Heima micro-brasserie in the Cher valley.

Arnaud Erhart at a wine tasting in the Jura

Chicago

Gusts of influence from both coasts helped bring natural wine to the Windy City in the last fifteen years. In 2008, Sean Krainik returned to the city after three years working at influential New York proto-natural wine shop UVA and partnered with his friend Nathan Adams to open Bucktown natural wine shop Red and White Wines. (Krainik later left the business, which Adams continues today.) A decade later, Oakland, California, natural wine restaurateur and retailer Bradford Taylor (of Ordinaire; see page 411) opened Diversey Wines in Logan Square, having relocated to the city with his family in 2016. Both establishments have since branched out, with Red and White opening an adjacent wine bar Noisette, and Diversey Wines supplying the wine list for the adjacent restaurant Cellar Door Provisions. Along with fellow travelers like Humboldt Park's Rootstock, these establishments today comprise a promising natural wine underground in the midwestern capital.

Red and White Wines

Roomy and timber-paneled, Red and White Wines has been bringing natural wine to its corner of Bucktown since 2008. Clients enjoy Saturday wine tastings with offerings that run the gamut from proto-natural icons like the Mosel's Ulli Stein to newcomers like Jura-based négociant project Les Valseuses.

Noisette

Opened as Red and White Wines' long-awaited adjacent wine bar in 2018, Noisette keeps the same hours as the wine shop, making it a quiet, chic destination for after-work drinks and afternoon indulgences. Its menu of refined bar snacks (olives, oysters, tinned fish, rillettes) takes inspiration from Parisian touchstones like Septime la Cave.

Diversey Wines

Diversey Wines' down-to-earth yet radical approach to natural wine retail will be familiar to fans of its sibling establishments, Oakland's Ordinaire and Los Angeles's Psychic Wines (see page 410). Its spartan shelving offers a Who's Who of legendary zero-zero vignerons ranging from the Sarthe's Jean-Pierre Robinot to Laureano Serres of Spain's Terra Alta.

Cellar Door Provisions

Ethan Pikas and Tony Bezsylko's stripped-down, farm-to-table café and dinner destination beside Diversey Wines boasts a laser-focused, well-priced wine selection from the latter's Bradford Taylor.

Rootstock

A cozy and unpretentious gastropub situated in Saul Bellow's old stomping ground, Rootstock focuses on "small production" wine and beer, a category that in practice includes a healthy smattering of offerings from natural wine mainstays like Olivier Cousin, Nicolas Vauthier, and Domaine Mosse (alongside many more conventional organic and biodynamic wines).

Diversey Wines in Logan Square. *From left*: Bradford Taylor, Ann Marie Meiers, and Mac Parsons.

Tokyo

Theories abound as to why Japan became, in the early 1990s, the biggest and earliest overseas market for French natural wine. Did the lightness and purity of the early French natural wines jibe with Japanese cuisine and Japanese constitutions? Or were the Japanese just culturally predisposed to accept the idea that wine, as a product of fermentation, should be alive?

Whatever the case, veteran natural vignerons from René-Jean Dard to Thierry Puzelat to Philippe Pacalet all credit Japanese importers with sustaining their winemaking careers at critical moments. Their ranks include Oeno Connexion's Yoshio Ito, a France-based former martial arts teacher who began importing the wines of Domaine du Prieuré Roch in 1993; Yasuko Goda, who began importing the wines of vignerons like Pierre Overnoy, Bruno Schueller, and Guy Breton at Racines Japan in 1997; Junko Arai of Cosmo Jun, who did much to help foster the natural winemaking community of the Cher valley; and François Dumas, a French expatriate who began importing natural wine in 1997 and later went on to found the Tokyo natural wine fair Festivin.

Today Japan's natural wine scene rivals that of France in its maturity and complexity, and even supports a nascent domestic natural winemaking culture. Here are some of the most influential Tokyo spots to start exploring natural wine in Japan.

Ahiru Store in Tomigaya, Tokyo

Ahiru Store

Siblings Teruhiko and Wakako Saito run this festive and unstoppably popular natural wine dining bar beside Yoyogi Park. It's known for its enormous natural wine selection, its astounding seasonal cuisine, and its lengthy lines for bar seats.

La Pioche

Founded in 2013, Shinya Hayashi's Nihonbashi-district French bistro is known among European vignerons for Hayashi's immense wine selection and the masterful grilled meats from its kitchen.

Bunon

Westerners exploring Tokyo's natural wine scene are often dismayed by the predominance of French cuisine. At its discreet address in Nishi-Azabu, Bunon artfully adopts European natural wine into the serene context of traditional Japanese farm-to-table cooking.

Shonzui

In an uninspiring wood-paneled dining room on the second floor of a building in Roppongi, you'll find Shonzui, Tokyo's first natural wine bar, which Shinsaku Katsuyama opened in 1993. Regulars feast on French-inspired terrines and Wagyu beef steaks washed down with the crème de la crème of French natural wines.

Winestand Waltz

Radical natural wine priest Yasuhiro Ooyama has run this confessional-size, standing-room-only natural wine bar since 2012. Waltz fills to capacity most nights with natural wine fans eager to taste Ooyama's latest obscure finds.

Beyond Natural Wine

Over the past four decades, natural wine has given rise to a passionate, international community of supporters. Retailers, chefs, restaurateurs, sommeliers, critics, writers, collectors, and amateurs alike are united in the conviction that naturalness matters. Natural wine, they understand, is more than a delicious beverage. It is the canary in the mine shaft, an indicator of the loss of culture that comes with agricultural industrialization.

The more you get to know natural wine, the more its power to delight and inebriate is eclipsed by its symbolic power.

In this way, natural wine is more than wine. It is a lens through which we can examine other aspects of the way we live today. Natural wine asks us to look at the cultural and social costs of technological progress. It questions the wisdom of the market.

Many other products, from bread, vegetables, and meat to building materials and fabric, are threatened by the same inexorable market forces that gave us manufactured conventional wines. We can't fight every battle. But if natural wine makes us question other aspects of contemporary life, it has gone to our heads in the best possible way.

Grains

Modern viticulture is dominated by cloned plant material (see page 48), but that crisis pales in comparison to the contemporary lack of biodiversity in grain agriculture. High-yield modern wheat strains with higher levels of gluten proteins have been implicated in the contemporary rise of gluten intolerance, while the mechanization of grain agriculture has accelerated rural depopulation. Meanwhile, synthetic chemical farming is almost ubiquitous

Gaillac vigneron, farmer, and distiller Laurent Cazottes with handfuls of his organically farmed Einkorn wheat—an ancient, low-yielding grain variety that produces bread rich in fiber, vitamins, and amino acids but low in gluten. Cazottes distills it to make an eau de vie.

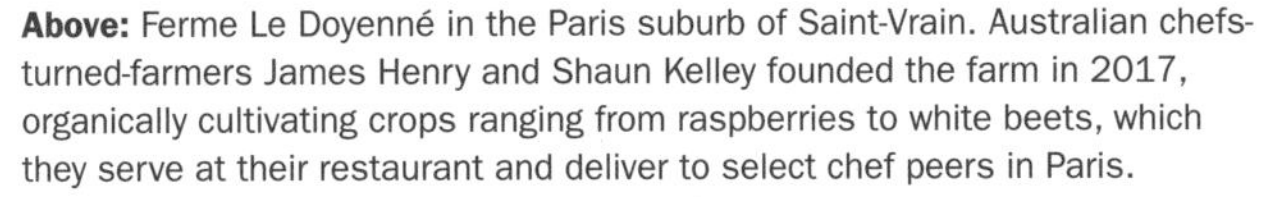

Above: Ferme Le Doyenné in the Paris suburb of Saint-Vrain. Australian chefs-turned-farmers James Henry and Shaun Kelley founded the farm in 2017, organically cultivating crops ranging from raspberries to white beets, which they serve at their restaurant and deliver to select chef peers in Paris.

Right: A cow in the herd of vigneron and mixed agriculture farmer Didier Barral in Faugères. Barral produces small quantities of veal, in addition to wine, which he supplies to restaurants including Nice's La Part des Anges and Antibes's Jeanne. Small-scale meat production within peasant modes of mixed agriculture is vanishingly rare even in Europe, and almost extinct in the US and the UK.

in grain agriculture. According to the 2018 Mercaris Acreage Report, in the US, just 4 to 5 percent of total acreage of grains including oats, barley, rye, and millet is farmed organically, while organic corn and soybean each amount to less than 1 percent of total production.

Fruits and Vegetables

Naturalness in the food we eat is arguably even more important than naturalness in the wine we drink. Reliance on high-yielding genetic strains of crops and the use of chemical fertilizers impoverish the quality of other fruits and vegetables just as they do wine grapes. In his 2004 analysis of nutrient data in the *Journal of the American College of Nutrition*, biochemist Donald Davis of the University of Texas demonstrated steady decline in the nutritional content of fruits and vegetables in the US over the last fifty years, which researchers blamed on crops genetically selected for high yields.

Meat

Contemporary industrial meat production is nothing less than a health, environmental, and humanitarian catastrophe. The production of low-cost meat at scale is a system reliant on hellish living conditions for animals, rendered possible only with the deployment of an ever-expanding arsenal of hormones, antibiotics, chemical sanitary treatments, and preservatives. (The industry is also notorious for creating intolerable working conditions for its human employees.) According to the FAIRR (Farm Animal Investment Risk and Return) initiative, cattle-ranching—and the grain production necessary to supply it—is responsible for 14.5 percent of all greenhouse gas emissions today.

Other systems of meat production are possible, but they require us to reevaluate the costs of meat production, its overly agglomerated structure in contemporary society, and our own level of entitlement to frequent meat consumption.

Alsace vigneron Christian Binner constructed his bioclimatic winery in 2012 with the aid of architect Mathieu Winter of Geishouse. Kept cool by its prairie roof, the winery requires no air-conditioning. Care was taken to ensure that the local timber in its construction was employed for structural support in the direction in which the wood grew.

Since 2009, Danish natural wine import company Rosforth & Rosforth has partnered with transport company Fairtransport and captain Anne-Flore Gannat to transport wine with a near-zero carbon footprint on a restored German sailing ship in a circuit between Denmark, the Caribbean, and the US.

Construction

Rapid economic expansion throughout the twentieth century and the advent of global supply chains has led to an impoverishment of building material and architectural practice. Air-conditioning allows modern buildings to ignore natural sources of insulation and temperature control, at great energy cost.

Transport

Among the unresolved dilemmas of the natural wine scene is its reliance on carbon-heavy international transport. Consumption of organic food and wine becomes more environmentally burdensome the farther they must travel to reach the market.

Labor

For centuries, technological innovation has been aimed at reducing human labor. Yet jobs sustain rural communities. The manual labor required by organic agriculture is honest, valuable work. Compared to synthetic chemical agriculture, it is better for the environment, better for the health of the worker, and better for the quality of the finished product.

Textiles

Just 2 percent of all cotton produced in the world, and 0.1 percent of cotton grown in the US, is organic. None is produced in France. The rest of what we wear is cultivated with many of the same chemicals natural vignerons avoid putting on their vines. According to the US Department of Agriculture, in 2019 the nation's farmers applied 986.6 million pounds of nitrogen fertilizer to cotton fields, 93 percent of which were also treated with herbicides.

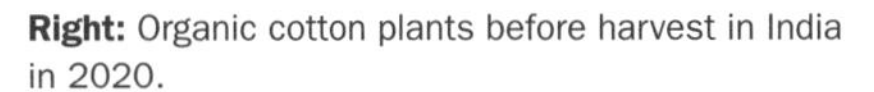

Right: Organic cotton plants before harvest in India in 2020.

Below: Stiekopf-Environment is an Alsace-based association that trains handicapped workers in the manual tasks needed for organic agricultural work. Here a student is pulling a vine's previous year's wood growth after winter pruning.

Further Reading

Thirsty for more natural wine knowledge? Here's a list of recommended resources to continue your explorations.

BLOGS, NEWSLETTERS, AND WEBSITES

The Pour by Eric Asimov of the *New York Times*
nytimes.com/column/the-pour

Not Drinking Poison by Aaron Ayscough
notdrinkingpoison.substack.com

Wine Terroirs by Bert Celce
wineterroirs.com

The Feiring Line by Alice Feiring
thefeiringline.com

No Wine Is Innocent by Antonin Iommi-Amenategui
nowineisinnocent.com

The Morning Claret by Simon J. Woolf
themorningclaret.com

BOOKS IN ENGLISH

Adventures on the Wine Route: Revised 25th Anniversary Edition by Kermit Lynch (Farrar, Strauss, and Giroux, 2013; first published 1988)

Agriculture Course: The Birth of the Biodynamic Method by Rudolf Steiner (Rudolf Steiner Press, 2016; first published 1958)

The Battle for Wine and Love: Or How I Saved the World from Parkerizaton by Alice Feiring (Mariner Books, 2009)

The Esthetics of Wine by Jules Chauvet (Editions Epure, 2020)

Jura Wine by Wink Lorch (Self-published, 2014)

Naked Wine: Letting Grapes Do What Comes Naturally by Alice Feiring (Da Capo Press, 2011)

Natural Wine by Isabelle Legeron (CICO Books, 2014)

Natural Wine for the People by Alice Feiring (Ten Speed Press, 2019)

The One-Straw Revolution by Masanobu Fukuoaka (New York Review of Books, 2009; first published 1978)

Pig Earth by John Berger (Writers and Readers Publishing Cooperative, 1979)

Wine in Question by Jules Chauvet (Editions Epure, 2020)

BOOKS IN FRENCH

Connaissance et travail du vin by Emile Peynaud (Dunod, 1975)

La Corne de Vache et la Microscope by Christelle Pineau (Editions de la découverte, 2019)

Entre Les Vignes #2 Avec Les Vignerons Natures d'Auvergne by Guillaume Laroche and Harry Annoni (Self-published, 2018)

Entre Les Vignes #3: 3 Generations de Vigneron-nes du Jura by Guillaume Laroche and Harry Annoni (Self-published, 2021)

Les Grands Vins Sans Sulfite by Arnaud Immélé (Vinédia, 2012)

Les méthodes biologiques appliquées à la vinification et à l'oenologie: Tome 1—Vinifications et Fermentations by Max Leglise (Courrier du Livre, 1994)

Réflexions d'un amateur de vins by Jacques Néauport (La Guilde, 1994)

Le Vin by Sylvie Augereau (Editions Tana, 2019)

Le Vin au Naturel by François Morel (Sang de la Terre, 2007)

Resources

What follows are lists of the key importers, wine shops, and dining establishments specializing in natural wine in Anglophone nations. The lists are not intended to be exhaustive—nor could they be, given the unpredictable pace of business openings and closures. Consider them a starting point for further explorations in natural wine, wherever you might find yourself in the world.

Most of the natural wine purveyors cited here are in major metropolitan areas. If you don't reside near these areas, try reaching out to importers in your region to see if they sell natural wines to any shops or dining establishments closer to home.

For importers as for wine shops and dining establishments, I've focused here on those who emphasize natural wine exclusively, or almost exclusively. There are two key exceptions. Among importers, Kermit Lynch Wine Merchant brings to the US such a plethora of foundational natural wine estates (Domaine Lapierre, Domaine Gramenon, etc.) that it seemed neglectful not to cite it here (along with its Berkeley, California, wine shop). Similarly, in New York City, Chambers Street Wines has long championed an artisanal, low-intervention wine production ethos, and stocks (and, in certain cases, directly imports) so much natural wine that its inclusion on these lists felt vital.

United States

IMPORTERS

Note: The US is an unusual market for wine importation due to the "three-tier system" that is the law in most, but not all, states. The three tiers are the importer (who stands in for the wine producers), the distributor, and the retailer. Importers may also distribute wines to retailers, but the two activities require different licenses. Some importers distribute nationally, while others limit their distribution to certain states and make agreements with other distributors to cover other states. Here I've focused on importers who either practice national distribution (whether alone or through agreements with other distributors), or who work within very large wine markets, such as California, Florida, or New York.

Arash Selects
instagram.com/arashselects

Bobo Selections
boboselections.com

Camille Rivière Selection
camillerivièreselection.com

F & R Wine Imports
frwineimports.com

Farm Wine Imports
farmwineimports.com

Fifi Imports
instagram.com/fifisimport

Goatboy Selections
instagram.com/goatboyselections

Jenny & François Selections
jennyandfrancois.com

Jose Pastor Selections
josepastorselections.com

Kermit Lynch Wine Merchant
kermitlynch.com

Louis/Dressner Selections
louisdressner.com

MFW Wine Co.
mfwwineco.com

Percy Selections
percyselections.com

SelectioNatural
selectionaturel.com

Selection Massale
selectionmassale.com

Super Glou
superglou.com

Terrell Wines
sidecar.terrellwines.com

Terrestrial Wine Co.
terrestrialwine.com

Tess Bryant Selections
tessbryant.wine

Thirsty Thirsty
instagram.com/thirstythirsty

Varda Selections
instagram.com/vardawine

Zev Rovine Selections
zrswines.com

BARS & RESTAURANTS

Atlanta

8arm
710 Ponce De Leon Avenue NE
8armatl.com

Austin

Lolo
1504 E 6th Street
lolo.wine

Boston

haley.henry
45 Province Street
haleyhenry.com

nathálie
186 Brookline Avenue
nathaliebar.com

Rebel Rebel
1 Bow Market Way
rebelrebelsomerville.com

Chicago

Cellar Door Provisions
3025 W Diversey Avenue
cellardoorprovisions.com

Noisette
1845 N Oakley Avenue
red-white-wines-chicago
.myshopify.com/pages/noisette

Rootstock
954 N California Avenue
rootstockbar.com

Houston

Light Years Wine
1304 W Alabama Street
lightyearswine.com

Vibrant
1931 Fairview Street
wearevibrant.com

Los Angeles

Bar Bandini
2150 Sunset Boulevard
barbandini.com

Botanica
1620 Silver Lake Boulevard
botanicarestaurant.com

Night + Market
Locations throughout Los Angeles
nightmarketsong.com

Tabula Rasa
5125 Hollywood Boulevard
tabularasabar.com

Voodoo Vins
713 N Virgil Avenue
voodoo-vin.myshopify.com

New York City

Contra
138 Orchard Street
contra.nyc

Ernesto's
259 E Broadway
ernestosnyc.com

The Four Horsemen
295 Grand Street
fourhorsemenbk.com

Frenchette
241 W Broadway
frenchettenyc.com

June
231 Court Street
junebk.com

Ops
346 Himrod Street
opsbk.com

Roberta's
261 Moore Street
robertaspizza.com

Ruffian
125 E 7th Street
ruffiannyc.com

The Ten Bells
247 Broome Street and 65 Irving Avenue
tenbellsnyc.com

Wildair
142 Orchard Street
wildair.nyc

Philadelphia

A.Kitchen and A.Bar
135 S 18th Street
akitchenandbar.com

Vedge
1221 Locust Street
vedgerestaurant.com

Portland

Bar Diane
Entrance on NW 21st Avenue through the gate next to Eb & Bean
bardiane.com

Bar Norman
2615 SE Clinton Street
barnorman.com

San Francisco Bay Area

Ordinaire
3354 Grand Avenue
ordinairewine.com

The Punchdown
1737 Broadway
punchdownwine.com

Ruby Wine
1419 18th Street
rubywinesf.com

Terroir
1116 Folsom Street
terroirsf.com

Verjus
528 Washington Street
verjuscAvenuecom

Seattle

Light Sleeper
1424 11th Avenue
lightsleeperseattle.com

Off Alley
1/2, 4903 Rainier Avenue S
offalleyseattle.com

WINE SHOPS

Austin

Lolo
1504 E 6th Street
lolo.wine

Sunrise Mini Mart
1809 W Anderson Lane
sunrisebottleshop.com

Boston

Rebel Rebel
1 Bow Market Way
rebelrebelsomerville.com

Social Wines
52 W Broadway
socialwinesbos.com

The Wine Bottega
341 Hanover Street
thewinebottega.com

Chicago

Diversey Wines
3023 W Diversey Avenue
diverseywine.com

Red & White Wines
1861 N Milwaukee Avenue
red-white-wines-chicago
.myshopify.com

Dallas

Bar & Garden
3314 Ross Avenue, Suite 150
barandgardendallas.com

Houston

Light Years Wine
1304 W Alabama Street
lightyearswine.com

Los Angeles

Domaine LA
6801 Melrose Avenue
domainela.com

Helen's Wines
4400 W Slauson Avenue
412 N Fairfax Avenue
11938 San Vicente Boulevard
helenswines.com

Lou Wine Shop
1911 Hillhurst Avenue
louwineshop.com

Psychic Wines
2825 Bellevue Avenue
psychicwinesla.com

Voodoo Vin
713 North Virgil Avenue
voodoo-vin.myshopify.com

New York City

Chambers Street Wines
148 Chambers Street A
chambersstwines.co

Dandelion Wines
153 Franklin Street
dandelionwineshop.com

Discovery Wines
16 Avenue B
discoverywines.com

Foret Wines
6838 Forest Avenue
foretwineshop.com

Henry's
69 Central Avenue
henrys.nyc

Peoples Wine
115 Delancey Street
peoples.wine

Thirst Wine Merchants
11 Greene Avenue
thirstmerchants.com

Uva
237 Bedford Avenue
uvawines.com

Wine Therapy
171 Elizabeth Street
winetherapynyc.com

Portland

Ardor Natural Wines
4243 SE Belmont Street,
Unit 400
ardornaturalwines.com

Bar Norman
2615 SE Clinton Street
barnorman.com

San Francisco Bay Area

Kermit Lynch Wine Merchant
1605 San Pablo Avenue
kermitlynch.com

Ordinaire
3354 Grand Avenue
ordinairewine.com

Ruby Wine
1419 18th Street
rubywinesf.com

Terroir
1116 Folsom Street
terroirsf.com

Vineyard Gate
238 Broadway
vineyardgate.com

Seattle

La Dive
721 E Pike Street
ladiveseattle.com

Molly's Bottle Shop
6406 32nd Avenue NW
mollysbottleshop.com

Rae Vino
321 NE 72nd Street
raevino.com

Canada

IMPORTERS

Ontario

Bogie's Best
bogiesbestimports.com

Context Wines
contextwines.com

Genuwine Imports
genuwineimports.com

Grape Witch Imports
grapewitches.com

The Living Vine
thelivingvine.ca

Parasol
parasolvin.com

Quebec

Bacchus 76
bacchus76.com

Boires
boires.ca

Dame-Jeane
vindamejeanne.com

La QV
laqv.ca

Le Vin dans les Voiles
levindanslesvoiles.com

Oenopole
oenopole.ca

Origines
origines.ca

RéZin
rezin.com

Vin i Vida
vinivida.ca

Vins Nomad
vinsnomad.com

Ward & Associés
wardetassocies.com

BARS & RESTAURANTS

Montreal

Alep and Le Petit Alep
Alep: 199 Rue Jean-Talon E
Le Petit Alep: 191 Rue Jean-Talon E
restaurantalep.com

Alma
1231 Avenue Lajoie
almamontreal.com

Bar Bara
4450 Notre-Dame Street W
barbaravin.com

Bar Henrietta
115 Avenue Laurier W
barhenrietta.com

Bar Suzanne
20 Avenue Duluth E
barsuzanne.ca

Candide
551 Rue Saint-Martin
restaurantcandide.com

Elena
5090 Notre-Dame Street W
coffeepizzawine.com

Foodlab
1201 Boulevard Saint-Laurent
linktr.ee/laboculinaire

Joe Beef
2491 Notre-Dame Street W
joebeef.com

Larry's
5201 Boulevard Saint-Laurent
lawrencemtl.com/larrys

Le Majestique
4105 Boulevard Saint-Laurent
restobarmajestique.com

Le Vin Papillon
2519 Notre-Dame Street W
vinpapillon.com

Loic
5001 Notre-Dame Street W
barloic.ca

Lundis au Soleil
801 Rue Jarry E
lundisausoleil.com

Mon Lapin
150 Rue Saint-Zotique E
vinmonlapin.com

Montreal Plaza
6230 Rue Saint-Hubert
montrealplaza.com

Nora Gray
1391 Rue Saint-Jacques
noragray.com

Pichai
5985 Rue Saint-Hubert
pichai.biz

Pullman
3424 Park Avenue
pullman-mtl.com

Toronto

Archive 909
909 Dundas Street W
archive909.com

The Comrade
758 Queen Street E
thecomraderestaurant.com

Donna's
827 Lansdowne Avenue
instagram.com/donnas.to

Dreyfus
96 Harbord Street
dreyfustoronto.com

The Federal
1438 Dundas Street W
thefed.ca

The Little Jerry
418 College Street
thelittlejerry.com

Midfield Wine Bar
1434 Dundas Street W
midfieldwine.com

Milou
1375 Dundas Street W
instagram.com/milou.to

Paradise Grapevine
841 Bloor Street W
paradisegrapevine.com

Paris Paris
146 Ossington Avenue
parisparis.ca

Union
72 Ossington Avenue
union72.ca

Wynona
819 Gerrard Street E
wynonatoronto.com

WINE SHOPS

Note: Independent wine retail is new to Ontario since the COVID-19 pandemic, when restaurants were permitted to function as retail shops during restaurant closures. Wine retail in Quebec is still a monopoly run by the Société des alcools du Québec (SAQ), but some natural wine establishments offer wine retail with the purchase of a takeout meal.

Toronto

Bodega Volo
608 College Street
birreriavolo.com

Boxcar Social
1208 Yonge Street
4 Boulton Avenue
boxcarsocial.ca

Good Cheese
614 Gerrard Street E
goodcheese.ca

Grape Witches
1247 Dundas Street W
grapewitches.com

Midfield Wine Bar
1434 Dundas Street W
midfieldwine.com

United Kingdom

IMPORTERS

Aubert & Mascoli
facebook.com/
aubertandmascoli

Basket Press Wines
basketpresswines.com

Beattie & Roberts
beattieandroberts.com

Gergovie Wines
gergovie-wines.com

Les Caves de Pyrène
lescaves.co.uk

Modal Wines
modalwines.com

Newcomer Wines
newcomerwines.com

Otros Vinos
otrosvinos.co.uk

Roland Wines
rolandwines.co.uk

Tutto Wines
tuttowines.com

266 Wines
266wines.co.uk

Uncharted Wines
unchartedwines.com

Wines Under the Bonnet
winesutb.com

Wright's Wines
store.wrightsfood.co.uk

BARS & RESTAURANTS

London

Antidote Wine Bar
12A Newburgh Street
antidotewinebar.com

Brawn
49 Columbia Road
brawn.co

Bright
1 Westgate Street
brightrestaurant.co.uk

Café Deco
43 Store Street
cafe-deco.co.uk

Duck Soup
41 Dean Street
ducksoupsoho.co.uk

40 Maltby Street
40 Maltby Street
40maltbystreet.com

Lyle's
Tea Building, 56 Shoreditch High Street
lyleslondon.com

Newcomer Wines
5 Dalston Lane
newcomerwines.com/
newcomerdalston

P.Franco
107 Lower Clapton Road
pfranco.co.uk

Sager + Wilde
193 Hackney Road
sagerandwilde.com

Silver Lining
13 Morning Lane
silverlininge9.com

Elsewhere in UK

Cave
286 Gloucester Road, Bristol
cavebristol.co.uk

Fitzroy
2 Fore Street, Fowley
fitzroycornwall.com

GlouGlou
17a Castle Gates, Shrewsbury
glouglou.uk

Kask
51 North Street, Bedminster
kaskwine.co.uk

Little Rascal
113D St John's Road, Corstorphine, Edinburgh
littlerascalwinebar.co.uk

Lovett's
1 The Coombe, Penzance
lovetts-newlyn.co.uk

Marmo
31 Baldwin Street, Bristol
marmo.restaurant

Petit Glou
Claremont Street, Shrewsbury
petitglou.uk

Two Belly
116 Whiteladies Road, Clifton
twobelly.co.uk

WINE SHOPS

London

Arch Rivals
361 Winchelsea Road
instagram.com/archrivals_e7

Dynamic Vines
Unit 5, Discovery Business Park
dynamicvines.com

Forest Wines
149 Forest Road
forestwines.com

40 Maltby Street
40 Maltby Street
40maltbystreet.com

Gnarly Vines
464 Hoe Street
gnarlyvines.co.uk

Newcomer Wines
5 Dalston Lane
newcomerwines.com/
newcomerdalston

Noble Fine Liquor
27 Broadway Market
noblefineliquor.co.uk

P.Franco
107 Lower Clapton Road, Lower Clapton
pfranco.co.uk

Provisions
167 Holloway Road
provisionslondon.co.uk

Shop Cuvée
189 Blackstock Road
shopcuvee.com

Silver Lining
13 Morning Lane
silverlininge9.com

Weino BIB
39 Balls Pond Road
weinobib.co.uk

Elsewhere in UK

Cave
286 Gloucester Road, Bristol
cavebristol.co.uk

GlouGlou
17a Castle Gates, Shrewsbury
glouglou.uk

Isca Wines
825 Stockport Road, Levenshulme, Manchester
iscawines.com

Kerb
04 Henry Street, Ancoats, Manchester
instagram.com/kerb.wine

Kork Deli
74 Whitley Road, Whitley Bay
korkwineanddeli.com

Made from Grapes
166-168 Nithsdale Road, Glasgow
madefromgrapes.shop

Monty Wines
Unit 8, Parkway Trading Estate, Street Werburgh's Road, St Werburgh's, Bristol
montywines.co.uk

Two Belly
116 Whiteladies Road, Clifton, Bristol
twobelly.co.uk

Wayward Wines
1C Regent Street, Chapel Allerton, Leeds
waywardwines.co.uk

Wine Freedom
28 Floodgate Street, Deritend, Birmingham

Ireland

IMPORTERS

Le Caveau
lecaveau.ie

BARS & RESTAURANTS

L'Atitude 51
1 Union Quay, Centre, Cork
latitude51.ie

Note
26 Fenian Street, Dublin
notedublin.com

Saint Francis Provisions
Short Quay, Sleveen, Kinsale, Co. Cork
instagram.com/stfranciskinsale

Table Wine
50 Pleasants Street, Saint Kevin's, Dublin
tablewine.ie

WINE SHOPS

Frank's Natural Wine Shop
Camden Street Lower, Saint Kevin's, Dublin
franksdublin.com

L'Atitude 51
1 Union Quay, Centre, Cork
latitude51.ie

Le Caveau
Market Yard, Gardens, Kilkenny
lecaveau.ie

New Zealand

IMPORTERS

Bare Wine
barewine.co.nz

The Last Drop
instagram.com/thelastdropnz

Salinity Imports
salinityimports.com

Wine Diamonds
winediamondstrade.co.nz

BARS & RESTAURANTS

Ascot
2/55 Ghuznee Street, Te Aro, Wellington
instagram.com/ascot.nz

Cave-a-Vin
146 Kitchener Road, Milford, Auckland
caveavin.co.nz

Celeste
146 Karangahape Road, Auckland CBD, Auckland
barceleste.com

Gatherings
5/2 Papanui Road, Merivale, Christchurch
gatherings.co.nz

Highwater
54 Cuba Street, Te Aro, Wellington
highwatereatery.co.nz

Mason
3 Wilson Street, Newtown, Wellington
barmason.co.nz

Omni
359 Dominion Road, Mount Eden, Auckland
atomni.co.nz

Puffin
60 Ghuznee Street, Te Aro, Wellington
puffinwinebar.com

WINE SHOPS

Everyday Wine
442 Karangahape Road, Newton, Auckland
177 Cuba Street, Te Aro, Wellington
everydaywine.co.nz

Star Superette
170 Karangahape Road, Auckland CBD, Auckland
starwines.co.nz

Australia

IMPORTERS

Andrew Guard
andrewguaRoadcom.au

Campbell Burton Wines
campbellburton.com.au

Giorgio de Maria Fun Wines
giorgiodemaria.com

Larkin Imports
larkinimports.com.au

Living Wines
livingwines.com.au

Lo-fi Wines
lofiwines.com

MESA Wines
mesa.wine

Vino Mito Wine Imports
vinomito.com

Vivant Selections
vivantselections.com

BARS & RESTAURANTS

Adelaide

Leigh Street Wine Room
9 Leigh Street, Adelaide
leighstreetwineroom.com

Loc Bottle Bar
6 Hindmarsh Square, Adelaide
locbottleshop.com

The Summertown Aristologist
1097 B26, Summertown
thesummertownaristologist.com

Hobart

Dier Makr
123 Collins Street
diermakr.com

Lucinda Wine
123 Collins Street
lucindawine.com

Sonny
120a Elizabeth Street
sonny.com.au

Melbourne

Bar Liberty
234 Johnston Street, Fitzroy
barliberty.com

Embla
122 Russell Street, Melbourne
embla.com.au

Hope Street Radio
35 Johnston Street, Collingwood
hopestradio.community

Napier Quarter
359 Napier Street, Fitzroy
napierquarter.com.au

Neighbourhood Wine
1 Reid Street, Fitzroy North
neighbourhoodwine.com

Old Palm Liquor
133B Lygon Street, Brunswick East
oldpalmliquor.com

Public Wine Shop
179 St Georges Road, Fitzroy North
publicwineshop.com.au

Perth

Casa
399 Oxford Street, Mount Hawthorn
casa-casa-casa.com

Si Paradiso
1/446 Beaufort Street, Highgate
si-paradiso.com

Wines of While
458 William Street, Perth
winesofwhile.com

Sydney

Dimitri's Pizzeria
215 Oxford Street, Darlinghurst
dimitrispizzeria.com

The Dolphin
412 Crown Street, Surry Hills
dolphinhotel.com.au

Ester
46-52 Meagher Street, Chippendale
ester-restaurant.com.au

Love, Tilly Devine
91 Crown Lane, Darlinghurst
instagram.com/lovetillydevine

Paski Vineria Popolare
239 Oxford Street, Darlinghurst
paski.com.au

10 William Street
10 William Street, Paddington
10williamStreetcom.au

Where's Nick
236 Marrickville Road, Marrickville
wheresnick.com.au

WINE SHOPS

Melbourne

Public Wine Shop
179 St Georges Road, Fitzroy North
publicwineshop.com.au

Perth

Casa
399 Oxford Street, Mount Hawthorn
casa-casa-casa.com

Wines of While
458 William Street, Perth
winesofwhile.com

Wise Child Wine Store
28 Thompson Road, North Fremantle
wcws.store

Sydney

DRNKS
Online only
drnks.com

P&V Merchants
64 Enmore Road, Newtown
pnvmerchants.com

Paski Vineria Popolare
239 Oxford Street, Darlinghurst
paski.com.au

Winona Wine
Shop 9/2-14 Pittwater Road, Manly
winonawine.com

Acknowledgments

My heartfelt thanks go out to Laura Nolan, my patient and encouraging agent, and to Judy Pray, Bridget Monroe Itkin, and the rest of the team at Artisan, for bringing this book into existence.

Its creation has occupied the better part of a decade, and I owe great debts of gratitude to many friends and loved ones. I'd like to offer special thanks to Jade Quintin, for tolerating many cold hours in damp, poorly lit cellars with vignerons discussing malolactic fermentation; to her mother, Marie Françoise Rigal, for housing us in the Loiret during COVID-19 lockdowns; to my mother and father, Elizabeth and Roger Ayscough, for never once pressuring me to return to the US and find a real job; to Charles Marlio, for critical support when I was first embarking on research for this book; to my dear landlords, Susannah Mowris and Guillaume Jeanson; to Jane Drotter, who kindly let me buy wines for her Paris restaurant back in the day; to Remo Hallauer and my former colleagues at Comme des Garçons, for indulging my work in the wine world; to Robert Dentice and Renée Patronik, for their support and their delightful company at so many expatriate Thanksgivings; to David Rosoff, my first mentor in wine; to Cliff Fong and Mark Siegel, for encouraging my move to France.

Throughout the research and production of this book, I've relied upon the kindness of many friends in restaurants and wine retail. I'd like to thank in particular Guy Jeu and Michel Moulherat, who first opened my eyes to natural wine; Pierre Jancou, a perennial inspiration; Olivier Camus, Jean-Pierre Robinot, François Morel, Raquel Careña, and Philippe Pinoteau, for their invaluable perspectives on natural wine history; Rodolphe Paquin of Repaire de Cartouche; Jean-Michel Wilmès of Aux Crieurs de Vin, Olivier Labarde of La Part des Anges, Richard and Pepita Grocat of Le Comptoir des Tontons, Olivier Roblin of Les Caves de Panthéon, and Michael Lemasle of Crus et Découvertes, for access to their extensive cellars; to Harry Lester, for his warm welcome in Auvergne; to Brice Fortunato and the team at Jeanne in Antibes, for enlivening the Côte d'Azur; to Oliver Lomeli of Chambre Noire, for being a great neighbor.

The travels necessary to research this book would have been significantly more burdensome were it not for the help of many friends in wine importation and regional administration. My thanks to Phil Sareil and Nick Gorevic of Jenny & François Imports, for introducing me to Austria and the Czech Republic; to Josh Eubank of Percy Selections, for introducing me to Catalonia; to Chris Terrell of Terrell Wines, for introducing me to Georgia; to Chris Santini of Kermit Lynch Wine Merchants, for making me feel at home in Burgundy; to Bill Fitch of Goatboy Selections, for introducing me to Barcelona; to Eric Narioo of Les Caves de Pyrène, for his kind welcome in Sicily; to Sune Rosforth of Rosforth & Rosforth and NOMA sommelier Mads Kleppe, for showing me the wonders of Copenhagen; to Barbara Repovs, for introducing me to Slovenia; to Josh Adler of Paris Wine Company, for getting me into the habit of visiting vignerons back in 2010.

Whether it be for help with photo sourcing, for inspiration, or for the pleasures of commiseration, I'd like to thank my fellow wine writers Bert Celce, Antonin Iommi-Amenategui, David Nelson, Sylvie Augereau, and Alice Feiring.

Most importantly, I'd like to thank all the vignerons who've shared their experiences, their savoir-faire, and their dinner tables with me in the past decade. With particular thanks to Jules and Yvon Métras, Julie Balagny, the entire Foillard family, Jean-Louis Dutraive, Ptit Max, Sylvain Chanudet, David and Michele Chapel, Romain des Grottes, Elisa Guerin, the whole Lapierre family, Emmanuel Lassaigne, Christian Binner, Kenji and Mai Hodgson, Agnès Mallet and Babass, Joe and Amandine Jefferies, the Andrieu siblings of Clos Fantine, Loïc Roure, Julien Altaber, Chiara Pepe, Jon Purcell, Stephen Roberts, Jeff Coutelou, Emma Bentley and the Maule family, Florien Klein-Snuverink, Phil Lardot, Thibault Pfifferling and Natalia Crozon, and Stephana Nicolescu and Andrea Calek. This book wouldn't exist without your insight and your kind support.

Index

Pages in *italics* indicate photographs. Pages followed by *m* indicate maps. Pages in **bold** indicate a full article or section of an article devoted to an individual.

Photography Credits

Jade Quentin: pp. 2, 12, 14, 23, 31 (bottom left and right), 33, 36 (left), 39 (bottom), 43–44, 54–56, 62, 65, 67–68, 72, 80 (top left), 89–92, 93 (top), 94 (bottom), 99, 105, 109 (top right), 110, 112, 115, 120, 130, 132, 144, 149, 150, 155, 159, 161, 164–165, 168, 169, 170, 174, 176, 180, 183, 186, 192, 196–198, 201–203, 206, 208, 210, 231, 232, 234, 237, 240, 242, 243, 253, 260, 263, 264, 272, 273, 275, 277, 278, 280, 282, 286, 290, 291, 293, 295, 297–298, 300–302, 306–307, 309, 313, 314, 316, 327, 329, 330, 337, 339–343, 347 (top and bottom), 349, 352 (bottom), 353, 354 (bottom), 361, 366, 369, 372, 374, 375 (top), 383 (top), 385, 387 (bottom), 390, 395 (left), 397 (left), 403–404; **Aaron Ayscough:** pp. 6–7, 10–11, 16–17, 19, 21, 35 (top right), 37–38, 39 (top), 40–42, 47, 49, 52–53, 57–61, 63, 64, 66, 69, 70, 73, 76, 77, 80 (bottom left and right), 93 (bottom), 94 (top and middle), 95–97, 100–104, 106–108, 109 (top left and bottom left), 111, 113, 114, 124, 127–131, 133–135, 137–139, 140–143, 147, 148, 152, 156, 162, 163, 167, 173, 175, 179, 181, 182, 184, 185, 187, 189, 190, 191, 195, 209, 211, 212, 215, 216, 218–226, 229, 233, 235, 236, 238, 239, 241, 244, 247, 249–252, 254–256, 259, 261, 265, 266, 269, 270, 271, 274, 276, 279, 284, 285, 289, 299, 305, 310, 312, 315, 319–323, 325, 326, 328, 333, 335, 336, 338, 342, 344, 346, 352 (top), 354 (top), 370, 371, 373, 375 (bottom), 379, 380, 382, 383 (bottom), 384, 386, 387 (top), 388–389, 392, 394, 395 (right), 396 (top right and bottom right), 397 (right), 398, 400–402, 413, 416–417, 418 (top); **History and Art Collection / Alamy Stock Photo:** p. 18; **Bruce Neyers. Photo courtesy of Jean-Jacques Robert:** p. 24; **Photo courtesy of Mathieu Lapierre (photographer unknown):** pp. 25, 32, 36 (right), and 75; **Yoshio Ito:** p. 26; **Photo courtesy of Aline Chauvet (photographer unknown):** p. 27; **Antillia Dufourmantelle. Photo courtesy of Mathieu Lapierre:** p. 28; **Courtesy of Bertrand Celce (original photographer unknown):** p. 30 (left); **Photo courtesy of Julien Guillot (photographer unknown):** p. 30 (right); **Photo courtesy of Jean-Baptiste Dutheil (photographer unknown):** p. 31 (top left); **Ilaria Bellotti:** p. 31 (top right); **Photo courtesy of Luc Desrousseaux (photographer unknown):** p. 34; **Photo courtesy of Olivier Camus (photographer unknown):** p. 35 (top left); **Photo courtesy of Jean-Pierre Robinot (photographer unknown):** p. 35 (bottom); **Bertrand Celce:** pp. 46, 48, 153–154, 157, and 217; **A. & G. Dolby. Photo courtesy of the Archives Départementales de la Côte d'Or:** p. 50; **Andre Arsicaud. Photo courtesy of the Archives Départementales d'Indre-et-Loir:** pp. 51 (top: left, middle, and right), 82 (top), and 368; **Photo courtesy of Smith Archive / Alamy (photographer unknown):** p. 51 (bottom); **Photo courtesy of the Archives Départementales d'Indre-et-Loir (photographer unknown):** p. 82 (bottom); **Photo courtesy of Keystone Press / Alamy (photographer unknown):** p. 83 (top); **AFP/ Stringer / Getty Images:** p. 83 (bottom); **Pere Karlsson/Alamy Stock Photo:** pp. 86 and 87; **David Rager:** pp. 166 and 172; **Photo courtesy of Michèle Aubery (photographer unknown):** p. 230; **Aurelien Avril:** p. 283; **Benoit Cortet:** p. 248; **Christophe Grilhe:** p. 324; **Petra Menclova:** p. 355; **Raef Hodgson:** p. 396 (top left); **Melanie Dunea:** p. 396 (bottom left); **Ditte Isager / Noma:** p. 405; **M Ramírez / Alamy Stock Photo:** p. 406; **Alessio Ceccotti:** p. 407; **Phil Bracey:** p. 408; **Dominique Lafond:** p. 409; **Yulia Zinshstein:** p. 410; **Aran Healy:** p. 411; **Damien Lafargue:** p. 412; **Eva Chaudoir:** p. 414; **World Discovery / Alamy Stock Photo:** p. 415; **Andreas Rosforth:** p. 418 (bottom); **Batuhan Toker / Alamy Stock Photo:** p. 419 (top); **Sophie Grieshaber:** p. 419 (bottom)

About the Author

Aaron Ayscough is a wine writer based in Paris. Since 2010, he has maintained a blog and newsletter about natural wine called Not Drinking Poison. He is the English translator of two works by the French winemaker-scientist Jules Chauvet: *Wine in Question* and *The Aesthetics of Wine*, and has worked extensively as a sommelier building wine selections for restaurants in the United States and France. His experience in wine production has included work at wine estates in the Beaujolais, Burgundy, and the Languedoc.